STUDENT'S SOLUTION MANUAL

LIAL • HORNSBY • MILLER

INTRODUCTORY ALGEBRA
FIFTH EDITION

USED BOOK

Prepared with the assistance of

August Zarcone
College of Dupage

Gerald Krusinski
College of DuPage

Abby Tanenbaum

Brian Hayes
Triton College

HarperCollins*CollegePublishers*

Student's Solution Manual to accompany Lial/Hornsby/Miller, *Introductory Algebra*, Fifth Edition.

Copyright © 1995 by HarperCollins College Publishers

All rights reserved. Printed in the United States of America. No part of this book may be reproduced in any manner whatsoever without written permission. For information, address HarperCollins College Publishers, 10 East 53rd Street, New York, NY 10022.

ISBN 0-673-99062-1

94 95 96 97 9 8 7 6 5 4 3 2 1

PREFACE

This book provides complete solutions for all margin exercises, section exercises numbered 3, 7, 11, ..., and all chapter review, chapter test, and cumulative review exercises in *Introductory Algebra*, fifth edition, by Margaret L. Lial, E. John Hornsby, Jr., and Charles D. Miller. Some solutions are presented in more detail than others. Thus, you may need to refer to a similar exercise to find a solution that is presented in sufficient detail. As needed, artwork is provided to clarify and illustrate solutions. Solutions are not provided for the exercises that involve open–response answers.

The following people have made valuable contributions to the production of this *Student's Solution Manual*: Abby Tanenbaum, editor; Judy Martinez, typist; Therese Brown and Charles Sullivan, artists; and Carmen Eldersveld, proofreader.

CONTENTS

R PREALGEBRA REVIEW

R.1 Fractions — 1
R.2 Decimals and Percents — 5

1 THE REAL NUMBER SYSTEM

1.1 Exponents, Order of Operations, and Inequality — 10
1.2 Variables, Expressions, and Equations — 12
1.3 Real Numbers and the Number Line — 16
1.4 Addition of Real Numbers — 18
1.5 Subtraction of Real Numbers — 22
1.6 Multiplication of Real Numbers — 24
1.7 Division of Real Numbers — 27
1.8 Properties of Addition and Multiplication — 29
1.9 Simplifying Expressions — 33
Chapter 1 Review Exercises — 35
Chapter 1 Test — 45

2 SOLVING EQUATIONS AND INEQUALITIES

2.1 The Addition Property of Equality — 48
2.2 The Multiplication Property of Equality — 50
2.3 More on Solving Linear Equations — 53
2.4 An Introduction to Applied Problems — 57
2.5 Formulas and Geometry Applications — 62
2.6 Percent; Ratio and Proportion — 65
2.7 The Addition and Multiplication Properties of Inequality — 69
Chapter 2 Review Exercises — 74
Chapter 2 Test — 85
Cumulative Review: Chapters R–2 — 89

3 EXPONENTS AND POLYNOMIALS

3.1 The Product Rule and Power Rules for Exponents	96
3.2 Integer Exponents and the Quotient Rule	99
3.3 An Application of Exponents: Scientific Notation	101
3.4 Addition and Subtraction of Polynomials	103
3.5 Multiplication of Polynomials	106
3.6 Products of Binomials	110
3.7 Division of a Polynomial by a Monomial	112
3.8 The Quotient of Two Polynomials	114
Chapter 3 Review Exercises	116
Chapter 3 Test	122
Cumulative Review: Chapters R–3	124

4 FACTORING

4.1 Factors; The Greatest Common Factor	129
4.2 Factoring Trinomials	132
4.3 More on Factoring Trinomials	136
4.4 Special Factorizations	142
Summary Exercises on Factoring	144
4.5 Solving Quadratic Equations by Factoring	145
4.6 Applications of Quadratic Equations	150
Chapter 4 Review Exercises	154
Chapter 4 Test	165
Cumulative Review: Chapters R–4	168

5 RATIONAL EXPRESSIONS

5.1 The Fundamental Property of Rational Expressions	173
5.2 Multiplication and Division of Rational Expressions	175
5.3 Least Common Denominators	178
5.4 Addition and Subtraction of Rational Expressions	181
5.5 Complex Fractions	186

5.6 Equations Involving Rational Expressions	188
Summary Exercises on Rational Expressions	194
5.7 Applications of Rational Expressions	195
Chapter 5 Review Exercises	200
Chapter 5 Test	211
Cumulative Review: Chapters R–5	214

6 GRAPHING LINEAR EQUATIONS

6.1 Linear Equations in Two Variables	219
6.2 Graphing Linear Equations in Two Variables	225
6.3 The Slope of a Line	229
6.4 Equations of a Line	233
6.5 Graphing Linear Inequalities in Two Variables	238
Chapter 6 Review Exercises	241
Chapter 6 Test	252
Cumulative Review: Chapters R–6	255

7 LINEAR SYSTEMS

7.1 Solving Systems of Linear Equations by Graphing	262
7.2 Solving Systems of Linear Equations by Addition	266
7.3 Solving Systems of Linear Equations by Substitution	273
7.4 Applications of Linear Systems	279
7.5 Solving Systems of Linear Inequalities	286
Chapter 7 Review Exercises	288
Chapter 7 Test	304
Cumulative Review: Chapters R–7	310

8 ROOTS AND RADICALS

8.1 Finding Roots	316
8.2 Multiplication and Division of Radicals	319
8.3 Addition and Subtraction of Radicals	323
8.4 Rationalizing the Denominator	324

8.5 Simplifying Radical Expressions	327
8.6 Equations with Radicals	330
Chapter 8 Review Exercises	335
Chapter 8 Test	344
Cumulative Review: Chapters R–8	346

9 QUADRATIC EQUATIONS

9.1 Solving Quadratic Equations by the Square Root Property	351
9.2 Solving Quadratic Equations by Completing the Square	353
9.3 Solving Quadratic Equations by the Quadratic Formula	357
Summary Exercises on Quadratic Equations	362
9.4 Graphing Quadratic Equations in Two Variables	364
Chapter 9 Review Exercises	367
Chapter 9 Test	378
Cumulative Review: Chapters R–9	381

CHAPTER R PREALGEBRA REVIEW

Section R.1 Fractions

R.1 Margin Exercises

1. **(a)** Since 12 can be divided by 2, it has more than two different factors, so it is a composite number.

   (b) 13 has exactly two different factors, 1 and 13, so it is a prime number.

   (c) Since 27 can be divided by 3, it has more than two different factors, so it is composite.

   (d) 59 has exactly two different factors, 1 and 59, so it is prime.

   (e) 1806 can be divided by 2, so it is composite.

2. **(a)** To write 70 in prime factored form, first divide by the smallest prime, 2, to get
   $$70 = 2 \cdot 35.$$
   Since 35 can be factored as $5 \cdot 7$, we have
   $$70 = 2 \cdot 5 \cdot 7,$$
   where all factors are prime.

   (b) $72 = 2 \cdot 36$
   $= 2 \cdot 2 \cdot 18$
   $= 2 \cdot 2 \cdot 2 \cdot 9$
   $72 = 2 \cdot 2 \cdot 2 \cdot 3 \cdot 3$

   (c) $693 = 3 \cdot 231$
   $= 3 \cdot 3 \cdot 77$
   $693 = 3 \cdot 3 \cdot 7 \cdot 11$

   (d) Since 97 is a prime number, its prime factored form is just 97.

3. **(a)** $\dfrac{8}{14} = \dfrac{4 \cdot 2}{7 \cdot 2} = \dfrac{4}{7}$

   (b) $\dfrac{35}{42} = \dfrac{5 \cdot 7}{6 \cdot 7} = \dfrac{5}{6}$

   (c) $\dfrac{120}{72} = \dfrac{5 \cdot 24}{3 \cdot 24} = \dfrac{5}{3}$

4. **(a)** $\dfrac{5}{8} \cdot \dfrac{2}{10} = \dfrac{5 \cdot 2}{8 \cdot 10}$ Multiply numerators; multiply denominators
   $= \dfrac{5 \cdot 2}{2 \cdot 4 \cdot 2 \cdot 5}$ Factor
   $= \dfrac{1}{8}$ Lowest terms

   (b) $\dfrac{1}{10} \cdot \dfrac{12}{5} = \dfrac{1 \cdot 12}{10 \cdot 5}$
   $= \dfrac{1 \cdot 2 \cdot 6}{2 \cdot 5 \cdot 5}$
   $= \dfrac{6}{25}$

   (c) $\dfrac{7}{9} \cdot \dfrac{12}{14} = \dfrac{7 \cdot 12}{9 \cdot 14}$
   $= \dfrac{7 \cdot 2 \cdot 2 \cdot 3}{3 \cdot 3 \cdot 2 \cdot 7}$
   $= \dfrac{2}{3}$

5. **(a)** $\dfrac{3}{10} \div \dfrac{2}{7} = \dfrac{3}{10} \cdot \dfrac{7}{2}$ Multiply by reciprocal of second fraction
   $= \dfrac{21}{20}$

   (b) $\dfrac{3}{4} \div \dfrac{7}{16} = \dfrac{3}{4} \cdot \dfrac{16}{7}$
   $= \dfrac{3 \cdot 4 \cdot 4}{7 \cdot 4} = \dfrac{12}{7}$

Chapter R Prealgebra Review

(c) $\dfrac{4}{3} \div 6 = \dfrac{4}{3} \div \dfrac{6}{1}$

$= \dfrac{4}{3} \cdot \dfrac{1}{6} = \dfrac{2 \cdot 2}{3 \cdot 2 \cdot 3} = \dfrac{2}{9}$

6. (a) $\dfrac{3}{5} + \dfrac{4}{5} = \dfrac{3+4}{5} = \dfrac{7}{5}$

(b) $\dfrac{5}{14} + \dfrac{3}{14} = \dfrac{5+3}{14}$

$= \dfrac{8}{14} = \dfrac{2 \cdot 4}{2 \cdot 7}$

$= \dfrac{4}{7}$

7. (a) $\dfrac{7}{30} + \dfrac{2}{45}$

Since $30 = 2 \cdot 3 \cdot 5$ and $45 = 3 \cdot 3 \cdot 5$, the least common denominator is

$2 \cdot 3 \cdot 3 \cdot 5 = 90.$

Write each fraction with a denominator of 90.

$\dfrac{7}{30} = \dfrac{7 \cdot 3}{30 \cdot 3} = \dfrac{21}{90}$

$\dfrac{2}{45} = \dfrac{2 \cdot 2}{45 \cdot 2} = \dfrac{4}{90}$

Now add.

$\dfrac{7}{30} + \dfrac{2}{45} = \dfrac{21}{90} + \dfrac{4}{90} = \dfrac{21+4}{90} = \dfrac{25}{90}$

25/90 can be written in lowest terms.

$\dfrac{25}{90} = \dfrac{5 \cdot 5}{5 \cdot 18} = \dfrac{5}{18}$

(b) $\dfrac{17}{10} + \dfrac{8}{27}$

Since $10 = 2 \cdot 5$ and $27 = 3 \cdot 3 \cdot 3$, the least common denominator is $2 \cdot 5 \cdot 3 \cdot 3 \cdot 3 = 270$. Write each fraction with a denominator of 270.

$\dfrac{17}{10} = \dfrac{17 \cdot 27}{10 \cdot 27} = \dfrac{459}{270}$

$\dfrac{8}{27} = \dfrac{8 \cdot 10}{27 \cdot 10} = \dfrac{80}{270}$

Now add.

$\dfrac{17}{10} + \dfrac{8}{27} = \dfrac{459}{270} + \dfrac{80}{270} = \dfrac{539}{270}$

(c) $2\dfrac{1}{8} + 1\dfrac{2}{3}$

Change both mixed numbers to improper fractions.

$2\dfrac{1}{8} = 2 + \dfrac{1}{8}$

$= \dfrac{16}{8} + \dfrac{1}{8} = \dfrac{17}{8}$

$1\dfrac{2}{3} = 1 + \dfrac{2}{3}$

$= \dfrac{3}{3} + \dfrac{2}{3} = \dfrac{5}{3}$

Thus,

$2\dfrac{1}{8} + 1\dfrac{2}{3} = \dfrac{17}{8} + \dfrac{5}{3}.$

The least common denominator is 24, so write each fraction with a denominator of 24.

$\dfrac{17}{8} = \dfrac{17 \cdot 3}{8 \cdot 3} = \dfrac{51}{24}$

$\dfrac{5}{3} = \dfrac{5 \cdot 8}{3 \cdot 8} = \dfrac{40}{24}$

Now add.

$\dfrac{17}{8} + \dfrac{5}{3} = \dfrac{51}{24} + \dfrac{40}{24} = \dfrac{51+40}{24}$

$= \dfrac{91}{24}$ or $3\dfrac{19}{24}$

8. (a) $\dfrac{9}{11} - \dfrac{3}{11} = \dfrac{9-3}{11} = \dfrac{6}{11}$

(b) $\dfrac{13}{15} - \dfrac{5}{6}$

Since $15 = 3 \cdot 5$ and $6 = 2 \cdot 3$, the least common denominator is $3 \cdot 5 \cdot 2 = 30$. Write each fraction with a denominator of 30.

$$\dfrac{13}{15} = \dfrac{13 \cdot 2}{15 \cdot 2} = \dfrac{26}{30}$$

$$\dfrac{5}{6} = \dfrac{5 \cdot 5}{6 \cdot 5} = \dfrac{25}{30}$$

Now subtract.

$$\dfrac{13}{15} - \dfrac{5}{6} = \dfrac{26}{30} - \dfrac{25}{30} = \dfrac{1}{30}$$

(c) $2\dfrac{3}{8} - 1\dfrac{1}{2}$

Change each mixed number into an improper fraction.

$$2\dfrac{3}{8} = 2 + \dfrac{3}{8}$$
$$= \dfrac{16}{8} + \dfrac{3}{8} = \dfrac{19}{8}$$

$$1\dfrac{1}{2} = 1 + \dfrac{1}{2}$$
$$= \dfrac{2}{2} + \dfrac{1}{2} = \dfrac{3}{2}$$

Thus,

$$2\dfrac{3}{8} - 1\dfrac{1}{2} = \dfrac{19}{8} - \dfrac{3}{2}.$$

The least common denominator is 8. Write each fraction with a denominator of 8. 19/8 remains unchanged, and

$$\dfrac{3}{2} = \dfrac{3 \cdot 4}{2 \cdot 4} = \dfrac{12}{8}.$$

Now subtract.

$$\dfrac{19}{8} - \dfrac{3}{2} = \dfrac{19}{8} - \dfrac{12}{8} = \dfrac{19 - 12}{8} = \dfrac{7}{8}$$

9. To find how many chairs can be upholstered, divide the total amount of fabric, 23 2/3 yards, by the amount needed for 1 chair, 2 1/4 yards.

$$23\dfrac{2}{3} \div 2\dfrac{1}{4} = \dfrac{71}{3} \div \dfrac{9}{4}$$
$$= \dfrac{71}{3} \cdot \dfrac{4}{9} = \dfrac{284}{27} = 10\dfrac{14}{27}$$

Since it is not possible to upholster 14/27 of a chair, only 10 chairs can be upholstered. There will be some fabric left over.

R.1 Section Exercises

3. $3\dfrac{4}{5} = 3 + \dfrac{4}{5}$
$$= \dfrac{3}{1} + \dfrac{4}{5}$$
$$= \dfrac{3 \cdot 5}{1 \cdot 5} + \dfrac{4}{5}$$
$$= \dfrac{15}{5} + \dfrac{4}{5}$$
$$= \dfrac{19}{5}$$

This fraction is in lowest terms since 19 and 5 have no common factors other than 1.

7. 17 is prime because it has exactly two different factors, 1 and 17.

11. Since 3458 can be divided by 2, it has more than two different factors, so it is composite.

15. To write 30 in prime factored form, first divide 30 by the smallest prime, 2.

$$30 = 2 \cdot 15$$

Since 15 can be factored as $3 \cdot 5$, we have

$$30 = 2 \cdot 3 \cdot 5,$$

where all factors are prime.

19. $124 = 2 \cdot 62$
$= 2 \cdot 2 \cdot 31$

23. $\frac{8}{16} = \frac{1 \cdot 8}{2 \cdot 8} = \frac{1}{2} \cdot \frac{8}{8} = \frac{1}{2} \cdot 1 = \frac{1}{2}$

27. $\frac{15}{45} = \frac{1 \cdot 15}{3 \cdot 15} = \frac{1}{3} \cdot \frac{15}{15} = \frac{1}{3} \cdot 1 = \frac{1}{3}$

31. $\frac{16}{24} = \frac{2 \cdot 8}{3 \cdot 8} = \frac{2}{3} \cdot \frac{8}{8} = \frac{2}{3} \cdot 1 = \frac{2}{3}$

Therefore, (c) is correct.

35. $\frac{1}{10} \cdot \frac{12}{5} = \frac{1 \cdot 12}{10 \cdot 5}$

$= \frac{6 \cdot 2}{2 \cdot 5 \cdot 5}$

$= \frac{6}{25}$

39. $2\frac{2}{3} \cdot 5\frac{4}{5}$

Change both mixed numbers to improper fractions.

$2\frac{2}{3} = 2 + \frac{2}{3} = \frac{6}{3} + \frac{2}{3} = \frac{8}{3}$

$5\frac{4}{5} = 5 + \frac{4}{5} = \frac{25}{5} + \frac{4}{5} = \frac{29}{5}$

$2\frac{2}{3} \cdot 5\frac{4}{5} = \frac{8}{3} \cdot \frac{29}{5}$

$= \frac{8 \cdot 29}{3 \cdot 5}$

$= \frac{232}{15}$ or $15\frac{7}{15}$

43. $\frac{32}{5} \div \frac{8}{15} = \frac{32}{5} \cdot \frac{15}{8}$ Multiply by reciprocal of second fraction

$= \frac{32 \cdot 15}{5 \cdot 8}$

$= \frac{8 \cdot 4 \cdot 3 \cdot 5}{5 \cdot 8}$

$= \frac{4 \cdot 3}{1} \cdot \frac{8 \cdot 5}{8 \cdot 5}$

$= 12 \cdot 1$

$= 12$

47. $2\frac{5}{8} \div 1\frac{15}{32}$

$2\frac{5}{8} = 2 + \frac{5}{8} = \frac{16}{8} + \frac{5}{8} = \frac{21}{8}$

$1\frac{15}{32} = 1 + \frac{15}{32} = \frac{32}{32} + \frac{15}{32} = \frac{47}{32}$

$2\frac{5}{8} \div 1\frac{15}{13} = \frac{21}{8} \div \frac{47}{32}$

$= \frac{21}{8} \cdot \frac{32}{47}$

$= \frac{21 \cdot 32}{8 \cdot 47}$

$= \frac{21 \cdot 8 \cdot 4}{8 \cdot 47}$

$= \frac{21 \cdot 4}{47} \cdot \frac{8}{8}$

$= \frac{84}{47} \cdot 1$

$= \frac{84}{47}$ or $1\frac{37}{47}$

51. $\dfrac{7}{12} + \dfrac{1}{12} = \dfrac{7+1}{12}$

$= \dfrac{8}{12}$

$= \dfrac{2 \cdot 4}{3 \cdot 4} = \dfrac{2}{3}$

55. $3\dfrac{1}{8} + \dfrac{1}{4}$

$3\dfrac{1}{8} = 3 + \dfrac{1}{8} = \dfrac{24}{8} + \dfrac{1}{8} = \dfrac{25}{8}$

$3\dfrac{1}{8} + \dfrac{1}{4} = \dfrac{25}{8} + \dfrac{1}{4}$

Since $8 = 2 \cdot 2 \cdot 2$ and $4 = 2 \cdot 2$, the LCD is $2 \cdot 2 \cdot 2$ or 8.

$3\dfrac{1}{8} + \dfrac{1}{4} = \dfrac{25}{8} + \dfrac{1 \cdot 2}{4 \cdot 2}$

$= \dfrac{25}{8} + \dfrac{2}{8}$

$= \dfrac{27}{8}$ or $3\dfrac{3}{8}$

59. $6\dfrac{1}{4} - 5\dfrac{1}{3}$

$6\dfrac{1}{4} = 6 + \dfrac{1}{4} = \dfrac{24}{4} + \dfrac{1}{4} = \dfrac{25}{4}$

$5\dfrac{1}{3} = 5 + \dfrac{1}{3} = \dfrac{15}{3} + \dfrac{1}{3} = \dfrac{16}{3}$

Since $4 = 2 \cdot 2$, the LCD is $2 \cdot 2 \cdot 3$ or 12.

$6\dfrac{1}{4} - 5\dfrac{1}{3} = \dfrac{25}{4} - \dfrac{16}{3}$

$= \dfrac{25 \cdot 3}{4 \cdot 3} - \dfrac{16 \cdot 4}{3 \cdot 4}$

$= \dfrac{75}{12} - \dfrac{64}{12}$

$= \dfrac{11}{12}$

63. Two dimes = \$.20 or $\dfrac{20}{100}$

Three dimes = \$.30 or $\dfrac{30}{100}$

One half-dollar = \$.50 or $\dfrac{50}{100}$

$2\left(\dfrac{10}{100}\right) + 3\left(\dfrac{10}{100}\right) = \dfrac{50}{100}$

67. The difference between the two measures is found by subtracting, using 16 as the LCD.

$\dfrac{3}{4} - \dfrac{3}{16} = \dfrac{3 \cdot 4}{4 \cdot 4} - \dfrac{3}{16}$

$= \dfrac{12}{16} - \dfrac{3}{16}$

$= \dfrac{9}{16}$

The difference between the two measures is 9/16 inch.

Section R.2   Decimals and Percents

R.2 Margin Exercises

1. (a) $.8 = \dfrac{8}{10}$

(b) $.431 = \dfrac{431}{1000}$

(c) $20.58 = \dfrac{2058}{100}$

2. (a) $\begin{array}{r} 68.9 \\ 42.72 \\ +\ 8.973 \\ \hline \end{array}$ becomes $\begin{array}{r} 68.900 \\ 42.720 \\ +\ 8.973 \\ \hline 120.593 \end{array}$.

(b) $\begin{array}{r} 32.5 \\ -\ 21.72 \\ \hline \end{array}$ becomes $\begin{array}{r} 32.50 \\ -\ 21.72 \\ \hline 10.78 \end{array}$.

Chapter R Prealgebra Review

(c) 42.83 + 71.629 + 3.074 becomes

$$\begin{array}{r} 42.830 \\ 71.629 \\ +\;3.074 \\ \hline 117.533 \end{array}.$$

(d) 351.8 − 2.706 becomes

$$\begin{array}{r} 351.800 \\ -\;\;\;2.706 \\ \hline 349.094 \end{array}.$$

3. (a) 2.13 × .05

$$\begin{array}{rl} 2.13 & 2 \text{ decimal places} \\ \times\;.05 & 2 \text{ decimal places} \\ \hline .1065 & 4 \text{ decimal places} \end{array}$$

(b) 69.32 × 1.4

$$\begin{array}{rl} 69.3\,2 & 2 \text{ decimal places} \\ \times\;\;\;\;1.4 & 1 \text{ decimal place} \\ \hline 27\,7\,2\,8 & 2 + 1 = 3 \\ 69\,3\,2\;\;\;\; & \\ \hline 97.0\,4\,8 & 3 \text{ decimal places} \end{array}$$

(c) 397.12 × .152

$$\begin{array}{rl} 397.12 & 2 \text{ decimal places} \\ \times\;\;\;\;.152 & 3 \text{ decimal places} \\ \hline 794\,24 & 2 + 3 = 5 \\ 19\,856\,0\;\; & \\ 39\,712\;\;\;\;\; & \\ \hline 60.362\,24 & 5 \text{ decimal places} \end{array}$$

(d) 42,980 × .012

$$\begin{array}{rl} 42{,}980 & 0 \text{ decimal places} \\ \times\;\;\;\;.012 & 3 \text{ decimal places} \\ \hline 85\,960 & 0 + 3 = 3 \\ 429\,80\;\;\; & \\ \hline 515.760 & 3 \text{ decimal places} \end{array}$$

or 515.76

4. (a) $32.3\,\overline{)481.27}$

To get the whole number 323 as the divisor, move the decimal point one place to the right. Then, move the decimal point the same number of places in 481.27 to get 4812.7.

$$32.3\underset{\curvearrowright}{\,.}\,\overline{)481.2\underset{\curvearrowright}{\,7\,}}$$

Bring the decimal point straight up, and divide as with whole numbers.

$$\begin{array}{r} 14.9 \\ 323\overline{)4812.7} \\ \underline{323\;\;\;\;\;} \\ 1582\;\; \\ \underline{1292\;\;} \\ 290\;7 \\ \underline{290\;7} \\ 0 \end{array}$$

(b) $.37\,\overline{)5.476}$

Move the decimal points two places to the right.

$$.37\underset{\curvearrowright}{.}\,\overline{)5.47\underset{\curvearrowright}{.}\,6}$$

Bring the decimal point straight up, and divide as with whole numbers.

$$\begin{array}{r} 14.8 \\ 37\overline{)547.6} \\ \underline{37\;\;\;\;} \\ 177\;\; \\ \underline{148\;\;} \\ 29\;6 \\ \underline{29\;6} \\ 0 \end{array}$$

(c) 375.125 ÷ 3.001 becomes

$$3.001\overline{)375.125}\,.$$

Move the decimal points three places to the right.

$$3.\underset{\curvearrowright}{001.}\,\overline{)375.\underset{\curvearrowright}{125.}}$$

Divide, bringing the decimal point straight up.

$$\begin{array}{r} 125. \\ 3001\overline{)375125.} \\ \underline{3001\;\;\;\;\;\;} \\ 7502\;\;\; \\ \underline{6002\;\;\;} \\ 15005 \\ \underline{15005} \\ 0 \end{array}$$

Thus,
$$375.125 \div 3.001 = 125.$$

5. **(a)** Divide the denominator, 9, into the numerator, 2. Attach zeros after the decimal point of the numerator as needed.

$$\begin{array}{r} .2222 \\ 9\overline{)2.0000...} \\ \underline{1\ 8} \\ 20 \\ \underline{18} \\ 20 \\ \underline{18} \\ 20 \\ \underline{18} \\ 2 \end{array}$$

The decimal will keep repeating the digit 2, so

$$\frac{2}{9} = .\overline{2}.$$

To the nearest thousandth,

$$\frac{2}{9} = .222.$$

(b)
$$\begin{array}{r} .85 \\ 20\overline{)17.00} \\ \underline{16\ 0} \\ 1\ 00 \\ \underline{1\ 00} \\ 0 \end{array}$$

$$\frac{17}{20} = .85$$

(c)
$$\begin{array}{r} .625 \\ 8\overline{)5.000} \\ \underline{4\ 8} \\ 20 \\ \underline{16} \\ 40 \\ \underline{40} \\ 0 \end{array}$$

$$\frac{5}{8} = .625$$

(d)
$$\begin{array}{r} .1428571 \\ 7\overline{)1.0000000...} \\ \underline{7} \\ 30 \\ \underline{28} \\ 20 \\ \underline{14} \\ 60 \\ \underline{56} \\ 40 \\ \underline{35} \\ 50 \\ \underline{49} \\ 10 \end{array}$$

The decimal has started to repeat.

$$\frac{1}{7} = .\overline{142857}$$

To the nearest thousandth,

$$\frac{1}{7} = .143.$$

6. **(a)** To convert 23% to a decimal, move the decimal point two places to the left and drop the percent sign.

$$23\% = .23$$

(b) $310\% = 3.10$

(c) To convert .71 to a percent, move the decimal point two places to the right and attach a percent sign.

$$.71 = 71\%$$

(d) $1.32 = 132\%$

Chapter R Prealgebra Review

R.2 Section Exercises

3. The best estimate for the sum 35.89 + 24.1 is (c) 60, since 35 + 24 is 59 and .89 + .1 is .99, which is almost 1.

7. $.4 = \dfrac{4}{10}$

11. $.138 = \dfrac{138}{1000}$

15. 25.32 + 10.92 + 85.74 + 29.826

$$\begin{array}{r} 25.32 \\ 10.92 \\ 85.74 \\ + \ 29.826 \\ \hline \end{array} \quad \text{becomes} \quad \begin{array}{r} 25.320 \\ 10.920 \\ 85.740 \\ + \ 29.826 \\ \hline 151.806 \end{array}$$

19. 43.5 − 28.17

$$\begin{array}{r} 43.5 \\ -\ 28.17 \\ \hline \end{array} \quad \text{becomes} \quad \begin{array}{r} 43.50 \\ -\ 28.17 \\ \hline 15.33 \end{array}$$

23.
$$\begin{array}{r} 7.56 \\ -\ 2.789 \\ \hline \end{array} \quad \text{becomes} \quad \begin{array}{r} 7.560 \\ -\ 2.789 \\ \hline 4.771 \end{array}$$

27. 34.045 × .56

$$\begin{array}{r} 34.045 \\ \times \quad .56 \\ \hline 2\ 04\ 270 \\ 17\ 02\ 25 \\ \hline 19.06\ 520 \end{array} \quad \begin{array}{l} 3\ decimal\ places \\ 2\ decimal\ places \\ \\ \\ 5\ decimal\ places \end{array}$$

or 19.0652

31. 24.837 ÷ 9.74

$$\begin{array}{r} 2.55 \\ 9.74.\overline{)24.83.70} \\ \underline{19\ 48} \\ 5\ 35\ 7 \\ \underline{4\ 87\ 0} \\ 48\ 70 \\ \underline{48\ 70} \\ 0 \end{array}$$

35. (a) 46.249 rounded to hundredths is 46.25.

(b) 46.249 rounded to tenths is 46.2.

(c) 46.249 rounded to ones or units is 46.

(d) 46.249 rounded to tens is 50.

39. $\dfrac{1}{4}$

$$\begin{array}{r} .25 \\ 4\overline{)1.00} \\ \underline{8} \\ 20 \\ \underline{20} \\ 0 \end{array}$$

$\dfrac{1}{4} = .25$

43. $\dfrac{3}{7}$

$$\begin{array}{r} .4285714 \\ 7\overline{)3.0000000} \\ \underline{2\ 8} \\ 20 \\ \underline{14} \\ 60 \\ \underline{56} \\ 40 \\ \underline{35} \\ 50 \\ \underline{49} \\ 10 \\ \underline{7} \\ 30 \end{array}$$

The decimal has started to repeat.

$\frac{3}{7} = .\overline{428571}$ or, rounding to the nearest thousandth, .429.

47. $\frac{1}{6}$

$$\begin{array}{r} .166 \\ 6\overline{)1.000} \\ \underline{6} \\ 40 \\ \underline{36} \\ 40 \\ \underline{36} \\ 4 \end{array}$$

$\frac{1}{6} = .1\overline{6}$ or, rounding to the nearest thousandth, .167.

51. $54\% = 54 \cdot 1\% = 54 \cdot .01 = .54$

55. $2.4\% = 2.4 \cdot 1\% = 2.4 \cdot .01 = .024$

59. $.75 = 75 \cdot .01 = 75 \cdot 1\% = 75\%$

63. $1.28 = 128 \cdot .01 = 128 \cdot 1\% = 128\%$

67. $\frac{3}{4}$

$$\begin{array}{r} .75 \\ 4\overline{)3.00} \\ \underline{28} \\ 20 \\ \underline{20} \\ 0 \end{array}$$

$\frac{3}{4} = .75 = 75 \cdot .01 = 75 \cdot 1\% = 75\%$

71. $\frac{5}{6}$

$$\begin{array}{r} .833 \\ 6\overline{)5.000} \\ \underline{48} \\ 20 \\ \underline{18} \\ 20 \\ \underline{18} \\ 2 \end{array}$$

$\frac{5}{6} = .8\overline{3} = 83.\overline{3} \cdot .01 = 83.\overline{3} \cdot 1\% = 83.\overline{3}\%$

CHAPTER 1   THE REAL NUMBER SYSTEM

Section 1.1   Exponents, Order of Operations, and Inequality

1.1 Margin Exercises

1. (a) $6^2 = 6 \cdot 6$ 6 used as a factor 2 times
   $= 36$

   (b) $3^5 = 3 \cdot 3 \cdot 3 \cdot 3 \cdot 3$ 3 used as a factor 5 times
   $= 243$

   (c) $(\frac{3}{4})^2 = \frac{3}{4} \cdot \frac{3}{4}$ 3/4 used as a factor 2 times
   $= \frac{9}{16}$

   (d) $(\frac{1}{2})^4 = \frac{1}{2} \cdot \frac{1}{2} \cdot \frac{1}{2} \cdot \frac{1}{2}$
   $= \frac{1}{16}$

   (e) $(.4)^3 = (.4)(.4)(.4)$
   $= .064$

2. (a) $3 \cdot 8 + 7 = 24 + 7$ Multiply
   $= 31$ Add

   (b) $9 + 12 \cdot 6 = 9 + 72$ Multiply
   $= 81$ Add

3. (a) $2 \cdot 9 + 7 \cdot 3 = 18 + 21$ Multiply
   $= 39$ Add

   (b) $7 \cdot 6 - 3(8 + 1) = 7 \cdot 6 - 3(9)$ Add inside parentheses
   $= 42 - 27$ Multiply
   $= 15$ Subtract

   (c) $\frac{2(7 + 8) + 2}{3 \cdot 5 + 1}$

   Simplify the numerator and denominator separately.

   $= \frac{2(15) + 2}{3 \cdot 5 + 1}$ Add inside parentheses

   $= \frac{30 + 2}{15 + 1}$ Multiply

   $= \frac{32}{16}$ Add

   $= 2$ Divide

   (d) $2 + 3^2 - 5 = 2 + 9 - 5$ Use exponent
   $= 11 - 5$ Add
   $= 6$ Subtract

4. (a) $4[7 + 3(6 + 1)]$
   $= 4[7 + 3(7)]$ Add inside parentheses
   $= 4[7 + 21]$ Multiply inside brackets
   $= 4[28]$ Add inside brackets
   $= 112$ Multiply

   (b) $9[(4 + 8) - 3]$
   $= 9[12 - 3]$ Add inside parentheses
   $= 9[9]$ Subtract inside brackets
   $= 81$ Multiply

5. (a) $7 < 5$ is written "Seven is less than five." This statement is false.

   (b) $12 > 6$ is written "Twelve is greater than six." This statement is true.

   (c) $4 \neq 10$ is written "Four is not equal to ten." This statement is true.

(d) 28 ≠ 4 · 7 is written "Twenty-eight is not equal to four times seven." Since 4 · 7 = 28, this statement is false.

6. (a) The statement 30 ≤ 40 is true since 30 < 40.

   (b) 25 ≥ 10 is a true statement since 25 > 10.

   (c) 40 ≤ 10 is false since 40 > 10.

   (d) 21 ≤ 21 is true since 21 = 21.

   (e) 3 ≥ 3 is true since 3 = 3.

7. (a) "Nine equals eleven minus two" is written 9 = 11 - 2.

   (b) "Seventeen is less than thirty" is written 17 < 30.

   (c) "Eight is not equal to ten" is written 8 ≠ 10.

   (d) "Fourteen is greater than twelve" is written 14 > 12.

   (e) "Thirty is less than or equal to fifty" is written 30 ≤ 50.

   (f) "Two is greater than or equal to two" is written 2 ≥ 2.

8. (a) 8 < 10 may be written as 10 > 8.

   (b) 3 > 1 may be written as 1 < 3.

   (c) 9 ≤ 15 may be written as 15 ≥ 9.

   (d) 6 ≥ 2 may be written as 2 ≤ 6.

1.1 Section Exercises

3. The inequality symbol should always point to the smaller number. The statement is true.

7. $12^2 = 12 \cdot 12 = 144$

11. $10^3 = 10 \cdot 10 \cdot 10 = 1000$

15. $4^5 = 4 \cdot 4 \cdot 4 \cdot 4 \cdot 4 = 1024$

19. $(.04)^3 = (.04)(.04)(.04) = .000064$

23. $9 \cdot 5 - 13 = 45 - 13$ *Subtract*
    $= 32$

27. $9 \cdot 4 - 8 \cdot 3 = 36 - 24$ *Multiply*
    $\qquad\qquad\quad\; = 12$ *Subtract*

31. $5[3 + 4(2^2)]$
    $= 5[3 + 4(4)]$ *Use the exponent*
    $= 5(3 + 16)$ *Multiply*
    $= 5(19)$ *Add*
    $= 95$ *Multiply*

35. $\dfrac{6(3^2 - 1) + 8}{3 \cdot 2 - 2}$

    Simplify the numerator and denominator separately.

    $\dfrac{6(3^2 - 1) + 8}{3 \cdot 2 - 2} = \dfrac{6(9 - 1) + 8}{6 - 2}$

    $= \dfrac{6(8) + 8}{4}$

    $= \dfrac{48 + 8}{4}$

    $= \dfrac{56}{4} = 14$

43. "$8 \geq 17$" means "8 is greater than or equal to 17." The statement is false since 8 is less than 17.

47. $6 \cdot 8 + 6 \cdot 6 \geq 0$
$48 + 36 \geq 0$
$84 \geq 0$

The statement is true since 84 is greater than zero.

51. $\frac{9(7-1) - 8 \cdot 2}{4(6-1)} > 3$

$\frac{9 \cdot 6 - 8 \cdot 2}{4(5)} > 3$

$\frac{54 - 16}{20} > 3$

$\frac{38}{20} > 3$

$\frac{19}{10} > 3$

The statement is false since 19/10 or 1 9/10 is less than 3.

55. "Fifteen is equal to five plus ten" is written

$15 = 5 + 10.$

59. "Sixteen is not equal to nineteen" is written

$16 \neq 19.$

63. "$7 < 19$" means "seven is less than nineteen."

The statement is true.

67. "$8 \geq 11$" means "eight is greater than or equal to eleven."

The statement is false.

71. $5 < 30$ becomes $30 > 5$ when the inequality symbol is reversed.

75. $(.53)^2 = (.53)(.53) = .2809$

Section 1.2 Variables, Expressions, and Equations

1.2 Margin Exercises

1. Replace p with 3 in each expression.

(a) $6p = 6(3)$
$= 18$

(b) $p + 12 = 3 + 12$
$= 15$

(c) $5p^2 = 5(3^2)$
$= 5 \cdot 9$
$= 45$

2. Replace x with 6 and y with 9 in each expression.

(a) $4x + 7y = 4(6) + 7(9)$
$= 24 + 63$ *Multiply*
$= 87$ *Add*

(b) $\frac{4x - 2y}{x + 1} = \frac{4(6) - 2(9)}{6 + 1}$

$= \frac{24 - 18}{7}$ *Multiply*

$= \frac{6}{7}$ *Subtract*

(c) $2x^2 + y^2 = 2 \cdot 6^2 + 9^2$
$= 2 \cdot 36 + 81$ *Use exponents*
$= 72 + 81$ *Multiply*
$= 153$ *Add*

3. (a) "The sum of 5 and a number" translates as 5 + x.

(b) "A number minus 4" translates as x − 4.

(c) "A number subtracted from 48" translates as 48 − x.

(d) "The product of 6 and a number" translates as 6 · x or 6x.

(e) "9 multiplied by the sum of a number and 5" translates as 9(x + 5).

4. (a) p − 1 = 3; 2
$\quad$ 2 − 1 = 3 $\quad$ *Replace p with 2*
$\quad\quad\quad$ 1 = 3 $\quad$ *False*

The number 2 is not a solution.

(b) $\quad$ 2k + 3 = 15; 7
$\quad$ 2 · 7 + 3 = 15 $\quad$ *Replace k with 7*
$\quad\quad$ 14 + 3 = 15
$\quad\quad\quad\quad$ 17 = 15 $\quad$ *False*

The number 7 is not a solution.

(c) $\quad$ 8p − 11 = 5; 2
$\quad$ 8 · 2 − 11 = 5 $\quad$ *Replace p with 2*
$\quad\quad$ 16 − 11 = 5
$\quad\quad\quad\quad$ 5 = 5 $\quad$ *True*

The number 2 is a solution.

5. (a) $\quad$ The sum of a number and 13 $\quad$ is $\quad$ 19.
$\quad\quad\quad\quad\downarrow\quad\quad\quad\downarrow\quad\quad\downarrow$
$\quad\quad\quad\quad$ x + 13 $\quad$ = $\quad$ 19

Try each number from the given domain {0, 2, 4, 6, 8, 10}. Since
$\quad\quad$ 6 + 13 = 19,
6 is a solution.

(b) "Three times a number is subtracted from 21, giving 15" translates

$\quad\quad$ 21 − 3x = 15.

Try each number from the domain {0, 2, 4, 6, 8, 10}. Since
$\quad\quad$ 21 − 3 · 2 = 21 − 6 = 15,
2 is a solution.

6. (a) 2x + 5y − 7 has no equals sign, so it is an expression.

(b) $\dfrac{3x - 1}{5}$ has no equals sign, so it is an expression.

(c) 2x + 5 = 7 has an equals sign, so it is an equation.

(d) $\dfrac{x}{y - 3}$ = 4x has an equals sign, so it is an equation.

1.2 Section Exercises

3. 7t + 2(t + 1) = 4

Because of the equals sign, this is an equation.

7. $2x^3 = 2 \cdot x \cdot x \cdot x$, which is not the same as $2x \cdot 2x \cdot 2x = 8x^3$.

11. 5x

(a) Replace x with 4 to get
$\quad\quad$ 5x = 5(4) = 20.

(b) Replace x with 6 to get
$\quad\quad$ 5x = 5(6) = 30.

14   Chapter 1   The Real Number System

15. $\dfrac{x + 1}{3}$

   (a) Replace x with 4 to get
   $$\dfrac{x + 1}{3} = \dfrac{4 + 1}{3} = \dfrac{5}{3}.$$

   (b) Replace x with 6 to get
   $$\dfrac{x + 1}{3} = \dfrac{6 + 1}{3} = \dfrac{7}{3}.$$

19. $3x^2 + x$

   (a) Replace x with 4 to get
   $$3x^2 + x = 3 \cdot 4^2 + 4$$
   $$= 3 \cdot 16 + 4$$
   $$= 48 + 4 = 52.$$

   (b) Replace x with 6 to get
   $$3x^2 + x = 3 \cdot 6^2 + 6$$
   $$= 3 \cdot 36 + 6$$
   $$= 108 + 6 = 114.$$

23. $8x + 3y + 5$

   (a) Replace x with 2 and y with 1 to get
   $$8x + 3y + 5 = 8(2) + 3(1) + 5$$
   $$= 16 + 3 + 5$$
   $$= 19 + 5 = 24.$$

   (b) Replace x with 1 and y with 5 to get
   $$8x + 3y + 5 = 8(1) + 3(5) + 5$$
   $$= 8 + 15 + 5$$
   $$= 23 + 5 = 28.$$

27. $x + \dfrac{4}{y}$

   (a) Replace x with 2 and y with 1 to get
   $$x + \dfrac{4}{y} = 2 + \dfrac{4}{1}$$
   $$= 2 + 4 = 6.$$

   (b) Replace x with 1 and y with 5 to get
   $$x + \dfrac{4}{y} = 1 + \dfrac{4}{5}$$
   $$= \dfrac{5}{5} + \dfrac{4}{5} = \dfrac{9}{5}.$$

31. $\dfrac{2x + 4y - 6}{5y + 2}$

   (a) Replace x with 2 and y with 1 to get
   $$\dfrac{2x + 4y - 6}{5y + 2} = \dfrac{2(2) + 4(1) - 6}{5(1) + 2}$$
   $$= \dfrac{4 + 4 - 6}{5 + 2}$$
   $$= \dfrac{8 - 6}{7}$$
   $$= \dfrac{2}{7}.$$

   (b) Replace x with 1 and y with 5 to get
   $$\dfrac{2x + 4y - 6}{5y + 2} = \dfrac{2(1) + 4(5) - 6}{5(5) + 2}$$
   $$= \dfrac{2 + 20 - 6}{25 + 2}$$
   $$= \dfrac{22 - 6}{27}$$
   $$= \dfrac{16}{27}.$$

1.2 Section Exercises

35. $\dfrac{3x + y^2}{2x + 3y}$

(a) Replace x with 2 and y with 1 to get

$$\dfrac{3x + y^2}{2x + 3y} = \dfrac{3(2) + 1^2}{2(2) + 3(1)}$$

$$= \dfrac{3(2) + 1}{4 + 3}$$

$$= \dfrac{6 + 1}{7}$$

$$= \dfrac{7}{7} = 1.$$

(b) Replace x with 1 and y with 5 to get

$$\dfrac{3x + y^2}{2x + 3y} = \dfrac{3(1) + 5^2}{2(1) + 3(5)}$$

$$= \dfrac{3(1) + 25}{2 + 15}$$

$$= \dfrac{3 + 25}{17} = \dfrac{28}{17}.$$

39. "Twelve times a number" translates as $12 \cdot x$ or $12x$.

43. "Two subtracted from a number" translates as $x - 2$.

47. "The difference between a number and 6" translates as $x - 6$.

51. "The product of 6 and four less than a number" translates as $6(x - 4)$.

55. $2x + y = 6$ is true for $x = 0$, $y = 6$ and for $x = 1$, $y = 4$. Other pairs are possible.

59. $5m + 2 = 7$; 1

$5(1) + 2 = 7$ Let $m = 1$
$5 + 2 = 7$
$7 = 7$ True

Because substituting 1 for m results in a true statement, 1 is a solution of the equation.

63. $6p + 4p + 9 = 11$; $\dfrac{1}{5}$

$6\left(\dfrac{1}{5}\right) + 4\left(\dfrac{1}{5}\right) + 9 = 11$ Let $p = 1/5$

$\dfrac{6}{5} + \dfrac{4}{5} + 9 = 11$

$\dfrac{10}{5} + 9 = 11$

$2 + 9 = 11$

$11 = 11$ True

The true result shows that 1/5 is a solution of the equation.

67. $\dfrac{z + 4}{2 - z} = \dfrac{13}{5}$; $\dfrac{1}{3}$

$\dfrac{\frac{1}{3} + 4}{2 - \frac{1}{3}} = \dfrac{13}{15}$ Let $z = 1/3$

$\dfrac{\frac{1}{3} + \frac{12}{3}}{\frac{6}{3} - \frac{1}{3}} = \dfrac{13}{15}$

$\dfrac{\frac{13}{3}}{\frac{5}{3}} = \dfrac{13}{15}$

$\dfrac{13}{3} \cdot \dfrac{3}{5} = \dfrac{13}{15}$

$\dfrac{13}{15} = \dfrac{13}{15}$ True

The true result shows that 1/3 is a solution of the equation.

16 Chapter 1 The Real Number System

71. "Sixteen minus three-fourths of a number is 13" translates as

$$16 - \frac{3}{4}x = 13.$$

Try each number from the given domain, $\{0, 2, 4, 6, 8, 10\}$ in turn.

$16 - \frac{3}{4}x = 13$ *Given equation*

$16 - \frac{3}{4}(0) = 13$ *False*

$16 - \frac{3}{4}(2) = 13$ *False*

$16 - \frac{3}{4}(4) = 13$ *True*

$16 - \frac{3}{4}(6) = 13$ *False*

$16 - \frac{3}{4}(8) = 13$ *False*

$16 - \frac{3}{4}(10) = 13$ *False*

The only solution is 4.

75. "Three times a number is equal to 8 more than twice the number" translates as

$$3x = 2x + 8.$$

Try each number from the given domain.

$3x = 2x + 8$ *Given equation*
$3(0) = 2(0) + 8$ *False*
$3(2) = 2(2) + 8$ *False*
$3(4) = 2(4) + 8$ *False*
$3(6) = 2(6) + 8$ *False*
$3(8) = 2(8) + 8$ *True*
$3(10) = 2(10) + 8$ *False*

The only solution is 8.

Section 1.3 Real Numbers and the Number Line

1.3 Margin Exercises

1. **(a)** Since Erin spent $53 more than she has in her checking account, her balance is -53.

   (b) Since the record high was 134° above zero, this temperature is expressed as 134.

   (c) Positive numbers are used to represent gains and negative numbers to represent losses, so a gain of 5 yards, followed by a loss of 10 yards, is expressed as 5, -10.

2. Graph -3, -2.75, -3/4, 1 1/2, and 17/8.

   To compare these numbers and locate them on the number line, rewrite -2.75 and 17/8 as mixed numbers.

   $$-2.75 = -2\frac{3}{4}$$

   $$\frac{17}{8} = 2\frac{1}{8}$$

   The order of the numbers from smallest to largest is

   -3, -2.75, -3/4, 1 1/2, 17/8.

   See the graph given as the answer for this margin exercise in the textbook.

3. **(a)** $-2 < 4$

   This statement is true because any negative number is smaller than any positive number.

(b) $6 > -3$

This statement is true because any positive number is greater than any negative number.

(c) $-9 < -12$

Since -9 is to the right of -12 on the number line, -9 *is greater than* -12. Therefore, the statement $-9 < -12$ is false.

(d) $-4 \geq -1$

Since -4 is to the left of -1 on the number line, -4 is smaller than -1. Therefore, the statement $-4 \geq -1$ is false.

(e) $-6 \leq 0$

This statement is true because any negative number is smaller than 0.

4.

	Number	*Opposite*
(a)	6	-6
(b)	15	-15
(c)	-9	$-(-9) = 9$
(d)	-12	$-(-12) = 12$
(e)	0	0

5. (a) $|-6| = 6$

(b) $|9| = 9$

(c) $-|15| = -(15)$ *Find absolute value first*

$ = -15$ *Find opposite*

(d) $-|-9| = -(9)$ *Find absolute value*

$ = -9$ *Find opposite*

(e) $-|32 - 2| = -|30|$ *Subtract inside absolute value bars*

$ = -(30)$ *Find absolute value*

$ = -30$ *Find opposite*

1.3 Section Exercises

3. Every rational number is a real number.

The real numbers are made up of all rational and irrational numbers, so the statement is true.

7. Some real numbers are not rational.

The real numbers include the irrational numbers as well as the rational numbers. Some examples of irrational numbers (real numbers that are not rational) are $\sqrt{2}$, $-\sqrt{5}$, and π. The statement is true.

11. A decrease of 1760 is represented by -1760.

15. A height of 8300 feet is represented by 8300.

19. $0, 3, -5, -6$

Place a dot on the number line at the point that corresponds to each number. The order of the numbers from smallest to largest is $-6, -5, 0, 3$. See the graph in the answer section of the textbook.

23. $\frac{1}{4}$, $2\frac{1}{2}$, $-3\frac{4}{5}$, -4, $-1\frac{5}{8}$

Place a dot on the number line at the point that corresponds to each number. The order of the numbers from smallest to largest is -4, $-3\ 4/5$, $-1\ 5/8$, $1/4$, $2\ 1/2$. See the graph in the answer section of the textbook.

27. -21, 1

Since -21 is to the left of 1 on the number line, -21 is the smaller number.

31. $-\frac{2}{3}$, $-\frac{1}{4}$

Since $-2/3$ is to the left of $-1/4$ on the number line, $-2/3$ is the smaller number.

35. $-3 < -2$

This statement is true since -3 is to the left of -2 on the number line.

39. (a) The opposite of 6 is -6.

(b) The distance between 0 and 6 on the number line is 6 units, so the absolute value of 6 is 6.

43. $|-7| = 7$

47. $-|12| = -12$

Notice that $-|12|$ and $|-12|$ are not the same.

51. $|13 - 4| = |9| = 9$

55. $\frac{3}{4} + \frac{9}{10} = \frac{3 \cdot 5}{4 \cdot 5} + \frac{9 \cdot 2}{10 \cdot 2}$

$= \frac{15}{20} + \frac{18}{20}$

$= \frac{33}{20}$

Section 1.4 Addition of Real Numbers
1.4 Margin Exercises

1. In this exercise, refer to the number lines included with the answers to the margin exercises in the textbook.

   (a) $1 + 4$
   Start at 0 on a number line. Draw an arrow 1 unit to the right to represent the addition of a positive number. From the right end of this arrow, draw a second arrow 4 units to the right. The number below the end of this second arrow is 5, so

   $$1 + 4 = 5.$$

   (b) $-2 + (-5)$

   Start at 0 on a number line. Draw an arrow 2 units to the left to represent the addition of a negative number. From the left end of this arrow, draw a second arrow 5 units to the left. The number below the end of this second arrow is −7, so

$$-2 + (-5) = -7.$$

2. **(a)** $-7 + (-3) = -(|-7| + |-3|)$
    $= -(7 + 3)$
    $= -10$

   (b) $-12 + (-18) = -(|-12| + |-18|)$
    $= -(12 + 18)$
    $= -30$

   (c) $-15 + (-4) = -19$ *Sum of two negative numbers is negative*

3. In this exercise, refer to the number lines included with the answers to the margin exercises in the textbook.

   (a) $6 + (-3)$

   Start at 0 on a number line, and draw an arrow 6 units to the right. From the right end of this arrow, draw a second arrow 3 units to the left. The number below the end of this second arrow is 3, so

   $$6 + (-3) = 3.$$

   (b) $-5 + 1$

   Start at 0 on a number line, and draw an arrow 5 units to the left. From the left end of this arrow, draw a second arrow 1 unit to the right. The number below the end of this second arrow is −4, so

   $$-5 + 1 = -4.$$

4. **(a)** $-8 + 2 = -6$

   The number with the greater absolute value is −8, so the sum will be negative. The answer of −6 is correct.

   (b) $-15 + 4 = -11$

   The answer is correct.

   (c) $17 + (-10) = 7$

   The number with the greater absolute value is 17, so the sum will be positive. The answer of 7 is correct.

   (d) $\frac{3}{4} + \left(-\frac{11}{8}\right) = -\frac{5}{8}$

   To check this answer, use a common denominator of 8.

   $$\frac{6}{8} + \left(-\frac{11}{8}\right) = -\frac{5}{8}$$

   The answer is correct.

   (e) $-9.5 + 3.8 = -5.7$

   The difference between the absolute values of −9.5 and 3.8 is 5.7. Since $|-9.5| > |3.8|$, the sum will be negative. The answer is correct.

5. **(a)** $2 + [7 + (-3)] = 2 + 4$ *Work inside brackets*
       $= 6$

   (b) $6 + [(-2 + 5) + 7]$
       $= 6 + [3 + 7]$ *Work inside parentheses*
       $= 6 + 10$ *Work inside brackets*
       $= 16$

(c) $-9 + [-4 + (-8 + 6)]$
    $= -9 + [-4 + (-2)]$ *Work inside parentheses*
    $= -9 + (-6)$ *Work inside brackets*
    $= -15$

6. (a) "4 more than -12" is written
    $-12 + 4 = -8.$

   (b) "The sum of 6 and -7" is written
    $6 + (-7) = -1.$

   (c) "-12 added to -31" is written
    $-31 + (-12) = -43.$

   (d) "7 increased by the sum of 8 and -3" is written
    $7 + [8 + (-3)] = 7 + 5$ *Work inside brackets*
    $= 12.$

7. Represent the losses by negative numbers and the gain by positive numbers.
    $-8 + (-5) + 7$
    $= [-8 + (-5)] + 7$ *Add from left to right*
    $= -13 + 7$
    $= -6$
   The team lost 6 yards.

1.4 Section Exercises

3. To simplify the expression $8 + [-2 + (-3 + 5)]$, I should begin by adding -3 and 5, according to the rules for order of operations.

7. $7 + (-10)$

   Because the numbers have different signs, find the difference between their absolute values:
    $10 - 7 = 3.$
   Because -10 has the larger absolute value, the sum is negative:
    $7 + (-10) = -3.$

11. $-10 + (-3)$

    Because the numbers have the same sign, add their absolute values:
    $10 + 3 = 13.$
    Because both numbers are negative, their sum is negative:
    $-10 + (-3) = -13.$

15. $-8 + 7 = -1$

19. $10 + [-3 + (-2)] = 10 + (-5) = 5$

23. $-8 + [3 + (-1) + (-2)]$
    $= -8 + [2 + (-2)]$
    $= -8 + 0$
    $= -8$

27. $\frac{5}{8} + \left(-\frac{17}{12}\right) = \frac{5 \cdot 3}{8 \cdot 3} + \left(-\frac{17 \cdot 2}{12 \cdot 2}\right)$
    $= \frac{15}{24} + \left(-\frac{34}{24}\right)$
    $= -\frac{19}{24}$

1.4 Section Exercises

31. $7.8 + (-9.4) = -1.6$

35. $[-8 + (-3)] + [-7 + (-7)]$
$= (-11) + (-14)$
$= -25$

39. $-5 + 0 = -5$

Since $-5 + 0$ is -5, the statement is true.

43. $-10 + 6 + 7 = -3$

Since $-10 + 6 + 7$ is 3, the statement is false.

47. $|-8 + 10| = -8 + (-10)$

Since $|-8 + 10|$ is 2 and $-8 + (-10)$ is -18, the statement is false.

51. $-7 + [-5 + (-3)] = [(-7) + (-5)] + 3$

Since
$$-7 + [-5 + (-3)] = -7 + (-8) = -15$$
and
$$[(-7) + (-5)] + 3 = -12 + 3 = -9,$$
the statement is false.

55. "14 added to the sum of -19 and -4" is written $[-19 + (-4)] + 14$.

$[-19 + (-4)] + 14 = (-23) + 14 = -9$

59. "4 more than the sum of 8 and -18" is written $[8 + (-18)] + 4$.

$[8 + (-18)] + 4 = (-10) + 4 = -6$

63. Distances below the surface are represented by negative numbers.

$0 + (-130) + (-54) = -184$

Their altitude is -184 meters.

67. Gains are represented by positive numbers and losses by negative numbers.

$6 + (-12) + 43 = -6 + 43 = 37$

The total net yardage is 37 yards.

71. Use negative numbers to represent amounts Kim owes and positive numbers to represent payments and credits.

$870.00 + 35.90 + 150.00 + (-.82.50)$
$+ (-10.00) + (-10.00) + 500.00$

Kim still owes $286.60.

In Exercises 75 and 79, mentally replace the variable with each number from the given domain until you find a number that makes the statement true.

75. $p + 4 = 1$

The solution is -3 since $-3 + 4 = 1$.

79. $b + (-6) = -6$

The solution is 0 since
$0 + (-6) = -6$.

83. $8[(9 + 2) - 4] = 8(11 - 4)$
$= 8(7) = 56$

Section 1.5 Subtraction of Real Numbers

1.5 Margin Exercises

1. In this exercise, refer to the number lines included with the answers to the margin exercises in the textbook.

    (a) $5 - 1$
    Begin at 0 on a number line, and draw an arrow 5 units to the right. From the right end of this arrow, draw an arrow 1 unit to the left. The number at the end of the second arrow is 4, so
    $$5 - 1 = 4.$$

    (b) $6 - 2$
    Begin at 0 on a number line, and draw an arrow 6 units to the right. From the right end of this arrow, draw an arrow 2 units to the left. The number at the end of the second arrow is 4, so
    $$6 - 2 = 4.$$

2. **(a)** $6 - 10 = 6 + (-10) = -4$
    (b) $-2 - 4 = -2 + (-4) = -6$
    (c) $3 - (-5) = 3 + 5 = 8$
    (d) $-8 - (-12) = -8 + 12 = 4$

3. **(a)** $2 - [(-3) - (4 + 6)]$
    $= 2 - [(-3) - 10]$
    $= 2 - [(-3) + (-10)]$
    $= 2 - (-13)$
    $= 2 + 13$
    $= 15$

    (b) $[(5 - 7) + 3] - 8$
    $= [(-2) + 3] - 8$
    $= 1 - 8$
    $= 1 + (-8)$
    $= -7$

    (c) $6 - [(-1 - 4) - 2]$
    $= 6 - [(-5) - 2]$
    $= 6 - (-7)$
    $= 6 + 7$
    $= 13$

4. **(a)** "The difference between -5 and -12" is written
    $$-5 - (-12) = -5 + 12$$
    $$= 7.$$

    (b) "-2 subtracted from the sum of 4 and -4" is written
    $$[4 + (-4)] - (-2) = 0 - (-2)$$
    $$= 0 + 2$$
    $$= 2.$$

    (c) "7 less than -2" is written
    $$-2 - 7 = -2 + (-7)$$
    $$= -9.$$

    (d) "9, decreased by 10 less than 7" is written
    $$9 - (7 - 10) = 9 - [7 + (-10)]$$
    $$= 9 - (-3)$$
    $$= 9 + 3$$
    $$= 12.$$

5. The difference between the highest and lowest elevations is given by
    $$6960 - (-40) = 6960 + 40$$
    $$= 7000.$$
    The difference is 7000 meters.

1.5 Section Exercises

3. In order to simplify

   $6 - [(7 - 8) - (8 - 12)]$

   according to the rules for the order of operations, we should begin by subtracting 8 from 7.

In Exercises 7-35, use the definition of subtraction to find the differences.

7. $4 - 7 = 4 + (-7) = -3$

11. $-7 - 3 = -7 + (-3) = -10$

15. $7 - (-4) = 7 + (4) = 11$

19. $-7 - (-3) = -7 + (3) = -4$

23. $-3 - (6 - 9) = -3 - [6 + (-9)]$
    $= -3 - (-3)$
    $= -3 + (3) = 0$

27. $-\frac{3}{4} - \frac{5}{8} = -\frac{3}{4} + (-\frac{5}{8})$
    $= -\frac{6}{8} + (-\frac{5}{8}) = -\frac{11}{8}$

31. $4.4 - (-9.2) = 4.4 + (9.2) = 13.6$

35. $-5.2 - (8.4 - 10.8)$
    $= -5.2 - [8.4 + (-10.8)]$
    $= -5.2 - (-2.4)$
    $= -5.2 + (2.4)$
    $= -2.8$

43. $-10 - [(5 - 4) - (-5 - 8)]$
    $= -10 - \{[5 + (-4)] - [-5 + (-8)]\}$
    $= -10 - [1 - (-13)]$
    $= -10 - [1 + 13]$
    $= -10 - (14)$
    $= -10 + (-14) = -24$

47. $(-\frac{3}{4} - \frac{5}{2}) - (-\frac{1}{8} - 1)$
    $= [-\frac{3}{4} + (-\frac{5}{2})] - [-\frac{1}{8} + (-1)]$
    $= [-\frac{3}{4} + (-\frac{10}{4})] - [-\frac{1}{8} + (-\frac{8}{8})]$
    $= (-\frac{13}{4}) - (-\frac{9}{8})$
    $= (-\frac{26}{8}) - (-\frac{9}{8})$
    $= (-\frac{26}{8}) + (\frac{9}{8}) = -\frac{17}{8}$

51. The difference between two negative numbers can be either positive or negative. In the following examples, the difference is a negative number.

    $-8 - (-2) = -8 + (2) = -6$
    $-6 - (-5) = -6 + (5) = -1$

55. "8 less than -2" is written

    $-2 - 8 = -2 + (-8) = -10.$

59. "12 less than the difference between 8 and -5" is written

    $[8 - (-5)] - 12 = [8 + (5)] - 12$
    $= 13 - 12$
    $= 13 + (-12)$
    $= 1.$

63. 14,494 − (−282) = 14,494 + (282)
    = 14,776

The difference between these two elevations is 14,776 feet.

67. 76,000 − (−29,000) = 76,000 + (29,000)
    = 105,000

The difference is $105,000.

71. $3 - x = 6$; $\{-3, -2, -1, 0, 1, 2, 3\}$

The solution is −3 since
$$3 - (-3) = 3 + (3) = 6.$$

75. $a - b = a + (-b)$

If b is negative, −b will be positive. Then $a - b = a + (-b)$ will be the sum of two positive numbers, which is positive.

79. $4x - 2y$
    $= 4(5) - 2(2)$ Let $x = 5$, $y = 2$
    $= 20 - 4$
    $= 16$

Section 1.6 Multiplication of Real Numbers

1.6 Margin Exercises

1. (a) $3(-3) = -3 + (-3) + (-3) = -9$
   (b) $3(-4) = -4 + (-4) + (-4) = -12$
   (c) $3(-5) = -5 + (-5) + (-5) = -15$

2. Use the following rule: The product of a positive number and a negative number is negative.

   (a) $2(-6) = -(2 \cdot 6) = -12$
   (b) $7(-8) = -(7 \cdot 8) = -56$
   (c) $(-9)(2) = -18$
   (d) $(-16)\left(\frac{5}{32}\right) = -\left(\frac{16}{1} \cdot \frac{5}{32}\right)$
       $= -\frac{16 \cdot 5}{2 \cdot 16}$
       $= -\frac{5}{2}$
   (e) $(4.56)(-10) = (4.56 \cdot 10)$
       $= -45.6$

3. Use the following rule: The product of two negative numbers is positive.

   (a) $(-5)(-6) = 5 \cdot 6 = 30$
   (b) $(-7)(-3) = 7 \cdot 3 = 21$
   (c) $(-8)(-5) = 8 \cdot 5 = 40$
   (d) $(-11)(-2) = 22$
   (e) $(-17)(-21) = 357$
   (f) $(-82)(-13) = 1066$

4. (a) $(-3)(4) - (2)(6) = -12 - 12$
       $= -12 + (-12)$
       $= -24$
   (b) $-7(-2 - 5) = -7[-2 + (-5)]$
       $= -7(-7)$
       $= 49$
   (c) $-8[-1 - (-4)(-5)] = -8[-1 - 20]$
       $= -8(-21)$
       $= 168$

5. (a) Replace x with −4 and y with 3.

$$2x - 7(y + 1) = 2(-4) - 7(3 + 1)$$
$$= -8 - 7(4)$$
$$= -8 - 28$$
$$= -36$$

(b) Replace x with 2 and y with −1.

$$(-3x)(4x - 2y) = (-3 \cdot 2)[4 \cdot 2 - 2(-1)]$$
$$= (-6)[8 - (-2)]$$
$$= (-6)[8 + 2]$$
$$= (-6)(10)$$
$$= -60$$

(c) Replace x with −2 and y with −3.

$$2x^2 - 4y^2 = 2(-2)^2 - 4(-3)^2$$
$$= 2(4) - 4(9)$$
$$= 8 - 36$$
$$= -28$$

6. (a) "The product of 6 and the sum of −5 and −4" is written

$$6[(-5) + (-4)] = 6(-9)$$
$$= -54.$$

(b) "Twice the difference between 8 and −4" is written

$$2[8 - (-4)] = 2[8 + 4]$$
$$= 2(12)$$
$$= 24.$$

(c) "Three-fifths of the sum of 2 and −7" is written

$$\tfrac{3}{5}[2 + (-7)] = \tfrac{3}{5}(-5)$$
$$= -3.$$

(d) "20% of the sum of 9 and −4" is written

$$.20[9 + (-4)] = .20(5)$$
$$= 1.$$

1.6 Section Exercises

3. "When the sum of two negative numbers is multiplied by a positive number, the product is a positive number" is a false statement because the sum of two negative numbers is a negative number, and the product of a negative number and a positive number is negative.

7. $(-4)(-5) = 20$

Note that the product of two negative numbers is positive.

11. $(-7)(4) = -(7 \cdot 4) = -28$

Note that the product of a negative number and a positive number is negative.

15. $(-8)(0) = 0$

19. $(-6.8)(.35) = -2.38$

In Exercises 23–39, use the order of operations.

23. $7 - 3 \cdot 6 = 7 - 18$
$$= -11$$

27. $15(8 - 12) = 15(-4)$
$$= -60$$

31. $(12 - 14)(1 - 4) = (-2)(-3)$
$$= 6$$

26 Chapter 1 The Real Number System

35. $(-2 - 8)(-6) + 7 = (-10)(-6) + 7$
$= 60 + 7$
$= 67$

39. $(-9 - 3)(-5) - (-4)$
$= (-12)(-5) - (-4)$
$= 60 - (-4)$
$= 60 + 4$
$= 64$

In Exercises 43–51, replace x with 6, y with -4, and a with 3. Then use the order of operations to evaluate each expression.

43. $5x - 2y + 3a = 5(6) - 2(-4) + 3(3)$
$= 30 - (-8) + 9$
$= 30 + 8 + 9$
$= 38 + 9$
$= 47$

47. $\left(\frac{1}{3}x - \frac{4}{5}y\right)\left(-\frac{1}{5}a\right)$
$= \left[\frac{1}{3}(6) - \frac{4}{5}(-4)\right]\left[-\frac{1}{5}(3)\right]$
$= \left[2 - \left(-\frac{16}{5}\right)\right]\left(-\frac{3}{5}\right)$
$= \left[2 + \frac{16}{5}\right]\left(-\frac{3}{5}\right)$
$= \left(\frac{10}{5} + \frac{16}{5}\right)\left(-\frac{3}{5}\right)$
$= \left(\frac{26}{5}\right)\left(-\frac{3}{5}\right)$
$= -\frac{78}{25}$

51. $-2y^2 + 3a = -2(-4)^2 + 3(3)$
$= -2(16) + 3(3)$
$= -32 + 9$
$= -23$

55. "The product of -9 and 2, added to 9" is written
$9 + (-9)(2) = 9 + (-18)$
$= -9.$

59. "Nine subtracted from the product of 7 and -12" is written
$(7)(-12) - 9 = -84 - 9$
$= -84 + (-9)$
$= -93.$

63. "Four-fifths of the sum of -8 and -2" is written
$\frac{4}{5}[-8 + (-2)] = \frac{4}{5}(-10)$
$= -\frac{40}{5} = -8.$

67. $-4y = 0$; $\{-3, -2, -1, 0, 1, 2, 3\}$
The solution is 0 since
$-4(0) = 0.$

71. $7x = -14$; $\{-3, -2, -1, 0, 1, 2, 3\}$
The solution is -2 since
$7(-2) = -14$

75. $6x + 10 = -8$; $\{-3, -2, -1, 0, 1, 2, 3\}$
The solution is -3 since
$6(-3) + 10 = -18 + 10 = -8.$

79. $\dfrac{2(x + 4y)}{2x + y} = \dfrac{2[2 + 4(10)]}{2(2) + 10}$

$= \dfrac{2(2 + 40)}{4 + 10}$

$= \dfrac{2(42)}{14}$

$= \dfrac{84}{14} = 6$

Section 1.7 Division of Real Numbers

1.7 Margin Exercises

1.

	Number	Reciprocal
(a)	6	$\dfrac{1}{6}$
(b)	-2	$\dfrac{1}{-2}$ or $-\dfrac{1}{2}$
(c)	$\dfrac{2}{3}$	$\dfrac{3}{2}$
(d)	$-\dfrac{1}{4}$	-4
(e)	0	None

2. (a) $\dfrac{42}{7} = 42 \cdot \dfrac{1}{7} = 6$

(b) $\dfrac{-36}{6} = -36 \cdot \dfrac{1}{6} = -6$

(c) $\dfrac{-12.56}{-.4} = -12.56 \cdot \dfrac{1}{-.4}$

$= -12.56(-2.5) = 31.4$

(d) $\dfrac{10}{7} \div \left(-\dfrac{24}{5}\right) = \dfrac{10}{7} \cdot \left(-\dfrac{5}{24}\right)$

$= -\dfrac{50}{168} = -\dfrac{25}{84}$

(e) $\dfrac{-3}{0}$ is undefined. Division by 0 is never permitted.

3. (a) $\dfrac{-8}{-2} = 4$

(b) $\dfrac{-16.4}{2.05} = -8$

$2.05\overline{)16.4}$ becomes $205\overline{)1640}$ with quotient 8, $\underline{1640}$, 0

(c) $\dfrac{1}{4} \div \left(-\dfrac{2}{3}\right) = \dfrac{1}{4}\left(-\dfrac{3}{2}\right)$ Multiply by reciprocal of $-2/3$

$= -\dfrac{3}{8}$

4. (a) $\dfrac{6(-4) - 2(5)}{3(2 - 7)} = \dfrac{-24 - 10}{3(-5)}$

$= \dfrac{-34}{-15}$

$= \dfrac{34}{15}$

(b) $\dfrac{-6(-8) + (-3)9}{(-2)[4 - (-3)]} = \dfrac{48 + (-27)}{(-2)[4 + (3)]}$

$= \dfrac{21}{(-2)(7)}$

$= \dfrac{21}{-14} = -\dfrac{3}{2}$

(c) $\dfrac{5^2 + 3^2}{3(-4) - 5} = \dfrac{25 + 9}{-12 - 5}$

$= \dfrac{34}{-17} = -2$

5. (a) "The quotient of 20 and the sum of 8 and -3" is written

$\dfrac{20}{8 + (-3)} = \dfrac{20}{5}$

$= 4.$

(b) "The product of -9 and 2, divided by the difference between 5 and -1" is written

$\dfrac{(-9)(2)}{5 - (-1)} = \dfrac{-18}{6}$

$= -3.$

6. (a) "Twice a number is −6" is written
$$2x = -6.$$
Since $2(-3) = -6$, the solution is −3.

(b) "The difference between −8 and a number is −11" is written
$$-8 - x = -11.$$
Since $-8 - 3 = -11$, the solution is 3.

(c) "The sum of 5 and a number is 8" is written
$$5 + x = 8.$$
Since $5 + 3 = 8$, the solution is 3.

(d) "The quotient of a number and −2 is 6" is written
$$\frac{x}{-2} = 6.$$
Since $\frac{-12}{-2} = 6$, the solution is −12.

1.7 Section Exercises

3. If two negative numbers are multiplied and their product is then divided by a negative number, the result is *less than* zero.

7. Since $11 \cdot \frac{1}{11} = 1$, the reciprocal of 11 is $\frac{1}{11}$.

11. Since $\frac{5}{6}\left(\frac{6}{5}\right) = 1$, the reciprocal of $\frac{5}{6}$ is $\frac{6}{5}$.

15. Since $\left(-\frac{8}{7}\right)\left(-\frac{7}{8}\right) = 1$, the reciprocal of $-\frac{8}{7}$ is $-\frac{7}{8}$.

19. The expression $\frac{5-5}{5-5}$ or $\frac{0}{0}$ is undefined. The correct response is (c).

23. $\frac{20}{-10} = -2$

Note that the quotient of two numbers having different signs is negative.

27. $\frac{0}{-3} = 0\left(-\frac{1}{3}\right) = 0$

31. $\left(-\frac{3}{4}\right) \div \left(-\frac{1}{2}\right) = \left(-\frac{3}{4}\right) \cdot \left(-\frac{2}{1}\right)$
$$= \frac{6}{4} = \frac{3}{2}$$

35. $\frac{18}{3-9} = \frac{18}{-6} = -3$

39. $\frac{-12 - 36}{-12} = \frac{-12 + (-36)}{-12}$
$$= \frac{-48}{-12}$$
$$= 4$$

43. $\frac{-21(3)}{-3 - 6} = \frac{-63}{-3 + (-6)}$
$$= \frac{-63}{-9}$$
$$= 7$$

47. $\dfrac{-27(-2) - (-12)(-2)}{-2(3) - 2(2)} = \dfrac{54 - 24}{-6 - 4}$

$= \dfrac{30}{-10} = -3$

51. $\dfrac{2^2 - 8^2}{6(-4 + 3)} = \dfrac{4 - 64}{6(-1)}$

$= \dfrac{-60}{-6} = 10$

55. "The quotient of -12 and the sum of -5 and -1"

$\dfrac{-12}{-5 + (-1)} = \dfrac{-12}{-6} = 2.$

59. "The product of -34 and 7, divided by -14" is written

$\dfrac{(-34)(7)}{-14} = \dfrac{-238}{-14} = 17.$

63. $\dfrac{n}{-2} = -2;$

$\{-8, -6, -4, -2, 0, 2, 4, 6, 8\}$

The solution is 4, since

$\dfrac{4}{-2} = -2.$

67. "Six times a number is -42" is written

$6x = -42.$

The solution is -7, since

$6(-7) = -42.$

71. "6 less than a number is 4" is written

$x - 6 = 4.$

The solution is 10, since

$10 - 6 = 4.$

75. $19 + (-19) = 0$

Section 1.8 Properties of Addition and Multiplication

1.8 Margin Exercises

1. (a) $x + 9 = 9 + x$

   (b) $(-12)(4) = 4(-12)$

   (c) $5x = x \cdot 5$

2. (a) $(9 + 10) + (-3)$
   $= 9 + [10 + (-3)]$

   (b) $-5 + (2 + 8) = (-5 + 2) + 8$

   (c) $10 \cdot [(-8) \cdot (-3)]$
   $= [10 \cdot (-8)] \cdot (-3)$

3. (a) $2(4 \cdot 6) = (2 \cdot 4)6$

   The order of the three numbers is the same on both sides of the equals sign. The only change is in the grouping of the numbers. Therefore, this is an example of an associative property.

   (b) $(2 \cdot 4)6 = (4 \cdot 2)6$

   While the same numbers are grouped inside the two pairs of parentheses, the order of the numbers has been changed. This illustrates a commutative property.

   (c) $(2 + 4) + 6 = 4 + (2 + 6)$

   Both the order and the grouping of the numbers have been changed. This is an example of both properties.

30   Chapter 1   The Real Number System

4. 8 + 4y + 10
    = (8 + 4y) + 10   Order of operations
    = (4y + 8) + 10   Commutative property
    = 4y + (8 + 10)   Associative property
    = 4y + 18   Add
    or 18 + 4y

5. (a) 9 + 0 = 9

    (b) 0 + (−7) = −7

    (c) $\frac{1}{4} \cdot \frac{3}{3} = \frac{3}{12}$ 3/3 = 1

    (d) 5 · 1 = 5

6. (a) $\frac{85}{105} = \frac{17 \cdot 5}{21 \cdot 5}$ Factor

    $= \frac{17}{21} \cdot \frac{5}{5}$ Write as a product

    $= \frac{17}{21} \cdot 1$ Property of 1

    $= \frac{17}{21}$ Identity property

    (b) $\frac{9}{10} - \frac{53}{50} = \frac{9}{10} \cdot 1 - \frac{53}{50}$ Identity property

    $= \frac{9}{10} \cdot \frac{5}{5} - \frac{53}{50}$ Property of 1

    $= \frac{45}{50} - \frac{53}{50}$ Multiply

    $= -\frac{8}{50}$ Subtract

    $= -\frac{4}{25}$ Lowest terms

7. (a) −6 + 6 = 0   Inverse property

    (b) $\frac{4}{3} \cdot \frac{3}{4} = 1$ Inverse property

    (c) $-\frac{1}{9} \cdot -9 = 1$ Inverse property

    (d) 275 + 0 = 275   Identity property

8. (a) 5m − 3 − 5m
    = (5m − 3) − 5m   Order of operations
    = (−3 + 5m) − 5m   Commutative property
    = −3 + (5m − 5m)   Associative property
    = −3 + 0   Inverse property
    = −3   Identity property

    (b) $\left(\frac{4}{3}\right)(-7)\left(\frac{3}{4}\right)$

    $= \left[\left(\frac{4}{3}\right)(-7)\right]\left(\frac{3}{4}\right)$ Order of operations

    $= \left[(-7)\left(\frac{4}{3}\right)\right]\left(\frac{3}{4}\right)$ Commutative property

    $= (-7)\left[\left(\frac{4}{3}\right)\left(\frac{3}{4}\right)\right]$ Asociative property

    $= (-7) \cdot 1$ Inverse property
    $= -7$ Identity property

9. (a) 2(p + 5) = 2 · p + 2 · 5
        = 2p + 10

    (b) −4(y + 7) = (−4)y + (−4)7
        = (−4y) + (−28)
        = −4y − 28

    (c) 5(m − 4) = 5 · m − 5 · 4
        = 5m − 20

    (d) 9 · k + 9 · 5 = 9(k + 5)

    (e) 3a − 3b = 3 · a − 3 · b
        = 3(a − b)

    (f) 7(2y + 7k − 9m)
        = 7(2y) + 7(7k) − 7(9m)
        = 14y + 49k − 63m

10. (a) −(3k − 5) = −1 · (3k − 5)
        = −1(3k) + (−1)(−5)
        = −3k + 5

(b) $\quad -(2 - r) = -1 \cdot (2 - r)$
$\qquad\qquad\quad = -1 \cdot 2 + (-1)(-r)$
$\qquad\qquad\quad = -2 + r$

(c) $\quad -(-5y + 8) = -1 \cdot (-5y) + (-1)(8)$
$\qquad\qquad\qquad = 5y - 8$

(d) $\quad -(-z + 4) = -1 \cdot (-z) + (-1)(4)$
$\qquad\qquad\qquad = z - 4$

11. (a) $4x + x = 4x + 1 \cdot x \quad$ *Identity property*
$\qquad\quad = (4 + 1)x \quad$ *Distributive property*
$\qquad\quad = 5x \qquad\qquad$ *Add*

(b) $a + a + a$
$\quad = 1 \cdot a + 1 \cdot a + 1 \cdot a \quad$ *Identity property*
$\quad = (1 + 1 + 1)a \quad$ *Distributive property*
$\quad = 3a \qquad\qquad\qquad$ *Add*

1.8 Section Exercises

3. "The additive inverse of a is 1/a" is a false statement since the additive inverse of a is −a, while 1/a is the *multiplicative* inverse of a.

7. "Every number has a multiplicative inverse" is a false statement since 0 does not have a multiplicative inverse.

11. $5(13 \cdot 7) = (5 \cdot 13) \cdot 7$

The numbers are in the same order but grouped differently, so this is an example of the associative property of multiplication.

15. $-6 + (12 + 7) = (-6 + 12) + 7$

This is an example of the associative property of addition.

19. $\left(\frac{2}{3}\right)\left(\frac{3}{2}\right) = 1$

This is an example of the multiplicative inverse property.

23. $(4 + 17) + 3 = 3 + (4 + 17)$

The order of the numbers has been changed, but not the grouping, so this is an example of the commutative property of addition.

27. $-\frac{5}{9} = -\frac{5}{9} \cdot \frac{3}{3} = -\frac{15}{27}$

This is an example of the identity property of addition.

31. Jack recognized the identity property of addition.

35. $r + 7$; commutative

$\qquad r + 7 = 7 + r$

39. $-6(x + 7)$; distributive

$\qquad -6(x + 7) = -6(x) + (-6)(7)$
$\qquad\qquad\qquad = -6x + (-42)$
$\qquad\qquad\qquad = -6x - 42$

43. $9 + 3x + 7$

$\quad = (9 + 3x) + 7 \quad$ Order of operations

$\quad = (3x + 9) + 7 \quad$ Commutative property

$\quad = 3x + (9 + 7) \quad$ Associative property

$\quad = 3x + 16 \quad$ Add

47. $-3w + 7 + 3w$

$\quad = (-3w + 7) + 3w \quad$ Order of operations

$\quad = [7 + (-3w)] + 3w \quad$ Commutative property

$\quad = 7 + [(-3w) + 3w] \quad$ Associative property

$\quad = 7 + 0 \quad$ Inverse property

$\quad = 7 \quad$ Identity property

51. $\left(-\frac{9}{7}\right)(-.38)\left(\frac{7}{9}\right)$

$\quad = \left[\left(-\frac{9}{7}\right)(-.38)\right]\left(\frac{7}{9}\right) \quad$ Order of operations

$\quad = \left[(-.38)\left(\frac{9}{7}\right)\right]\left(\frac{7}{9}\right) \quad$ Commutative property

$\quad = (-.38)\left[\left(\frac{9}{7}\right)\left(\frac{7}{9}\right)\right] \quad$ Associative property

$\quad = (-.38)(1) \quad$ Inverse property

$\quad = -.38 \quad$ Identity property

55. $5x + x = 5x + 1x$

$\quad\quad\quad\quad = (5 + 1)x$

$\quad\quad\quad\quad = 6x$

59. $-8(r + 3) = -8(r) + (-8)(3)$

$\quad\quad\quad\quad\quad = -8r + (-24)$

$\quad\quad\quad\quad\quad = -8r - 24$

63. $-\frac{4}{3}(12y + 15z)$

$\quad = -\frac{4}{3}(12y) + \left(-\frac{4}{3}\right)(15z)$

$\quad = \left[\left(-\frac{4}{3}\right) \cdot 12\right]y + \left[\left(-\frac{4}{3}\right) \cdot 15\right]z$

$\quad = -16y + (-20)z$

$\quad = -16y - 20z$

67. $7(2v) + 7(5r) = 7(2v + 5r)$

71. $q + q + q = 1 \cdot q + 1 \cdot q + 1 \cdot q$

$\quad\quad\quad\quad\quad = (1 + 1 + 1)q$

$\quad\quad\quad\quad\quad = 3q$

75. $-(4t + 5m)$

$\quad = -1(4t + 5m) \quad$ Identity property

$\quad = -1(4t) + (-1)(5m) \quad$ Distributive property

$\quad = -4t - 5m \quad$ Multiply

79. $-(-3q + 5r - 8s)$

$\quad = -1(-3q + 5r - 8s)$

$\quad = -1(-3q) + (-1)(5r) + (-1)(-8s)$

$\quad = 3q - 5r + 8s$

83. $(-12) + 26 + 19 + (-2)$

$\quad = \{[(-12) + 26] + 19\} + (-2)$

$\quad = (14 + 19) + (-2)$

$\quad = 33 + (-2)$

$\quad = 31$

Section 1.9 Simplifying Expressions

1.9 Margin Exercises

1. **(a)** $9k + 12 - 5 = 9k + (12 - 5)$
   $= 9k + 7$

   (b) $7(3p + 2q) = 7(3p) + 7(2q)$
   $= 21p + 14q$

   (c) $2 + 5(3z - 1) = 2 + 5(3z) - 5(1)$
   $= 2 + 15z - 5$
   $= 15z - 3$

   Note: By the order of operations, 5 is multiplied by $(3z - 1)$ first; then 2 is added to the product.

   (d) $-3 - (2 + 5y) = -3 - 1(2 + 5y)$
   $= -3 - 1(2) - 1(5y)$
   $= -3 - 2 - 5y$
   $= -5 - 5y$

2. The numerical coefficient is the number in front of the variable or variables.

Term	Numerical coefficient
(a) $15q$	15
(b) $-2m^3$	-2
(c) $-18m^7q^4$	-18
(d) $-r$	$-1 \quad -r = -1 \cdot r$

3. **(a)** $9x$ and $4x$ have the same variable, x, and the same exponent, which is understood to be 1, so they are like terms.

   (b) $-8y^3$ and $12y^2$ have the same variable but different exponents on y, so they are unlike terms.

   (c) $7x^2y^4$ and $-7x^2y^4$ have the same variables and exponents, so they are like terms.

   (d) $13kt$ and $4tk = 4kt$ have the same variables and exponents, so they are like terms.

4. **(a)** $4k + 7k = (4 + 7)k = 11k$

   (b) $4r - r = 4r - 1r = (4 - 1)r$
   $= 3r$

   (c) $5z + 9z - 4z = (5 + 9 - 4)z$
   $= 10z$

   (d) $8p + 8p^2$ cannot be simplified. $8p$ and $8p^2$ are unlike terms and cannot be combined.

5. **(a)** $10p + 3(5 + 2p)$
   $= 10p + 3(5) + 3(2p)$
   $= 10p + 15 + 6p$
   $= 16p + 15$

   (b) $7z - 2 - 4(1 + z)$
   $= 7z - 2 - 4 - 4z$
   $= 3z - 6$

   (c) $-(3 + 5k) + 7k$
   $= -1(3 + 5k) + 7k$
   $= -3 - 5k + 7k$
   $= -3 + 2k \quad \text{or} \quad 2k - 3$

6. **(a)** "Three times a number is subtracted from the sum of the number and 8" is written
   $(x + 8) - 3x = x + 8 - 3x$
   $= -2x + 8.$

34   Chapter 1   The Real Number System

(b) "Twice a number added to the sum of 6 and the number" is written

$$2x + (6 + x) = (2x + x) + 6$$
$$= 3x + 6.$$

1.9 Section Exercises

3. The numerical coefficient of $5x^3y^7$ is 5. Therefore, the correct response is (a).

7. $8(4q - 3t) = 8(4q) - 8(3t)$
   $= 32q - 24t$

11. $-2 - (5 - 3p) = -2 - 1(5 - 3p)$
    $= -2 - 1(5) - 1(-3p)$
    $= -2 - 5 + 3p$
    $= -7 + 3p$

15. The numerical coefficient of the term $-12k$ is -12.

19. Because xw can be written as $1 \cdot xw$, the numerical coefficient of the term xw is 1.

23. The numerical coefficient of the term 74 is 74.

27. $8r$ and $-13r$ are like terms since they have the same variable with the same exponent (which is understood to be 1).

31. 4, 9, and -24 are like terms since all numerical terms without variables are like terms.

39. $-\frac{4}{3} + 2t + \frac{1}{3}t - 8 - \frac{8}{3}t$

   $= \left(2t + \frac{1}{3}t - \frac{8}{3}t\right) + \left(-\frac{4}{3} - 8\right)$

   Group like terms

   $= \left(2 + \frac{1}{3} - \frac{8}{3}\right)t + \left(-\frac{4}{3} - 8\right)$

   Distributive property

   $= \left(\frac{6}{3} + \frac{1}{3} - \frac{8}{3}\right)t + \left(-\frac{4}{3} - \frac{24}{3}\right)$

   LCD = 3

   $= -\frac{1}{3}t + \left(-\frac{28}{3}\right)$

   $= -\frac{1}{3}t - \frac{28}{3}$

43. $2y^2 - 7y^3 - 4y^2 + 10y^3$

   $= (2y^2 - 4y^2) + (-7y^3 + 10y^3)$

   Group like terms

   $= (2 - 4)y^2 + (-7 + 10)y^3$

   Distributive property

   $= -2y^3 + 3y^3$

47. $-4(y - 7) - 6 = -4(y) - (-4)(7) - 6$

   $= -4y - (-28) - 6$

   $= -4y + 28 - 6$

   $= -4y + 22$

51. $-4(-3k + 3) - (6k - 4) - 2k + 1$

   $= -4(-3k + 3) - 1(6k - 4) - 2k + 1$

   $= 12k - 12 - 6k + 4 - 2k + 1$

   Distributive property

   $= (12k - 6k - 2k) + (-12 + 4 + 1)$

   Group like terms

   $= 4k - 7$ *Combine like terms*

55. "Five times a number, added to the sum of the number and three" is written

$$(x + 3) + 5x = x + 3 + 5x$$
$$= x + 5x + 3$$
$$= 6x + 3.$$

59. "Six times a number added to -4, subtracted from twice the sum of three times the number and 4" is written

$$2(3x + 4) - (-4 + 6x)$$
$$= 2(3x + 4) - 1(-4 + 6x)$$
$$= 6x + 8 + 4 - 6x$$
$$= 6x + (-6x) + 8 + 4$$
$$= 0 + 12 = 12.$$

63. The opposite or additive inverse of 5 is -5.

67. $(x - 2) + 2 = x - 2 + 2 = x$,

so the number to be added is 2.

Chapter 1 Review Exercises

1. $5^4 = 5 \cdot 5 \cdot 5 \cdot 5 = 625$

2. $\left(\frac{3}{5}\right)^3 = \frac{3}{5} \cdot \frac{3}{5} \cdot \frac{3}{5} = \frac{27}{125}$

3. $(.02)^5 = (.02)(.02)(.02)(.02)(.02)$
$= .0000000032$

4. $(.001)^3 = (.001)(.001)(.001)$
$= .000000001$

5. $8 \cdot 5 - 13 = 40 - 13 = 27$

6. $7[3 + 6(3^2)] = 7[3 + 6(9)]$
$= 7(3 + 54)$
$= 7(57)$
$= 399$

7. $\frac{9(4^2 - 3)}{4 \cdot 5 - 17} = \frac{9(16 - 3)}{20 - 17}$
$= \frac{9(13)}{3}$
$= \frac{117}{3} = 39$

8. $\frac{6(5 - 4) + 2(4 - 2)}{3^2 - (4 + 3)} = \frac{6(1) + 2(2)}{9 - (4 + 3)}$
$= \frac{6 + 4}{9 - 7}$
$= \frac{10}{2} = 5$

9. $12 \cdot 3 - 6 \cdot 6 = 36 - 36 = 0$

Therefore, the statement "$12 \cdot 3 - 6 \cdot 6 \leq 0$" is true.

10. $3[5(2) - 3] = 3(10 - 3) = 3(7) = 21$

Therefore, the statement "$3[5(2) - 3] > 20$" is true.

11. $4^2 - 8 = 16 - 8 = 8$

Therefore, the statement "$9 \leq 4^2 - 8$" is false.

12. $9 \cdot 2 - 6 \cdot 3 = 18 - 18 = 0$

Therefore, the statement "$9 \cdot 2 - 6 \cdot 3 \geq 0$" is true.

13. "Thirteen is less than seventeen" is written $13 < 17$.

14. "Five plus two is not equal to 10" is written $5 + 2 \neq 10$.

In part (a) of Exercises 15–18, replace x with 3. In part (b), replace x with 15.

15. (a) $4x = 4(3) = 12$

(b) $4x = 4(15) = 60$

16. (a) $\dfrac{x + 2}{x + 1} = \dfrac{3 + 2}{3 + 1} = \dfrac{5}{4}$

(b) $\dfrac{x + 2}{x + 1} = \dfrac{15 + 2}{15 + 1} = \dfrac{17}{16}$

17. (a) $3x + x^2 = 3(3) + 3^2$
    $= 3(3) + 9$
    $= 9 + 9 = 18$

(b) $3x + x^2 = 3(15) + 15^2$
    $= 3(15) + 225$
    $= 45 + 225 = 270$

18. (a) $7.45(x + 1) = 7.45(3 + 1)$
    $= 7.45(4) = 29.8$

(b) $7.45(x + 1) = 7.45(15 + 1)$
    $= 7.45(16) = 119.2$

In Exercises 19–22, replace x with 6 and y with 3.

19. $2x + 6y = 2(6) + 6(3)$
    $= 12 + 18 = 30$

20. $4(3x - y) = 4[3(6) - 3]$
    $= 4(18 - 3)$
    $= 4(15) = 60$

21. $\dfrac{x}{3} + 4y = \dfrac{6}{3} + 4(3)$
    $= 2 + 12 = 14$

22. $\dfrac{x^2 + 3}{3y - x} = \dfrac{6^2 + 3}{3(3) - 6}$
    $= \dfrac{36 + 3}{9 - 6}$
    $= \dfrac{39}{3} = 13$

23. "Six added to a number" translates as $x + 6$.

24. "A number subtracted from eight" translates as $8 - x$.

25. "Nine subtracted from six times a number" translates as $6x - 9$.

26. "Three-fifths of a number added to 12" translates as $12 + \dfrac{3}{5}x$.

27. $5x + 3(x + 2) = 22$; 2

$5x + 3(x + 2) = 5(2) + 3(2 + 2)$ Let $x = 2$
$= 5(2) + 3(4)$
$= 10 + 12 = 22$

Therefore, 2 is a solution of the given equation.

28. $\dfrac{t + 5}{3t} = 1$; 6

$\dfrac{t + 5}{3t} = \dfrac{6 + 5}{3(6)}$ Let $t = 6$
$= \dfrac{11}{18}$

Therefore, 6 is not a solution of the equation.

29. "Six less than twice a number is 10" is written

$2x - 6 = 10.$

Since $2(8) - 6 = 16 - 6 = 10$, the solution is 8.

30. "The product of a number and 4 is 8" is written

$4x = 8.$

Since $4(2) = 8$, the solution is 2.

For Exercises 31–34, see the graphs in the answer section of the textbook.

31. $-4, -\dfrac{1}{2}, 0, 2.5, 5$

Graph these numbers on a number line. They are already arranged in order from smallest to largest.

32. $-2, -3, |-3|, |-1|$

Recall that $|-3| = 3$ and $|-1| = 1$. From smallest to largest, the numbers are $-3, -2, |-1|, |-3|$.

33. $-3\dfrac{1}{4}, 2\dfrac{4}{5}, -1\dfrac{1}{8}, \dfrac{5}{6}$

From smallest to largest, the numbers are $-3\dfrac{1}{4}, -1\dfrac{1}{8}, \dfrac{5}{6}, 2\dfrac{4}{5}$.

34. $|-4|, -|-3|, -|-5|, -6$

Recall that $|-4| = 4$, $-|-3| = -3$, and $-|-5| = -5$. From smallest to largest, the numbers are $-6, -|-5|, -|-3|$, and $|-4|$.

35. $-10, 5$

Since any negative number is smaller than any positive number, -10 is the smaller number.

36. $-8, -9$

Since -9 is to the left of -8 on the number line, -9 is the smaller number.

37. $-\dfrac{2}{3}, -\dfrac{3}{4}$

To compare these fractions, use a common denominator

$-\dfrac{2}{3} = -\dfrac{8}{12}, \quad -\dfrac{3}{4} = -\dfrac{9}{12}$

Since $-9/12$ is to the left of $-8/12$ on the number line, $-3/4$ is the smaller number.

38. $0, -|23|$

Since $-|23| = -23$ and $-23 < 0$, $-|23|$ is the smaller number.

39. $12 > -13$

This statement is true since 12 is to the right of -13 on a number line.

40. $0 > -5$

This statement is true since 0 is to the right of -5 on a number line.

41. $-9 < -7$

This statement is true since -9 is to the left of -7 on the number line.

42. $-13 > -13$

This is a false statement since $-13 = -13$.

43. -9

(a) $-(-9) = 9$

(b) $|-9| = 9$

44. 0

(a) $-0 = 0$

(b) $|0| = 0$

45. 6

(a) $-(6) = -6$

(b) $|6| = 6$

46. $-\dfrac{5}{7}$

(a) $-\left(-\dfrac{5}{7}\right) = \dfrac{5}{7}$

(b) $\left|-\dfrac{5}{7}\right| = \dfrac{5}{7}$

47. $|-12| = 12$

48. $-|3| = -3$

49. $-|-19| = -19$

50. $-|9 - 2| = -|7| = -7$

51. $-10 + 4 = -6$

52. $14 + (-18) = -4$

53. $-8 + (-9) = -17$

54. $\dfrac{4}{9} + \left(-\dfrac{5}{4}\right) = \dfrac{4 \cdot 4}{9 \cdot 4} + \left(-\dfrac{5 \cdot 9}{4 \cdot 9}\right)$

$= \dfrac{16}{36} + \left(-\dfrac{45}{36}\right)$

$= -\dfrac{29}{36}$

55. $-13.5 + (-8.3) = -21.8$

56. $(-10 + 7) + (-11) = (-3) + (-11)$
$= -14$

57. $[-6 + (-8) + 8] + [9 + (-13)]$
$= \{[-6 + (-8)] + 8\} + (-4)$
$= [(-14) + 8] + (-4)$
$= (-6) + (-4) = -10$

58. $(-4 + 7) + (-11 + 3) + (-15 + 1)$
$= (3) + (-8) + (-14)$
$= [3 + (-8)] + (-14)$
$= (-5) + (-14) = -19$

59. "19 added to the sum of -31 and 12" is written

$(-31 + 12) + 19 = (-19) + 19$
$= 0.$

60. "13 more than the sum of -4 and -8" is written

$[-4 + (-8)] + 13 = -12 + 13$
$= 1.$

61. $18 + (-26) = -8$

Tri's balance is -$8.

62. $93 - 6 = 93 + (-6)$
$= 87$

The new temperature is 87° F.

63. $x + (-2) = -4$

Because

$(-2) + (-2) = -4,$

the solution is -2.

64. $12 + x = 11$

Because

$12 + (-1) = 11,$

the solution is -1.

65. $-7 - 4 = -7 + (-4) = -11$

66. $-12 - (-11) = -12 + (11) = -1$

67. $5 - (-2) = 5 + (2) = 7$

68. $-\frac{3}{7} - \frac{4}{5} = -\frac{3 \cdot 5}{7 \cdot 5} - \frac{4 \cdot 7}{5 \cdot 7}$
$= -\frac{15}{35} - \frac{28}{35}$
$= -\frac{15}{35} + \left(-\frac{28}{35}\right)$
$= -\frac{43}{35}$

69. $2.56 - (-7.75) = 2.56 + (7.75)$
$= 10.31$

70. $(-10 - 4) - (-2) = [-10 + (-4)] + 2$
$= (-14) + (2)$
$= -12$

71. $(-3 + 4) - (-1) = (-3 + 4) + 1$
$= 1 + 1$
$= 2$

72. $-(-5 + 6) - 2 = -(1) + (-2)$
$= -1 + (-2)$
$= -3$

40   Chapter 1   The Real Number System

73. "The difference between -4 and -6" is written

$$-4 - (-6) = -4 + 6$$
$$= 2.$$

74. "Five less than the sum of 4 and -8" is written

$$[4 + (-8)] - 5 = (-4) + (-5)$$
$$= -9.$$

75. $-28 + 13 - 14 = (-28 + 13) - 14$
$$= (-28 + 13) + (-14)$$
$$= -15 + (-14)$$
$$= -29$$

His present financial status is $-\$29$.

76. $-3 - 7 = -3 + (-7)$
$$= -10$$

The new temperature is $-10°$.

78. Yes, the difference between two negative numbers can be positive. This will happen whenever the first number is larger. For example,

$$-8 - (-12) = -8 + 12 = 4.$$

79. $(-12)(-3) = 36$

80. $15(-7) = -(15 \cdot 7)$
$$= -105$$

81. $\left(-\dfrac{4}{3}\right)\left(-\dfrac{3}{8}\right) = \dfrac{4}{3} \cdot \dfrac{3}{8}$
$$= \dfrac{4 \cdot 3}{3 \cdot 8}$$
$$= \dfrac{4}{8} = \dfrac{1}{2}$$

82. $(-4.8)(-2.1) = 10.08$

83. $5(8 - 12) = 5[8 + (-12)]$
$$= 5(-4) = -20$$

84. $(5 - 7)(8 - 3)$
$$= [5 + (-7)][8 + (-3)]$$
$$= (-2)(5) = -10$$

85. $2(-6) - (-4)(-3) = -12 - (12)$
$$= -12 + (-12)$$
$$= -24$$

86. $3(-10) - 5 = -30 + (-5) = -35$

In Exercises 87–90, replace x with -5, y with 4, and z with -3.

87. $6x - 4z = 6(-5) - 4(-3)$
$$= -30 - (-12)$$
$$= -30 + 12 = -18$$

88. $5x + y - z = 5(-5) + (4) - (-3)$
$$= (-25 + 4) + 3$$
$$= -21 + 3 = -18$$

89. $5x^2 = 5(-5)^2$
$$= 5(25)$$
$$= 125$$

90. $z^2(3x - 8y) = (-3)^2[3(-5) - 8(4)]$
$= 9(-15 - 32)$
$= 9[-15 + (-32)]$
$= 9(-47) = -423$

91. "Nine less than the product of -4 and 5" is written
$(-4)(5) - 9 = -20 + (-9)$
$= -29.$

92. "Five-sixths of the sum of 12 and -6" is written
$\frac{5}{6}[12 + (-6)] = \frac{5}{6}(6)$
$= 5.$

93. $\frac{-36}{-9} = 4$

94. $\frac{220}{-11} = -20$

95. $-\frac{1}{2} \div \frac{2}{3} = -\frac{1}{2} \cdot \frac{3}{2} = -\frac{3}{4}$

96. $-33.9 \div (-3) = \frac{-33.9}{-3} = 11.3$

97. $\frac{-5(3) - 1}{8 - 4(-2)} = \frac{-15 + (-1)}{8 - (-8)}$
$= \frac{-16}{8 + 8}$
$= \frac{-16}{16} = -1$

98. $\frac{5(-2) - 3(4)}{-2[3 - (-2)] - 1} = \frac{-10 - 12}{-2(3 + 2) - 1}$
$= \frac{-10 + (-12)}{-2(5) - 1}$
$= \frac{-22}{-10 + (-1)}$
$= \frac{-22}{-11} = 2$

99. $\frac{10^2 - 5^2}{8^2 + 3^2 - (-2)} = \frac{100 - 25}{64 + 9 + 2}$
$= \frac{75}{75} = 1$

100. $\frac{(.6)^2 + (.8)^2}{(-1.2)^2 - (-.56)} = \frac{.36 + .64}{1.44 + .56}$
$= \frac{1.00}{2.00} = .5$

101. "The quotient of 12 and the sum of 8 and -4" is written
$\frac{12}{8 + (-4)} = \frac{12}{4} = 3.$

102. "The product of -20 and 12, divided by the difference between 15 and -15" is written
$\frac{(-20)(12)}{15 - (-15)} = \frac{-240}{15 + 15}$
$= \frac{-240}{30} = -8.$

103. "8 times a number is -24" is written
$8x = -24.$
If $x = -3$,
$8x = 8(-3) = -24.$
The solution is -3.

104. "The quotient of a number and 3 is −2" is written

$$\frac{x}{3} = -2.$$

If $x = -6$,

$$\frac{x}{3} = \frac{-6}{3} = -2.$$

The solution is −6.

105. "3 less than a number is −7" is written

$$x - 3 = -7.$$

If $x = -4$,

$$x - 3 = -4 - 3 = -4 + (-3) = -7.$$

The solution is −4.

106. "The sum of a number and 5 is −6" is written

$$x + 5 = -6.$$

If $x = -11$,

$$x + 5 = (-11) + 5 = -6.$$

The solution is −11.

107. $6 + 0 = 6$

This is an example of an identity property.

108. $5 \cdot 1 = 5$

This is an example of an identity property.

109. $-\frac{2}{3}\left(-\frac{3}{2}\right) = 1$

This is an example of an inverse property.

110. $17 + (-17) = 0$

This is an example of an inverse property.

111. $5 + (-9 + 2) = [5 + (-9)] + 2$

This is an example of an associative property.

112. $w(xy) = (wx)y$

This is an example of an associative property.

113. $3x + 3y = 3(x + y)$

This is an example of the distributive property.

114. $(1 + 2) + 3 = 3 + (1 + 2)$

This is an example of a commutative property.

115. $7y + y = 7y + 1y = (7 + 1)y = 8y$

116. $-12(4 - t) = -12 \cdot 4 - (-12)(t)$
$$= -48 + 12t$$

117. $3(2s) + 3(4y) = 3(2s + 4y)$
$$= 6s + 12y$$

118. $-(-4r + 5s) = -1(-4r + 5s)$
$= (-1)(-4r) + (-1)(5s)$
$= 4r - (1)(5s)$
$= 4r - 5s$

121. $16p^2 - 8p^2 + 9p^2 = (16 - 8 + 9)p^2$
$= 17p^2$

122. $4r^2 - 3r + 10r + 12r^2$
$= (4r^2 + 12r^2) + (-3r + 10r)$
$= (4 + 12)r^2 + (-3 + 10)r$
$= 16r^2 + 7r$

123. $-8(5k - 6) + 3(7k + 2)$
$= (-8)(5k) - (-8)(6) + 3(7k) + 3(2)$
$= -40k - (-48) + 21k + 6$
$= -40k + 48 + 21k + 6$
$= (-40 + 21)k + (48 + 6)$
$= -19k + 54$

124. $2s - (-3s + 6) = 2s - 1(-3s + 6)$
$= 2s + 3s - 6$
$= 5s - 6$

125. $-7(2t - 4) - 4(3t + 8) - 19(t + 1)$
$= -14t + 28 - 12t - 32 - 19t - 19$
$= (-14t - 12t - 19t) + (28 - 32 - 1)$
$= -45t - 23$

126. $3.6t^2 + 9t - 8.1(6t^2 + 4t)$
$= 3.6t^2 + 9t - (8.1)(6t^2)$
$\quad - (8.1)(4t)$
$= 3.6t^2 + 9t - 48.6t^2 - 32.4t$
$= (3.6t^2 - 48.6t^2) + (9t - 32.4t)$
$= -45t^2 - 23.4t$

127. "Seven times a number subtracted from the product of -2 and three times the number" is written
$(-2)(3x) - 7x = -6x - 7x$
$= -13x.$

128. "The quotient of 9 more than a number and 6 less than the number" is written
$$\frac{x + 9}{x - 6}.$$

129. In Exercises 127, the word *and* does not signify addition. The phrase "the product of -2 and ..." means the same thing as "-2 times" and signifies multiplication. Here the word *and* cannot be considered in isolation.

130. Answers may vary. Two ways to write the expression $3(4x - 6)$ in words are "Three times the difference between four times a number and 6" and "The product of 3 and 6 less than four times a number."

131. $[(-2) + 7 - (-5)] + [-4 - (-10)]$
$= \{[(-2) + 7] - (-5)\} + (-4 + 10)$
$= (5 + 5) + 6$
$= 10 + 6 = 16$

132. $\left(-\frac{5}{6}\right)^2 = \left(-\frac{5}{6}\right)\left(-\frac{5}{6}\right)$
$= \frac{25}{36}$

133. $-|(-7)(-4)| - (-2) = -|28| + 2$
$= -28 + 2 = -26$

134. $\dfrac{6(-4) + 2(-12)}{5(-3) + (-3)} = \dfrac{-24 + (-24)}{-15 + (-3)}$
$= \dfrac{-48}{-18} = \dfrac{8}{3}$

135. $\dfrac{3}{8} - \dfrac{5}{12} = \dfrac{3 \cdot 3}{8 \cdot 3} - \dfrac{5 \cdot 2}{12 \cdot 2}$
$= \dfrac{9}{24} - \dfrac{10}{24}$
$= \dfrac{9}{24} + \left(-\dfrac{10}{24}\right)$
$= -\dfrac{1}{24}$

136. $\dfrac{12^2 + 2^2 - 8}{10^2 - (-4)(-15)} = \dfrac{144 + 4 - 8}{100 - (-4)(-15)}$
$= \dfrac{148 - 8}{100 - 60}$
$= \dfrac{140}{40} = \dfrac{7}{2}$

137. $\dfrac{8^2 + 6^2}{7^2 + 1^2} = \dfrac{64 + 36}{49 + 1}$
$= \dfrac{100}{50} = 2$

138. $-16(-3.5) - 7.2(-3)$
$= 56 - [-(7.2)(3)]$
$= 56 - (-21.6)$
$= 56 + 21.6$
$= 77.6$

139. $2\dfrac{5}{6} - 4\dfrac{1}{3} = \dfrac{17}{6} - \dfrac{13}{3}$
$= \dfrac{17}{6} - \dfrac{13 \cdot 2}{3 \cdot 2}$
$= \dfrac{17}{6} - \dfrac{26}{6}$
$= \dfrac{17}{6} + \left(-\dfrac{26}{6}\right)$
$= -\dfrac{9}{6} = -\dfrac{3}{2} = -1\dfrac{1}{2}$

140. $-8 + [(-4 + 17) - (-3 - 3)]$
$= -8 + \{(13) - [-3 + (-3)]\}$
$= -8 + [13 - (-6)]$
$= -8 + (13 + 6)$
$= -8 + 19 = 11$

141. $-\dfrac{12}{5} \div \dfrac{9}{7} = -\dfrac{12}{5} \cdot \dfrac{7}{9}$
$= -\dfrac{12 \cdot 7}{5 \cdot 9}$
$= -\dfrac{84}{45} = -\dfrac{28 \cdot 3}{15 \cdot 3}$
$= -\dfrac{28}{15}$

142. $(-8 - 3) - 5(2 - 9)$
$= [-8 + (-3)] - 5[2 + (-9)]$
$= -11 - 5(-7)$
$= -11 - (-35)$
$= -11 + 35 = 24$

143. $[-7 + (-2) - (-3)] + [8 + (-13)]$
$= \{[-7 + (-2)] + 3\} + (-5)$
$= (-9 + 3) + (-5)$
$= -6 + (-5) = -11$

144. $\dfrac{15}{2} \cdot \left(-\dfrac{4}{5}\right) = -\dfrac{15 \cdot 4}{2 \cdot 5}$
$= -\dfrac{60}{10} = -6$

145. $13{,}600 - 1400 = 13{,}600 + (-1400)$
$= 12{,}200$

$12{,}200 was spent on advertising in 1994.

146. "The quotient of a number and 14 less than three times the number" is written

$$\frac{x}{3x - 14}.$$

Chapter 1 Test

1. $4[-20 + 7(-2)] = 4[-20 + (-14)]$
    $= 4(-34) = -136$

    Since $-136 \leq 135$, the statement "$4[-20 + 7(-2)] \leq 135$" is true.

2. $(-3)^2 + 2^2 = 9 + 4 = 13$
    $5^2 = 25$

    Since $13 \neq 25$, the statement "$(-3)^2 + 2^2 = 5^2$" is false.

3. $-1, -3, |-4|, |-1|$

    Recall that $|-4| = 4$ and $|-1| = 1$. From smallest to largest, the numbers are $-3, -1, |-1|, |-4|$. See the graph in the answer section of the textbook.

4. $6, -|-8|$

    $-|-8| = -(8) = -8$

    Since $-8 < 6$, $-|-8|$ (or -8) is the smaller number.

5. $-.742, -1.277$

    Since -1.277 is to the left of $-.742$ on the number line, -1.277 is the smaller number.

6. "The quotient of -6 and the sum of 2 and -8" is written

    $$\frac{-6}{2 + (-8)}, \text{ and } \frac{-6}{2 + (-8)} = \frac{-6}{-6} = 1.$$

7. $\dfrac{a + b}{a \cdot b}$

    If a and b are both negative, $a + b$ will be negative, and $a \cdot b$ will be positive. Because the quotient of a negative number and a positive number is negative, $\dfrac{a + b}{a \cdot b}$ would be negative.

8. $-2 - (5 - 17) + (-6)$
    $= -2 - [5 + (-17)] + (-6)$
    $= -2 - (-12) + (-6)$
    $= (-2 + 12) + (-6)$
    $= 10 + (-6) = 4$

9. $-5\frac{1}{2} + 2\frac{2}{3} = -\frac{11}{2} + \frac{8}{3}$
    $= -\frac{11 \cdot 3}{2 \cdot 3} + \frac{8 \cdot 2}{3 \cdot 2}$
    $= -\frac{33}{6} + \frac{16}{6}$
    $= -\frac{17}{6} = -2\frac{5}{6}$

10. $-6 - [-7 + (2 - 3)]$
    $= -6 - [-7 + (-1)]$
    $= -6 - (-8)$
    $= -6 + 8 = 2$

11. $4^2 + (-8) - (2^3 - 6)$
    $= 16 + (-8) - (8 - 6)$
    $= 16 + (-8) - 2$
    $= 8 - 2 = 6$

12. $(-5)(-12) + 4(-4) + (-8)^2$
    $= (-5)(-12) + 4(-4) + 64$
    $= 60 + (-16) + 64$
    $= 44 + 64 = 108$

13. $\dfrac{-7 - (-6 + 2)}{-5 - (-4)} = \dfrac{-7 - (-4)}{-5 + 4}$
    $= \dfrac{-7 + 4}{-1}$
    $= \dfrac{-3}{-1} = 3$

14. $\dfrac{30(-1 - 2)}{-9[3 - (-2)] - 12(-2)}$
    $= \dfrac{30(-3)}{-9(5) - (-24)}$
    $= \dfrac{-90}{-45 + 24}$
    $= \dfrac{-90}{-21} = \dfrac{30}{7}$

15. $-3x = -12$
    If $x = 4$,
    $-3x = -3(4) = -12$.
    Therefore, the solution is 4.

16. $\dfrac{x}{-4} = -1$
    If $x = 4$,
    $\dfrac{x}{-4} = \dfrac{4}{-4} = -1$.
    Therefore, the solution is 4.

17. $3x - 4y^2 = 3(-2) - 4(4^2)$ Let $x = -2$, $y = 4$
    $= 3(-2) - 4(16)$
    $= -6 - 64 = -70$

18. $\dfrac{5x + 7y}{3(x + 4)} = \dfrac{5(-2) + 7(4)}{3(-2 + 4)}$ Let $x = -2$, $y = 4$
    $= \dfrac{-10 + 28}{3(2)}$
    $= \dfrac{18}{6} = 3$

19. $118 - (-60) = 118 + 60$
    $= 178$

    The difference between the highest and lowest temperatures is 178° F.

20. Commutative

    $(5 + 2) + 8 = 8 + (5 + 2)$
    illustrates a commutative property because the order of the numbers is changed, but not the grouping.
    The correct response is B.

21. Associative

    $-5 + (3 + 2) = (-5 + 3) + 2$,
    illustrates an associative property because the grouping of the numbers is changed, but not the order.
    The correct response is D.

22. Inverse

    $-\dfrac{5}{3}\left(-\dfrac{3}{5}\right) = 1$ illustrates an inverse property.
    The correct response is E.

23. Identity

    $3x + 0 = 3x$ illustrates an identity property.
    The correct response is A.

24. Distributive

   $-3(x + y) = -3x + (-3y)$ illustrates the distributive property.
   The correct response is C.

25. $-2(3x^2 + 4) - 3(x^2 + 2x)$
   $= -2(3x^2) + (-2)(4) - 3(x^2) - 3(2x)$
   $= -6x^2 + (-8) - 3x^2 - 6x$
   $= -9x^2 - 6x - 8$

CHAPTER 2 SOLVING EQUATIONS AND INEQUALITIES

Section 2.1 The Addition Property of Equality

2.1 Margin Exercises

1. (a) $\quad m - 2.9 = -6.4$
   $\quad m - 2.9 + 2.9 = -6.4 + 2.9$
   $\quad\quad\quad m = -3.5$

   (b) $\quad y - 4.1 = 6.3$
   $\quad y - 4.1 + 4.1 = 6.3 + 4.1$
   $\quad\quad\quad y = 10.4$

2. (a) $\quad a + 2 = -3$
   $\quad a + 2 - 2 = -3 - 2$
   $\quad\quad\quad a = -5$

   (b) $\quad r + 16 = 22$
   $\quad r + 16 - 16 = 22 - 16$
   $\quad\quad\quad r = 6$

3. $\quad \frac{7}{2}m + 1 = \frac{9}{2}m$
   $\quad \frac{7}{2}m + 1 - \frac{7}{2}m = \frac{9}{2}m - \frac{7}{2}m \quad$ *Subtract $(7/2)m$*
   $\quad\quad\quad 1 = \frac{2}{2}m \quad$ *Combine terms*
   $\quad\quad\quad 1 = m \quad\quad 2/2 = 1$

4. (a) $\quad -(5 - 3r) + 4(-r + 1) = 1$
   $\quad -1(5 - 3r) + 4(-r + 1) = 1$
   $\quad -5 + 3r - 4r + 4 = 1 \quad$ *Distributive property*
   $\quad -1 - r = 1 \quad$ *Combine terms*
   $\quad -1 - r + r = 1 + r \quad$ *Add r*
   $\quad -1 = 1 + r \quad$ *Combine terms*
   $\quad -1 - 1 = 1 + r - 1 \quad$ *Subtract 1*
   $\quad -2 = r$

   (b) $\quad -3(m - 4) + 2(5 + 2m) = 29$
   $\quad -3m + 12 + 10 + 4m = 29 \quad$ *Distributive property*
   $\quad m + 22 = 29 \quad$ *Combine terms*
   $\quad m + 22 - 22 = 29 - 22 \quad$ *Subtract 22*
   $\quad m = 7$

5. (a) $\quad 2(x - 6) = 2x - 12$
   $\quad 2x - 12 = 2x - 12$
   $\quad 2x - 12 + 12 = 2x - 12 + 12 \quad$ *Add 12*
   $\quad 2x = 2x$
   $\quad 2x - 2x = 2x - 2x \quad$ *Subtract $2x$*
   $\quad 0 = 0 \quad$ *True*

   The variable x has "disappeared," and a true statement has resulted. This means that for every real number value of x, the equation is true. Thus, all real numbers are solutions of the equation.

   (b) $\quad 3x + 6(x + 1) = 9x - 4$
   $\quad 3x + 6x + 6 = 9x - 4$
   $\quad 9x + 6 = 9x - 4$
   $\quad 9x + 6 - 9x = 9x - 4 - 9x \quad$ *Subtract $9x$*
   $\quad 6 = -4 \quad$ *False*

   The variable x has "disappeared," and a false statement has resulted. This means that for every real number value of x, the equation is false. Thus, the equation has no solution.

2.1 Section Exercises

For Exercises 3–43, all solutions should be checked by substituting into the original equation. Checks will be shown here for only a few of the exercises.

3. $7 + r = -3$

If the left side of this equation were just x, the solution would be known. Get x alone by subtracting 7 from each side.

$$7 + r = -3$$
$$7 + r - 7 = -3 - 7$$
$$r + 7 - 7 = -3 - 7$$
$$r + 0 = -10$$
$$r = -10$$

Check this solution by replacing r with -10 in the original equation.

$$7 + r = -3$$
$$7 + (-10) = -3 \quad ? \quad \textit{Let } r = -10$$
$$-3 = -3 \quad \textit{True}$$

Because the final statement is true, -10 checks as the solution.

7.
$$x - 6.5 = -2.3$$
$$x - 6.5 + 6.5 = -2.3 + 6.5$$
$$x + 0 = 4.2$$
$$x = 4.2$$

11.
$$5.6x + 2 = 4.6x$$
$$5.6x + 2 - 4.6x = 4.6x - 4.6x$$
$$1.0x + 2 = 0$$
$$x + 2 - 2 = 0 - 2$$
$$x = -2$$

Check:
$$5.6x + 2 = 4.6x$$
$$5.6(-2) + 2 = 4.6(-2) \quad ? \quad \textit{Let } x = -2$$
$$-11.2 + 2 = -9.2 \quad ?$$
$$-9.2 = -9.2 \quad \textit{True}$$

15.
$$1.2y - 4 = .2y - 4$$
$$1.2y - 4 - .2y = .2y - 4 - .2y$$
$$1.0y - 4 = -4$$
$$y - 4 + 4 = -4 + 4$$
$$y = 0$$

19.
$$8x + 1 = 1 + 8x$$
$$8x + 1 - 8x = 1 + 8x - 8x$$
$$1 = 1 \quad \textit{True}$$

The variable has disappeared and a true statement resulted. This is a signal that any real number is a solution. We indicate the solution as "all real numbers."

23.
$$5t + 3 + 2t - 6t = 4 + 12$$
$$t + 3 = 16 \quad \textit{Combine like terms}$$
$$t + 3 - 3 = 16 - 3$$
$$t = 13$$

27.
$$6x + 5 - 7x + 3 = 5x - 6x - 4$$
$$-x + 8 = -x - 4$$
$$-x + x + 8 = -x + x - 4$$
$$8 = 4 \quad \textit{False}$$

Because this is a false statement, the equation has no solution.

31. $\frac{5}{7}x + \frac{1}{3} = \frac{2}{5} - \frac{2}{7}x + \frac{2}{5}$

$\frac{5}{7}x + \frac{1}{3} = \frac{4}{5} - \frac{2}{7}x$

$\frac{5}{7}x + \frac{2}{7}x + \frac{1}{3} = \frac{4}{5} - \frac{2}{7}x + \frac{2}{7}x$
                      Add (2/7)s

$\frac{7}{7}x + \frac{1}{3} = \frac{4}{5}$ *Combine like terms*

$1x + \frac{1}{3} = \frac{4}{5}$

$x + \frac{1}{3} - \frac{1}{3} = \frac{4}{5} - \frac{1}{3}$ *Subtract 1/3*

$x = \frac{12}{15} - \frac{5}{15}$ *LCD = 15*

$x = \frac{7}{15}$

35. $2(p + 5) - (9 + p) = -3$

$2(p + 5) - 1(9 + p) = -3$

$2p + 10 - 9 - p = -3$

$p + 1 = -3$

$p + 1 - 1 = -3 - 1$

$p = -4$

Check

$2(p + 5) - (9 + p) = -3$

$2(-4 + 5) - [9 + (-4)] = -3$?
                    Let p = -4

$2(1) - (5) = -3$?

$2 - 5 = -3$?

$-3 = -3$
     True

39. $10(-2x + 1) = -14(x + 2) + 38 - 6x$

$-20x + 10 = -14x - 28 + 38 - 6x$

$-20x + 10 = -20x + 10$

$-20x + 20x + 10 = -20x + 20x + 10$

$10 = 10$ *True*

All real numbers are solutions.

43. $4(7x - 1) + 3(2 - 5x) = 4(3x + 5) - 6$

$28x - 4 + 6 - 15x = 12x + 20 - 6$

$13x + 2 = 12x + 14$

$13x - 12x + 2 = 12x - 12x + 14$

$x + 2 = 14$

$x + 2 - 2 = 14 - 2$

$x = 12$

47. $4\left(\frac{1}{4}m\right) = \left(4 \cdot \frac{1}{4}\right)m$ *Associative property*

$= 1 \cdot m$ *Inverse property*

$= m$ *Identity property*

51. 7x must be multiplied by 1/7 to give a result of just x.

$\frac{1}{7}(7x) = \left(\frac{1}{7} \cdot 7\right)x$

$= 1 \cdot x$

$= x$

Section 2.2 The Multiplication Property of Equality

2.2 Margin Exercises

1. To check that 5 is the solution of $3x = 15$, replace x with 5 in the given equation.

$3x = 15$ *Given equation*

$3(5) = 15$ *Let x = 5*

$15 = 15$ *True*

The solution 5 is correct.

2. (a) $-6p = -14$

$\frac{-6p}{-6} = \frac{-14}{-6}$ *Divide by -6*

$p = \frac{7}{3}$ *Lowest terms*

(b) $3r = -12$

$\dfrac{3r}{3} = \dfrac{-12}{3}$ *Divide by 3*

$r = -4$

(c) $-2m = 16$

$\dfrac{-2m}{-2} = \dfrac{16}{-2}$ *Divide by -2*

$m = -8$

3. (a) $\dfrac{y}{5} = 5$

$\dfrac{1}{5}y = 5$

$5 \cdot \dfrac{1}{5}y = 5 \cdot 5$ *Multiply by 5, the reciprocal of 1/5*

$y = 25$

(b) $\dfrac{p}{4} = -6$

$\dfrac{1}{4}p = -6$

$4 \cdot \dfrac{1}{4}p = 4(-6)$ *Multiply by 4, the reciprocal of 1/4*

$p = -24$

4. (a) $-\dfrac{5}{6}t = -15$

$-\dfrac{6}{5}\left(-\dfrac{5}{6}t\right) = -\dfrac{6}{5}(-15)$ *Multiply by -6/5*

$t = 18$

(b) $\dfrac{3}{4}k = -21$

$\dfrac{4}{3} \cdot \dfrac{3}{4}k = \dfrac{4}{3}(-21)$ *Multiply by 4/3*

$k = -28$

5. (a) $-.7m = -5.04$

$\dfrac{-.7m}{-.7} = \dfrac{-5.04}{-.7}$ *Divide by -.7*

$m = 7.2$

2.2 Section Exercises

(b) $12.5k = -63.75$

$\dfrac{12.5k}{12.5} = \dfrac{-63.75}{12.5}$ *Divide by 12.5*

$k = -5.1$

6. (a) $4r - 9r = 20$

$-5r = 20$ *Combine terms*

$\dfrac{-5r}{-5} = \dfrac{20}{-5}$ *Divide by -5*

$r = -4$

(b) $7m - 5m = -12$

$2m = -12$ *Combine terms*

$\dfrac{2m}{2} = \dfrac{-12}{2}$ *Divide by 2*

$m = -6$

7. (a) $-m = 2$

$-1 \cdot m = 2$ $-m = -1 \cdot m$

$(-1)(-1 \cdot m) = -1 \cdot 2$ *Multiply by -1*

$1 \cdot m = -2$

$m = -2$

(b) $-p = -7$

$-1 \cdot p = -7$

$(-1)(-1) \cdot p = (-1)(-7)$ *Multiply by -1*

$p = 7$

2.2 Section Exercises

3. $.1x = 2$

We must multiply both sides by the reciprocal of .1 or 1/10, which is 10.

52 Chapter 2 Solving Equations and Inequalities

7. $-x = .45$

This equation is equivalent to $-1x = .45$. We must multiply both sides by the reciprocal of -1, which is -1.

11. $-4x = 10$

We must divide both sides by -4.

15. $-x = 32$

This equation is equivalent to $-1x = 32$. We must divide both sides by -1.

In Exercises 19–51, all solutions should be checked by substituting into the original equation. Checks will be shown here for only a few of the exercises.

19. $2m = 15$

$\dfrac{2m}{2} = \dfrac{15}{2}$ Divide by 2

$m = \dfrac{15}{2}$

23. $10t = -36$

$\dfrac{10t}{10} = \dfrac{-36}{10}$ Divide by 10

$t = -\dfrac{18}{5}$ Lowest terms

Check

$10t = -36$

$10\left(-\dfrac{18}{5}\right) = -36$? Let $t = -18/5$

$-\dfrac{180}{5} = -36$?

$-36 = -36$ True

27. $2r = 0$

$\dfrac{2r}{2} = \dfrac{0}{2}$

$r = 0$

31. $-y = 12$

$-1y = 12$

$\dfrac{-1y}{-1} = \dfrac{12}{-1}$

$y = -12$

35. $.2t = 8$

$\dfrac{.2t}{.2} = \dfrac{8}{.2}$

$t = 40$

39. $5m + 6m - 2m = 63$

$9m = 63$

$\dfrac{9m}{9} = \dfrac{63}{9}$

$m = 7$

43. $\dfrac{x}{7} = -5$

$\dfrac{1}{7}x = -5$

$7\left(\dfrac{1}{7}x\right) = 7(-5)$

$x = -35$

47. $-\dfrac{2}{7}p = -5$

$-\dfrac{7}{2}\left(-\dfrac{2}{7}p\right) = -\dfrac{7}{2}(-5)$

$p = \dfrac{35}{2}$

Check

$-\dfrac{2}{7}p = -5$

$-\dfrac{2}{7}\left(\dfrac{35}{2}\right) = -5$? Let $p = 35/2$

$-\dfrac{70}{14} = -5$?

$-5 = -5$ True

51. $-2.1m = 25.62$

$$\frac{-2.1m}{-2.1} = \frac{25.62}{-2.1}$$

$$m = -12.2$$

Check

$$-2.1m = 25.62$$
$$-2.1(-12.2) = 25.62 \quad ? \quad \textit{Let } m = -12.2$$
$$25.62 = 25.62 \quad \textit{True}$$

55. $8(3q + 4) = 8(3q) + 8(4)$
$$= 24q + 32$$

59. $6 - 7(2 - 8p) = 6 - 7(2) + 7(8p)$
$$= 6 - 14 + 56p$$
$$= -8 + 56p$$

Section 2.3 More on Solving Linear Equations

2.3 Margin Exercises

1. (a) $5y - 7y + 6y - 9 = 3 + 2y$

$$4y - 9 = 3 + 2y$$
$$\textit{Combine terms}$$

$$4y - 9 + 9 = 3 + 2y + 9$$
$$\textit{Add 9}$$

$$4y = 12 + 2y$$

$$4y - 2y = 12 + 2y - 2y$$
$$\textit{Subtract 2y}$$

$$2y = 12$$

$$\frac{2y}{2} = \frac{12}{2}$$
$$\textit{Divide by 2}$$

$$y = 6$$

(b) $-3k - 5k - 6 + 11 = 2k - 5$

$$-8k + 5 = 2k - 5$$
$$\textit{Combine terms}$$

$$-8k + 5 + 5 = 2k - 5 + 5$$
$$\textit{Add 5}$$

$$-8k + 10 = 2k$$

$$-8k + 10 + 8k = 2k + 8k$$
$$\textit{Add 8k}$$

$$10 = 10k$$

$$\frac{10}{10} = \frac{10k}{10}$$
$$\textit{Divide by 10}$$

$$1 = k$$

2. (a) $7(p - 2) + p = 2p + 4$

$$7p - 14 + p = 2p + 4$$
$$\textit{Distributive property}$$

$$8p - 14 = 2p + 4$$
$$\textit{Combine terms}$$

$$8p - 14 + 14 = 2p + 4 + 14$$
$$\textit{Add 14}$$

$$8p = 2p + 18$$

$$8p - 2p = 2p + 18 - 2p$$
$$\textit{Subtract 2p}$$

$$6p = 18$$

$$\frac{6p}{6} = \frac{18}{6}$$
$$\textit{Divide by 6}$$

$$p = 3$$

(b) $3(m + 5) - 1 + 2m = 5(m + 2)$

$$3m + 15 - 1 + 2m = 5m + 10$$
$$\textit{Distributive property}$$

$$5m + 14 = 5m + 10$$

$$5m + 14 - 5m = 5m + 10 - 5m$$
$$\textit{Subtract 5m}$$

$$14 = 10$$
$$\textit{False}$$

The equation has no solution.

3. (a) $7m - (2m - 9) = 39$
   $7m - 2m + 9 = 39$ *Distributive property*
   $5m + 9 = 39$
   $5m + 9 - 9 = 39 - 9$ *Subtract 9*
   $5m = 30$
   $\dfrac{5m}{5} = \dfrac{30}{5}$ *Divide by 5*
   $m = 6$

   (b) $4x + 2(3 - 2x) = 6$
   $4x + 6 - 4x = 6$ *Distributive property*
   $6 = 6$ *True*

   Since this is a true statement for every real number value of x, all real numbers are solutions of the equation.

4. (a) $2(4 + 3r) = 3(r + 1) + 11$
   $8 + 6r = 3r + 3 + 11$ *Distributive property*
   $8 + 6r = 3r + 14$
   $8 + 6r - 8 = 3r + 14 - 8$ *Subtract 8*
   $6r = 3r + 6$
   $6r - 3r = 3r + 6 - 3r$ *Subtract 3r*
   $3r = 6$
   $\dfrac{3r}{3} = \dfrac{6}{3}$ *Divide by 3*
   $r = 2$

(b) $2 - 3(2 + 6z) = 4(z + 1) + 18$
   $2 - 6 - 18z = 4z + 4 + 18$ *Distributive property*
   $-4 - 18z = 4z + 22$
   $-4 - 18z + 4 = 4z + 22 + 4$ *Add 4*
   $-18z = 4z + 26$
   $-18z - 4z = 4z + 26 - 4z$ *Subtract 4z*
   $-22z = 26$
   $\dfrac{-22z}{-22} = \dfrac{26}{-22}$ *Divide by -22*
   $z = -\dfrac{13}{11}$

5. $\dfrac{1}{4}x - 4 = \dfrac{3}{2}x + \dfrac{3}{4}x$
   $4\left(\dfrac{1}{4}x - 4\right) = 4\left(\dfrac{3}{2}x + \dfrac{3}{4}x\right)$ *Multiply by LCD = 4*
   $4\left(\dfrac{1}{4}x\right) - 4(4) = 4\left(\dfrac{3}{2}x\right) + 4\left(\dfrac{3}{4}x\right)$ *Distributive property*
   $x - 16 = 6x + 3x$
   $x - 16 = 9x$ *Combine terms*
   $x - x - 16 = 9x - x$ *Subtract x*
   $-16 = 8x$
   $\dfrac{-16}{8} = \dfrac{8x}{8}$ *Divide by 8*
   $-2 = x$

6. $.06(100 - y) + .04y = .05(92)$

   To clear decimals, multiply both sides by 100.

   $100[.06(100 - y) + .04y] = 100[.05(92)]$
   $6(100 - y) + 4y = 5(92)$
   $600 - 6y + 4y = 460$
   $600 - 2y = 460$
   $-2y = -140$
   $y = 70$

2.3 Section Exercises

In Exercises 3-15, use the four-step method for solving linear equations given in the text. The details of these steps, including the check, will only be shown for a few of the exercises.

3.
$$5m + 8 = 7 + 4m$$
$$5m - 4m + 8 = 7 + 4m - 4m$$
$$m + 8 = 7$$
$$m + 8 - 8 = 7 - 8$$
$$m = -1$$

Check
$$5m + 8 = 7 + 4m$$
$$5(-1) + 8 = 7 + 4(-1) \quad ? \quad \textit{Let } m = -1$$
$$-5 + 8 = 7 + (-4) \quad ?$$
$$3 = 3 \quad \textit{True}$$

The solution of the given equation is -1.

7. $\quad x + 3 = -(2x + 2)$

Step 1 $\quad x + 3 = -1(2x + 2)$
$\qquad x + 3 = -2x - 2 \quad \textit{Distributive property}$

Step 2 $\quad x + 2x + 3 = -2x + 2x - 2$
$\qquad\qquad\qquad\qquad\qquad \textit{Add } 2x$
$\qquad 3x + 3 = -2$
$\qquad 3x + 3 - 3 = -2 - 3$
$\qquad\qquad\qquad\qquad \textit{Subtract } 3$
$\qquad 3x = -5$

Step 3 $\quad \dfrac{3x}{3} = \dfrac{-5}{3}$
$\qquad\qquad x = -\dfrac{5}{3}$

Step 4 Substitute $-5/3$ for x in the original equation.

$$x + 3 = -(2x + 2)$$
$$-\tfrac{5}{3} + 3 = -\left[2\left(-\tfrac{5}{3}\right) + 2\right] \quad ? \quad \begin{array}{l}\textit{Let}\\ x = -5/3\end{array}$$
$$-\tfrac{5}{3} + \tfrac{9}{3} = -\left(-\tfrac{10}{3} + \tfrac{6}{3}\right) \quad ?$$
$$\tfrac{4}{3} = -\left(-\tfrac{4}{3}\right) \quad ?$$
$$\tfrac{4}{3} = \tfrac{4}{3} \quad \textit{True}$$

The solution of the given equation is $-5/3$.

11.
$$6(4x - 1) = 12(2x + 3)$$
$$24x - 6 = 24x + 36$$
$$24x - 24x - 6 = 24x - 24x + 36$$
$$-6 = 36 \quad \textit{False}$$

Because this is a false statement, the equation has no solution.

15.
$$7r - 5r + 2 = 5r - r$$
$$2r + 2 = 4r$$
$$2 = 2r$$
$$1 = r$$

19. $\quad \dfrac{3}{5}t - \dfrac{1}{10}t = t - \dfrac{5}{2}$

The least common denominator of all the fractions in the equation is 10.

$$10\left(\tfrac{3}{5}t - \tfrac{1}{10}t\right) = 10\left(t - \tfrac{5}{2}\right)$$
$$\textit{Multiply both sides by } 10$$
$$10\left(\tfrac{3}{5}t\right) + 10\left(-\tfrac{1}{10}t\right) = 10t + 10\left(-\tfrac{5}{2}\right)$$
$$\textit{Distributive property}$$
$$6t - t = 10t - 25$$
$$5t = 10t - 25$$

$5t - 10t = 10t - 25 - 10t$
  Subtract 10t from both sides

$-5t = -25$

$\dfrac{-5t}{-5} = \dfrac{-25}{-5}$ Divide both sides by -5

$t = 5$

23. $\dfrac{2}{3}k - \left(k + \dfrac{1}{4}\right) = \dfrac{1}{12}(k + 4)$

The least common denominator of all the fractions in the equation is 12.

$12\left[\dfrac{2}{3}k - \left(k + \dfrac{1}{4}\right)\right] = 12\left[\dfrac{1}{12}(k + 4)\right]$
  Multiply both sides by 12

$12\left(\dfrac{2}{3}k\right) + 12\left[-\left(k + \dfrac{1}{4}\right)\right] = 12\left[\dfrac{1}{12}(k + 4)\right]$
  Distributive property

$8k + 12(-k) + 12\left(-\dfrac{1}{4}\right) = 1(k + 4)$

$8k - 12k - 3 = k + 4$

$-4k - 3 = k + 4$

$-4k - 3 - k = k + 4 - k$
  Subtract k from both sides

$-5k - 3 = 4$

$-5k - 3 + 3 = 4 + 3$

$-5k = 7$

$\dfrac{-5k}{-5} = \dfrac{7}{-5}$
  Divide both sides by -5

$k = -\dfrac{7}{5}$

27. $1.00x + .05(12 - x) = .10(63)$

To eliminate the decimals, we multiply both sides by 100.

$100[1.00x + .05(12 - x)] = 100[.10(63)]$

$100(1.00x) + 100[.05(12 - x)] = (100)(.10)(63)$

$100x + 5(12 - x) = 10(63)$

$100x + 60 - 5x = 630$

$95x + 60 = 630$

$95x = 570$

$\dfrac{95x}{95} = \dfrac{570}{95}$

$x = 6$

In Exercises 31–43, all solutions should be checked by substituting in the original equation. Here, only the check for Exercise 31 will be shown.

31. $10(2x - 1) = 8(2x + 1) + 14$

$20x - 10 = 16x + 8 + 14$

$20x - 10 = 16x + 22$

$4x - 10 = 22$

$4x = 32$

$x = 8$

Check

$10(2x - 1) = 8(2x + 1) + 14$

$10[2(8) - 1] = 8[2(8) + 1] + 14$?
  Let $x = 8$

$10(16 - 1) = 8(16 + 1) + 14$?

$10(15) = 8(17) + 14$?

$150 = 136 + 14$?

$150 = 150$ True

The solution is 8.

35. $-(4y + 2) - (-3y - 5) = 3$

$-1(4y + 2) - 1(-3y - 5) = 3$

$-4y - 2 + 3y + 5 = 3$

$-y + 3 = 3$

$-y = 0$

$y = 0$

39. $.10(x + 80) + .20x = 14$

To eliminate the decimals, we multiply both sides by 100.

$$100[.10(x + 80) + .20x] = 100(14)$$
$$10(x + 80) + 20x = 1400$$
$$10x + 800 + 20x = 1400$$
$$30x + 800 = 1400$$
$$30x = 600$$
$$x = 20$$

43. $9(v + 1) - 3v = 2(3v + 1) - 8$
$$9v + 9 - 3v = 6v + 2 - 8$$
$$6v + 9 = 6v - 6$$
$$9 = -6 \quad \textit{False}$$

Because this is a false statement, the equation has no solution.

47. A number added to -6 is written

$$-6 + x.$$

51. The quotient of -6 and a nonzero number is written

$$\frac{-6}{x}.$$

Section 2.4 An Introduction to Applied Problems

2.4 Margin Exercises

1. Let x = the number.

5 added to	the product of 9 and a number	is	19 less than the number.
↓ ↓	↓	↓	↓
5 +	9x	=	x − 19

Solve the equation.

$$5 + 9x = x - 19$$
$$5 + 9x - 5 = x - 19 - 5 \quad \text{Subtract 5}$$
$$9x = x - 24$$
$$9x - x = x - 24 - x \quad \text{Subtract } x$$
$$8x = -24$$
$$\frac{8x}{8} = \frac{-24}{8} \quad \text{Divide by 8}$$
$$x = -3$$

The number is -3. A check shows that

$$5 + 9(-3) = 5 + (-27) = -22$$

and

$$-3 - 19 = -22.$$

Both results are -22, so -3 is the correct answer.

2. Let x = the number of miles Jim drove;

  3x = the number of miles Annie drove.

Number of miles Jim drove	plus	number of miles Annie drove	is	84 miles altogether.
↓	↓	↓	↓	↓
x	+	3x	=	84

Solve this equation.

$$x + 3x = 84$$
$$4x = 84$$
$$\frac{4x}{4} = \frac{84}{4} \quad \text{Divide by 4}$$
$$x = 21$$

Jim drove 21 miles and Annie drove $3 \cdot 21 = 63$ miles. Since $21 + 63 = 84$, the check shows they drove 84 miles altogether. The answers are correct.

3. Let x = the length of the middle-sized piece;

   $x + 10$ = the length of the longest piece;

   $x - 5$ = the length of the shortest piece.

Length of shortest piece	plus	length of middle-sized piece	plus
↓	↓	↓	↓
$x - 5$	+	x	+

length of longest piece	is	total length of pipe.
↓	↓	↓
$x + 10$	=	50

Solve this equation.

$$(x - 5) + x + (x + 10) = 50$$
$$3x + 5 = 50$$
$$3x + 5 - 5 = 50 - 5$$
$$3x = 45$$
$$\frac{3x}{3} = \frac{45}{3}$$
$$x = 15$$

The middle-sized piece is 15 inches long, the longest piece is $15 + 10 = 25$ inches long, and the shortest piece is $15 - 5 = 10$ inches long. Since $15 + 25 + 10 = 50$ inches, the length of the pipe, the answers are correct.

4. Let x = the number of members;

   $2x$ = the number of nonmembers.

   (If each member brought two nonmembers, there would be twice as many nonmembers as members).

Members	plus	nonmembers
↓	↓	↓
x	+	$2x$

is	the total in attendance.
↓	↓
=	27

Solve the equation.

$$x + 2x = 27$$
$$3x = 27$$
$$\frac{3x}{3} = \frac{27}{3}$$
$$x = 9$$

There were 9 members and $2 \cdot 9 = 18$ nonmembers. Since $9 + 18 = 27$, the total number of people, the answers are correct.

5. Let x = the degree measure of the angle.

   Then $90 - x$ = the degree measure of its complement

   and $180 - x$ = the degree measure of its supplement.

```
Twice the      is    the        less  30°.
complement           supplement
    ↓          ↓      ↓           ↓    ↓
 2(90 - x)    =    (180 - x)   -   30
```

Solve the equation.

$$180 - 2x = 180 - x - 30$$
$$180 - 2x = 150 - x$$
$$180 - 2x + 2x = 150 - x + 2x$$
$$180 = 150 + x$$
$$180 - 150 = 150 - 150 + x$$
$$30 = x$$

The measure of the angle is 30°.

2.4 Section Exercises

The applied problems in Exercises 7–35 should be solved by using the six-step method shown in the text. These steps will only be listed in a few of the solutions, but all of the solutions are based on this method.

7. *Step 1* Let x = the unknown number

   Step 2 $2x + 3$ = the sum of 3 and twice the number

   $4(2x + 3)$ = the product of 4 with this sum

   $7x + 8$ = the sum of 8 and seven times the number

   Step 3 $4(2x + 3) = 7x + 8$

   Step 4
   $$8x + 12 = 7x + 8$$
   $$8x - 7x + 12 = 7x - 7x + 8$$
   $$x + 12 = 8$$
   $$x + 12 - 12 = 8 - 12$$
   $$x = -4$$

   Step 5 The number is -4.

   Step 6 Check that -4 is the correct answer by substituting this result into the words of the original problem.

   Twice the number is $2(-4) = -8$.

   Three added to twice the number is $-8 + 3 = -5$.

   This sum multiplied by 4 is $4(-5) = -20$.

   The number multiplied by 7 is $7(-4) = -28$.

   Eight added to this product is $-28 + 8 = -20$.

   Because both results are -20, the answer, -4, checks.

11. Let x = the number of wins.
    Then $x + 558$ = the number of losses.

    The total number of games was 1516, so
    $$x + (x + 558) = 1516.$$

    Solve this equation.
    $$x + x + 558 = 1516$$
    $$2x + 558 = 1516$$
    $$2x = 958$$
    $$x = 479$$
    $$x + 558 = 1037$$

    Auerbach has 1037 wins. Because $479 + 1037 = 1516$, this answer checks.

15. Let x = the number of packages delivered by United Parcel Service.

Then $x + 2$ = the number of packages delivered by Airborne Express

and $3(x + 2)$ = the number of packages delivered by Federal Express.

$$x + (x + 2) + 3(x + 2) = 13$$
$$x + x + 2 + 3x + 6 = 13$$
$$5x + 8 = 13$$
$$5x = 5$$
$$x = 1$$

One package was delivered by United Parcel Service, 3 were delivered by Airborne Express, and 9 were delivered by Federal Express.

19. Let x = the number of hits Cabrera had,

$47x$ = number of hits Gant had,

$47x - 17$ = number of hits Justice had.

$$x + 47x + (47x - 17) = 268$$
$$x + 47x + 47x - 17 = 268$$
$$95x - 17 = 268$$
$$95x = 285$$
$$x = 3$$

$$47x = 47(3) = 141$$
$$47x - 17 = 141 - 17 = 124$$

Cabrera had 3 hits, Gant had 141 hits, and Justice had 124 hits.

$3 + 141 + 124 = 268$, so this answer checks.

23. Let x = the suggested list price of the power door locks;

$\frac{10}{3}x$ = the suggested list price for the antilock brake system.

$$x + \frac{10}{3}x = 1040$$

$$3(x + \frac{10}{3}x) = 3(1040)$$
$$3x + 10x = 3120$$
$$13x = 3120$$
$$x = 240$$

The power door locks cost $240 and the antilock brake system costs $800. The combined cost is $240 + $800 = $1040, so this answer checks.

27. *Step 1* Let x = the measure of the angle.

Step 2 $90 - x$ = the measure of its complement;

$180 - x$ = the measure of its supplement.

Step 3 Write an equation.

The measure of the supplement	is	10 times	the measure of the complement.
↓	↓	↓	↓
$180 - x$	=	$10 \cdot$	$(90 - x)$

Step 4 Solve the equation.

$$180 - x = 10(90 - x)$$
$$180 - x = 900 - 10x$$
$$180 - x - 180 = 900 - 10x - 180$$
Subtract 180
$$-x = 720 - 10x$$
$$-x + 10x = 720 - 10x + 10x$$
Add 10x

$9x = 720$

$\dfrac{9x}{9} = \dfrac{720}{9}$ *Divide by 9*

$x = 80$

Step 5 The measure of the angle is 80°.

Step 6 The measure of the supplement (180 − 80 = 100) is 10 times the complement (90 − 80 = 10).

31. *Step 1* Let x = the measure of the angle.

    Step 2 $90 - x$ = the measure of its complement;

    $180 - x$ = the measure of its supplement.

    Step 3 Write an equation.

Measure of complement		Measure of supplement		Sum
↓		↓		↓
$(90 - x)$	$+$	$(180 - x)$	$=$	160

    Step 4 Solve the equation.

    $(90 - x) + (180 - x) = 160$

    $-2x + 270 = 160$

    $-2x + 270 - 270 = 160 - 270$ *Subtract 270*

    $-2x = -110$

    $\dfrac{-2x}{-2} = \dfrac{-110}{-2}$ *Divide by 2*

    $x = 55$

    Step 5 The measure of the angle is 55°.

Step 6 The sum of the measures of its complement (90° − 55° = 35°) and its supplement (180° − 55° = 125°) is 160° (35° + 125° = 160°).

35. *Step 1* Let x = an even integer.

    Step 2 $x + 2$ = the next larger even integer.

    Step 3 Write an equation.

The smaller integer	added to	3 times	The larger integer
↓	↓	↓	↓
x	$+$	$3 \cdot$	$(x + 2)$
totals		86.	
↓		↓	
$=$		86	

    Step 4 Solve the equation.

    $x + 3(x + 2) = 86$

    $x + 3x + 6 = 86$

    $4x + 6 = 86$

    $4x + 6 - 6 = 86 - 6$ *Subtract 6*

    $4x = 80$

    $\dfrac{4x}{4} = \dfrac{80}{4}$ *Divide by 4*

    $x = 20$

    Step 5 The first even integer is 20; the next larger even integer is 20 + 2 = 22.

    Step 6 The smaller added to three times the larger (20 + 66) is 86.

39. prt; $p = 4000$, $r = .04$, $t = 2$

    $prt = (4000)(.04)(2)$
    $ = (160)(2)$
    $ = 320$

Section 2.5 Formulas and Geometry Applications

2.5 Margin Exercises

1. **(a)** Given the formula $I = prt$, the problem is to find p when $I = \$246$, $r = .06$, and $t = 2$. Substitute these values into the formula, and solve for p.

$$I = prt$$
$$\$246 = p(.06)(2)$$
$$\$246 = .12p$$
$$\frac{\$246}{.12} = \frac{.12p}{.12} \quad \text{Divide by .12}$$
$$\$2050 = p$$

(b) Using $P = 2L + 2W$ as the formula, find L when $P = 126$ and $W = 25$. Substitute these values into the formula, and solve for L.

$$P = 2L + 2W$$
$$126 = 2L + 2(25)$$
$$126 = 2L + 50$$
$$126 - 50 = 2L + 50 - 50 \quad \text{Subtract 50}$$
$$76 = 2L$$
$$\frac{76}{2} = \frac{2L}{2} \quad \text{Divide by 2}$$
$$38 = L$$

2. The fence will enclose the perimeter of the rectangular field, so use the formula for the perimeter of a rectangle. Find the length of the field by substituting $P = 800$ and $W = 175$ into the formula and solving for L.

$$P = 2L + 2W$$
$$800 = 2L + 2(175)$$
$$800 = 2L + 350$$
$$800 - 350 = 2L + 350 - 350$$
$$450 = 2L$$
$$\frac{450}{2} = \frac{2L}{2}$$
$$225 = L$$

The length of the field is 225 meters.

3. Use the formula for the area of a triangle.

$$A = \tfrac{1}{2}bh$$
$$120 = \tfrac{1}{2}b(24) \quad \text{Let } A = 120, \; h = 24$$
$$120 = 12b$$
$$\frac{120}{12} = \frac{12b}{12}$$
$$10 = b$$

The length of the base is 10 meters.

4. **(a)** Because the angles are vertical angles, they have the same measure.

$$2x + 24 = 4x - 40$$
$$24 = 2x - 40$$
$$64 = 2x$$
$$32 = x$$

If $x = 32$,
$$2x + 24 = 2(32) + 24$$
$$= 64 + 24 = 88,$$
and
$$4x - 40 = 4(32) - 40$$
$$= 128 - 40 = 88.$$

Both angles measure 88°.

(b) The sum of the measures of the two angles is 180°.

$$(5x + 12) + 3x = 180$$
$$5x + 12 + 3x = 180$$
$$8x + 12 = 180$$
$$8x = 168$$
$$x = 21$$

If $x = 21$,

$$5x + 12 = 5(21) + 12$$
$$= 105 + 12 = 117$$

and

$$3x = 3(21) = 63.$$

The measures of the angles are 117° and 63°.

5. (a) Solve $I = prt$ for t.

$$I = prt$$
$$\frac{I}{pr} = \frac{prt}{pr} \quad \text{Divide by } pr$$
$$\frac{I}{pr} = t, \text{ or } t = \frac{I}{pr}$$

(b) Solve $P = a + b + c$ for a.

$$P = a + b + c$$
$$P - (b + c) = a + b + c - (b + c)$$
$$\quad\quad\quad\quad\quad \text{Subtract } (b + c)$$
$$P - b - c = a, \text{ or } a = P - b - c$$

6. (a) Solve $A = p + prt$ for t.

$$A = p + prt$$
$$A - p = p + prt - p$$
$$\quad\quad\quad\quad \text{Subtract } p$$
$$A - p = prt$$
$$\frac{A - p}{pr} = \frac{prt}{pr} \quad \text{Divide by } pr$$
$$\frac{A - p}{pr} = t, \text{ or } t = \frac{A - p}{pr}$$

(b) Solve $y = mx + b$ for x.

$$y = mx + b$$
$$y - b = mx + b - b$$
$$\quad\quad\quad\quad \text{Subtract } b$$
$$y - b = mx$$
$$\frac{y - b}{m} = \frac{mx}{m} \quad \text{Divide by } m$$
$$\frac{y - b}{m} = x, \text{ or } x = \frac{y - b}{m}$$

2.5 Section Exercises

3. Carpeting for a bedroom covers the surface of the bedroom floor, so area would be used.

7. Tile for a bathroom covers the surface of the bathroom floor, so area would be used.

In Exercises 11-29, substitute the give value into the formula and then solve for the remaining variable.

11. $P = 2L + 2W$; $L = 6$, $W = 4$

$$P = 2L + 2W$$
$$= 2(6) + 2(4)$$
$$= 12 + 8$$
$$P = 20$$

15. $A = \frac{1}{2}bh$; $b = 10$, $h = 14$

$$A = \frac{1}{2}bh$$
$$A = \frac{1}{2}(10)(14)$$
$$= (5)(14)$$
$$A = 70$$

64 Chapter 2 Solving Equations and Inequalities

19. $d = rt$; $d = 100$, $t = 2.5$

$$d = rt$$
$$100 = r(2.5)$$
$$100 = 2.5r$$
$$\frac{100}{2.5} = \frac{2.5r}{2.5}$$
$$40 = r$$

23. $A = \frac{1}{2}h(b + B)$; $h = 7$, $b = 12$, $B = 14$

$$A = \frac{1}{2}bh$$
$$= \frac{1}{2}(7)(12 + 14)$$
$$= \frac{1}{2}(7)(26)$$
$$= \frac{1}{2}(182)$$
$$A = 91$$

27. $A = \pi r^2$; $r = 12$, $\pi = 3.14$

$$A = \pi r^2$$
$$A = 3.14(12)^2$$
$$= 3.14(144)$$
$$A = 452.16$$

31. $V = \frac{1}{3}Bh$; $B = 36$, $h = 4$

$$V = \frac{1}{3}Bh$$
$$= \frac{1}{3}(36)(4)$$
$$= (12)(4)$$
$$V = 48$$

39. A page of the newspaper is a rectangle with length 51 inches and width 35 inches, so use the formulas for the perimeter and area of a rectangle.

$$P = 2L + 2W$$
$$= 2(51) + 2(35)$$
$$= 102 + 70$$
$$P = 172$$

The perimeter was 172 inches.

$$A = LW$$
$$= (51)(35)$$
$$A = 1785$$

The area was 1785 square inches.

43. Use the formula for the area of a trapezoid with $B = 115.80$, $b = 171.00$, and $h = 165.97$.

$$A = \frac{1}{2}(B + b)h$$
$$= \frac{1}{2}(115.80 + 171.00)(165.97)$$
$$= \frac{1}{2}(286.80)(165.97)$$
$$= 23,800.098$$

To the nearest hundredth of a square foot, the combined area of the two lots is 23,800.10 square feet.

47. The angles are opposite each other, so they are vertical angles. Set their measures equal to each other and solve for x.

$$7x + 5 = 3x + 45$$
$$4x + 5 = 45 \quad \textit{Subtract 3x}$$
$$4x = 40 \quad \textit{Subtract 5}$$
$$x = 10 \quad \textit{Divide by 10}$$

The measure of the first angle is $7(10) + 5 = 75°$; the measure of the second angle is $3(10) + 45$, which is also 75°.

51. $A = LW$ for L

$$A = LW$$
$$\frac{A}{W} = \frac{LW}{W} \quad \text{Divide by } W$$
$$\frac{A}{W} = L$$
$$\text{or} \quad L = \frac{A}{W}$$

55. $I = prt$ for p

$$\frac{I}{rt} = \frac{prt}{rt} \quad \text{Divide by } rt$$
$$\frac{I}{rt} = p$$
$$\text{or} \quad p = \frac{I}{rt}$$

59. $A = \frac{1}{2}bh$ for b

$$2A = 2\left(\frac{1}{2}bh\right) \quad \text{Multiply by 2}$$
$$2A = bh$$
$$\frac{2A}{h} = \frac{bh}{h} \quad \text{Divide by } h$$
$$\frac{2A}{h} = b$$
$$\text{or} \quad b = \frac{2A}{h}$$

63. $V = \pi r^2 h$ for h

$$\frac{V}{\pi r^2} = \frac{\pi r^2 h}{\pi r^2} \quad \text{Divide by } \pi r^2$$
$$\frac{V}{\pi r^2} = h$$
$$\text{or} \quad h = \frac{V}{\pi r^2}$$

67. To convert .05 to a percent, move the decimal point two places to the right and attach a percent sign.

$$.05 = 5\%$$

71. $\frac{3}{4}y = 21$

$$\frac{4}{3}\left(\frac{3}{4}y\right) = \frac{4}{3}(21) \quad \begin{array}{l}\text{Multiply by 4/3, the}\\\text{reciprocal of 3/4}\end{array}$$
$$y = \frac{84}{3} = 28$$

2.6 Percent; Ratio and Proportion

2.6 Margin Exercises

1. (a) To find 20% of 70, multiply the decimal equivalent of 20% by 70.

$$(.20)(70) = 14$$

(b) Find 25% of $270 by muliplying.

$$(25)(270) = 67.50$$

The discount is $67.50. The sale price of the set is $270 - $67.50 = $202.50.

2. (a) Let x represent the percent in decimal form.

90 is what percent of 270?
↓ ↓ ↓ ↓ ↓
90 = x · 270

$$90 = 270x$$
$$\frac{90}{270} = \frac{270x}{270}$$
$$\frac{1}{3} = x$$
$$x = \frac{1}{x} = .\overline{3}$$

$.\overline{3} = 33\frac{1}{3}\%$, so 90 is 33 1/3% of 270.

(b) Let x represent the percent in decimal form.

682 is what percent of $11,000 ?

682 = x · 11,000

$$682 = 11{,}000x$$

$$\frac{682}{11{,}000} = \frac{11{,}000x}{11{,}000}$$

$$.062 = x$$

.062 = 6.2%, so $682 is 6.2% of $11,000.

3. (a) The ratio of 9 women to 5 women is

$$\frac{9}{5}.$$

(b) To find the ratio of 4 inches to 1 foot, first convert 1 foot to 12 inches. The ratio of 4 inches to 1 foot is then

$$\frac{4}{12} = \frac{1}{3}.$$

4. (a) $\frac{y}{6} = \frac{35}{42}$

Find the cross products.

$$42y = 6 \cdot 35$$

$$y = \frac{6 \cdot 35}{42}$$

$$= \frac{6 \cdot 5 \cdot 7}{6 \cdot 7} = 5$$

(b) $\frac{a}{24} = \frac{15}{16}$

Find the cross products.

$$16a = 24 \cdot 15$$

$$a = \frac{24 \cdot 15}{16}$$

$$= \frac{3 \cdot 8 \cdot 15}{2 \cdot 8}$$

$$= \frac{45}{2}$$

5. (a) $\frac{z}{2} = \frac{z+1}{3}$

Find the cross products.

$$3z = 2(z + 1)$$

$$3z = 2z + 2$$

$$z = 2$$

(b) $\frac{p+3}{3} = \frac{p-5}{4}$

Find the cross products.

$$4(p + 3) = 3(p - 5)$$

$$4p + 12 = 3p - 15$$

$$p = -27$$

6. Let x = the number of miles represented by 30 inches.

Inches Miles

$$\frac{12}{30} = \frac{500}{x}$$

$$12x = 30 \cdot 500 \quad \text{Cross products}$$

$$12x = 15{,}000$$

$$x = 1250 \quad \text{Divide by 12}$$

1250 miles would be represented by 30 inches.

7.

Size	Unit cost (dollars per square foot)
1 roll	$\frac{\$.85}{63} = \$.0135$
3-roll package	$\frac{\$2.89}{3(63)} = \$.0153$

The best buy is 1 roll for $.85.

2.6 Section Exercises

3. What is 26% of 480?

26% of 480 = (.26)(480) = 124.8

7. 25% of what number is 150?

Let x represent the unknown number. Change 25% to a decimal and translate the question into symbols.

$$.25x = 150$$
$$\frac{.25x}{.25} = \frac{150}{.25}$$
$$x = 600$$

25% of 600 is 150.

11. The problem may be stated as follows: "What is 16.4% of $3250?"

16.4% of 3250 = (.164)(3250)
= 533.000,

so 16.4% of $3250 is $533.

15. Let x represent the percent discount in decimal form. The amount of discount is

$$\$180 - \$150 = \$30,$$

so the problem may be stated "What percent of 180 is 30?"

$$(x)(180) = 30$$
$$180x = 30$$
$$\frac{180x}{180} = \frac{30}{180}$$
$$x = \frac{1}{6} = .1\overline{6}$$

$.1\overline{6} = 16.\overline{6}$% or 16 2/3%.
Thus, the percent discount is $16.\overline{6}$% or 16 2/3%.

19. Let x represent the percent increase in decimal form.

The increase in value is

$$\$1750 - \$625 = \$1125.$$
$$(x)(625) = 1125$$
$$625x = 1125$$
$$\frac{625x}{625} = \frac{1125}{625}$$
$$x = 1.8.$$

1.8 = 180%, so the percent increase is 180%.

23. New York

$$\text{winning percentage} = \frac{\text{number of wins}}{\text{number of games}}$$
$$= \frac{52}{52 + 63}$$
$$= \frac{52}{115}$$
$$= .452$$

27. The ratio of 40 miles to 30 miles is

$$\frac{40}{30} = \frac{4}{3}.$$

31. To find the ratio of 20 yards to 8 feet, first convert 20 yards to feet.

$$20 \text{ yards} = 20 \cdot 3 = 60 \text{ feet}$$

The ratio of 20 yards to 8 feet is thus

$$\frac{60}{8} = \frac{15 \cdot 4}{2 \cdot 4} = \frac{15}{2}.$$

35. To convert 8 days to 40 hours, first convert 8 days to hours.

$$8 \text{ days} = 8 \cdot 24 = 192 \text{ hours}$$

The ratio of 8 days to 40 hours is thus

$$\frac{192}{40} = \frac{8 \cdot 24}{8 \cdot 5} = \frac{24}{5}.$$

39. $\quad \dfrac{k}{4} = \dfrac{175}{20}$

$\quad 20k = 4(175) \quad$ *Cross products are equal*

$\quad 20k = 700$

$\quad \dfrac{20k}{20} = \dfrac{700}{20} \quad$ *Divide by 20*

$\quad k = 35$

43. $\quad \dfrac{3y - 2}{5} = \dfrac{6y - 5}{11}$

$\quad 11(3y - 2) = 5(6y - 5)$
$\quad\quad\quad\quad\quad\quad\quad\;$ *Cross products are equal*

$\quad 33y - 22 = 30y - 25$
$\quad\quad\quad\quad\quad\quad\quad\;$ *Distributive property*

$\quad 3y - 22 = -25 \quad$ *Subtract 30y*

$\quad 3y = -3 \quad$ *Add 22*

$\quad y = -1 \quad$ *Divide by 3*

47. Let x = the number of calories in 12 slices.

Set up a proportion with number of calories in one ratio and number of slices in the other.

$$\frac{x \text{ calories}}{85 \text{ calories}} = \frac{12 \text{ slices}}{2 \text{ slices}}$$

$$\frac{x}{85} = \frac{12}{2}$$

$(x)(2) = (85)(12) \quad$ *Cross products*

$\quad 2x = 1020$

$\quad \dfrac{2x}{2} = \dfrac{1020}{2} \quad$ *Divide by 2*

$\quad x = 510$

Twelve slices of bacon provide 510 calories.

51. Let x = the number of inches between Mexico City and Cairo on the map.

Set up a proportion.

$$\frac{11 \text{ inches}}{x \text{ inches}} = \frac{3300 \text{ miles}}{7700 \text{ miles}}$$

$$\frac{11}{x} = \frac{3300}{7700}$$

$3300x = 11(7700) \quad$ *Cross products*

$3300x = 84{,}700$

$\quad x = \dfrac{84{,}700}{3300}$

$\quad\quad = \dfrac{847}{33} = 25\dfrac{2}{3} \quad$ *Lowest terms*

Mexico City and Cairo are 25 2/3 inches apart on the map.

In Exercises 55 and 59, to find the best buy, divide the price by the number of units to get the unit cost. Each result was found by using a calculator and rounding the answers to three decimal places.

55.

<u>Size   Unit Cost (dollars per bag)</u>

20-count $\frac{\$2.49}{20} = \$.125$

30-count $\frac{\$4.29}{30} = \$.143$

The 20-count size is the best buy.

59.

<u>Size   Unit Cost (dollars per ounce)</u>

14-ounce size $\frac{\$.93}{14} = \$.066$

32-ounce size $\frac{\$1.19}{32} = \$.037$

44-ounce size $\frac{\$2.19}{44} = \$.050$

The 32-ounce size is the best buy.

63. $-13 < -6$ because -13 is to the left of -6 on the number line.

Section 2.7 The Addition and Multiplication Properties of Inequality

2.7 Margin Exercises

In Exercises 1-3, refer to the graphs included with the answers to the margin exercises in the textbook.

1. **(a)** $x \leq 3$

The statement $x \leq 3$ includes all real numbers less than or equal to 3. To graph this, place a solid dot at 3 on a number line and draw an arrow extending from the dot to the left.

(b) $x > -4$

x can be any real number greater than -4 but not equal to -4. To graph this, place an open circle at -4 on a number line and draw an arrow to the right.

(c) $-4 \geq x$

$-4 \geq x$ is the same as $x \leq -4$. Graph this by placing a solid dot at -4 on a number line and drawing an arrow to the left.

(d) $0 < x$

$0 < x$ is the same as $x > 0$. Graph this by placing an open circle at 0 on a number line and drawing an arrow to the right.

Chapter 2 Solving Equations and Inequalities

2. (a) $-7 < x < -2$

To graph this inequality, place open circles at -7 and -2 on a number line. Then draw a line segment between the two circles.

(b) $-6 < x \leq -4$

To graph this inequality, draw an open circle at -6 and a solid dot at -4 on a number line. Then draw a line segment between the circle and the dot.

3. (a)
$$-1 + 8r < 7r + 2$$
$$-1 + 8r + 1 < 7r + 2 + 1 \quad \text{Add 1}$$
$$8r < 7r + 3$$
$$8r - 7r < 7r + 3 - 7r \quad \text{Subtract } 7r$$
$$r < 3$$

To graph this, place an open circle at 3 on a number line and draw an arrow to the left.

(b)
$$4m \geq 5m - \frac{4}{3}$$
$$3(4m) \geq 3\left(5m - \frac{4}{3}\right) \quad \text{Multiply by 3}$$
$$12m \geq 15m - 4 \quad \text{Distributive property}$$
$$-3m \geq -4 \quad \text{Subtract 15}$$
$$\frac{-3m}{-3} \leq \frac{-4}{-3} \quad \text{Divide by } -3; \text{ reverse the symbol}$$
$$m \leq \frac{4}{3}$$

To graph this, place a solid dot at 4/3 and draw an arrow to the left.

4. (a)
$$-2 < 8$$
$$6(-2) < 6(8) \quad \text{Multiply by 6}$$
$$-12 < 48$$

$$-2 < 8$$
$$-5(-2) > -5(8) \quad \text{Multiply by } -5; \text{ reverse symbol}$$
$$10 > -40$$

(b)
$$-4 > -9$$
$$2(-4) > 2(-9) \quad \text{Multiply by 2}$$
$$-8 > -18$$

$$-4 > -9$$
$$-8(-4) < -8(-9) \quad \text{Multiply by } -8; \text{ reverse symbol}$$
$$32 < 72$$

In Exercises 5–7, refer to the graphs included with the answers to the margin exercises in the textbook.

5. (a)
$$9y < -18$$
$$\frac{9y}{9} < \frac{-18}{9} \quad \text{Divide by 9}$$
$$y < -2$$

(b)
$$-2r > -12$$
$$\frac{-2r}{-2} < \frac{-12}{-2} \quad \text{Divide by } -2; \text{ reverse symbol}$$
$$r < 6$$

(c)
$$-5p \leq 0$$
$$\frac{-5p}{-5} \geq \frac{0}{-5} \quad \text{Divide by } -5; \text{ reverse symbol}$$
$$p \geq 0$$

6.
$$5r - r + 2 < 7r - 5$$
$$4r + 2 < 7r - 5$$
$$4r + 2 - 7r < 7r - 5 - 7r \quad \text{Subtract } 7r$$
$$-3r + 2 < -5$$
$$-3r + 2 - 2 < -5 - 2 \quad \text{Subtract 2}$$

$$-3r < -7$$
$$\frac{-3r}{-3} > \frac{-7}{-3}$$
Divide by -3; reverse symbol
$$r > \frac{7}{3}$$

7. $4(y - 1) - 3y > -15 - (2y + 1)$
    $4y - 4 - 3y > -15 - 2y - 1$ *Distributive property*
    $y - 4 > -16 - 2y$
    $y - 4 + 2y > -16 - 2y + 2y$ *Add 2y*
    $3y - 4 > -16$
    $3y - 4 + 4 > -16 + 4$ *Add 4*
    $3y > -12$
    $\frac{3y}{3} > \frac{-12}{3}$ *Divide by 3*
    $y > -4$

8. Let x = Maggie's score on the fourth test.

The average	is at least	90.
↓	↓	↓
$\frac{1}{4}(98 + 86 + 88 + x)$	≥	90

    Solve the inequality.
    $$\frac{1}{4}(98 + 86 + 88 + x) \geq 90$$
    $$\frac{1}{4}(272 + x) \geq 90$$
    $$4 \cdot \frac{1}{4}(272 + x) \geq 4 \cdot 90 \quad \text{Multiply by 4}$$
    $$272 + x \geq 360$$
    $$272 + x - 272 \geq 360 - 272$$
    $$x \geq 88$$

    She should score 88 or more.

2.7 Section Exercises

For Exercises 3 and 7, see the number line graphs in the answer section of the textbook.

3. $x < -3$

    This statement says that x can take any value less than -3. Place an open circle at -3 (because -3 is not part of the graph) and draw an arrow extending to the left.

7. $8 \leq x \leq 10$

    Place solid dots at 8 and 10 (because both endpoints are included in the graph) and draw a line segment between the two dots.

11. An open circle is used at any endpoint where the inequality symbol is < or > to show that the endpoint is *not* included in the graph. A closed circle (also called a "solid dot") is used where the inequality symbol is ≤ or ≥ to show that the endpoint *is* included in the graph.

For Exercises 15 and 19, see the number line graphs in the answer section of the textbook.

15. $z - 8 \geq -7$
    $z - 8 + 8 \geq -7 + 8$ *Add 8*
    $z \geq 1$

    Place a solid dot at 1 and draw an arrow extending to the right.

19.
$$3n + 5 < 2n - 6$$
$$3n - 2n + 5 < 2n - 2n - 6 \quad \text{Subtract } 2n$$
$$n + 5 < -6$$
$$n + 5 - 5 < -6 - 5 \quad \text{Subtract } 5$$
$$n < -11$$

Place an open circle at -11 and draw an arrow extending to the left.

23. To solve the inequality $6x < -42$, you would divide both sides by 6, a positive number, so you would not reverse the direction of the inequality. The direction of the inequality is reversed only when both sides are multiplied or divided by a negative number.

For Exercises 27–47, see the number line graphs in the answer section of the textbook.

27.
$$2y \geq -20$$
$$\frac{2y}{2} \geq \frac{-20}{2}$$
$$y \geq -10$$

Place a solid dot at -10 and draw an arrow extending to the right.

31.
$$-x \geq 0$$
$$-1x \geq 0$$
$$\frac{-1x}{-1} \leq \frac{0}{-1} \quad \text{Divide by } -1; \text{ reverse the symbol}$$
$$x \leq 0$$

Place a solid dot at 0 and draw an arrow extending to the left.

35.
$$-.02x \leq .06$$
$$\frac{-.02x}{-.02} \geq \frac{.06}{-.02} \quad \text{Divide by } -.02; \text{ reverse the symbol}$$
$$x \geq -3$$

Place a solid dot at -3 and draw an arrow extending to the right.

39.
$$6x + 3 + x < 2 + 4x + 4$$
$$7x + 3 < 4x + 6$$
$$7x - 4x + 3 < 4x - 4x + 6$$
$$3x + 3 < 6$$
$$3x + 3 - 3 < 6 - 3$$
$$3x < 3$$
$$\frac{3x}{3} < \frac{3}{3}$$
$$x < 1$$

Place an open circle at 1 and draw an arrow extending to the left.

43.
$$5(x + 3) - 6x \leq 3(2x + 1) - 4x$$
$$5x + 15 - 6x \leq 6x + 3 - 4x \quad \text{Distributive property}$$
$$-x + 15 \leq 2x + 3$$
$$-x - 2x + 15 \leq 2x - 2x + 3 \quad \text{Subtract } 2x$$
$$-3x + 15 \leq 3$$
$$-3x + 15 - 15 \leq 3 - 15 \quad \text{Subtract } 15$$
$$-3x \leq -12$$
$$\frac{-3x}{-3} \geq \frac{-12}{-3} \quad \text{Divide by } -3; \text{ reverse the symbol}$$
$$x \geq 4$$

Place a solid dot at 4 and draw an arrow extending to the right.

47.
$$4x - (6x + 1) \le 8x + 2(x - 3)$$
$$4x - 1(6x + 1) \le 8x + 2(x - 3)$$
$$4x - 6x - 1 \le 8x + 2x - 6 \quad \textit{Distributive property}$$
$$-2x - 1 \le 10x - 6 \quad \textit{Combine like terms}$$
$$-2x - 10x - 1 \le 10x - 10x - 6 \quad \textit{Subtract 10x}$$
$$-12x - 1 \le -6$$
$$-12x - 1 + 1 \le -6 + 1 \quad \textit{Add 1}$$
$$-12x \le -5$$
$$\frac{-12x}{-12} \ge \frac{-5}{-12} \quad \textit{Divide by -12; reverse the symbol}$$
$$x \ge \frac{5}{12}$$

Place a solid dot at 5/12 and draw an arrow extending to the right.

51. Let x = the number.

Translate the statement given in the exercise into symbols to obtain the inequality

$$(3x + 6) - 8 < x + 4.$$

Solve the inequality.
$$3x + 6 - 8 < x + 8$$
$$3x - 2 < x + 4$$
$$2x - 2 < 4$$
$$2x < 6$$
$$x < 3$$

All numbers less than 3 satisfy the given condition.

55. Let F = the required Fahrenheit temperature.

The Fahrenheit temperature must correspond to a Celsius temperature that is less than or equal to 30 degrees.

$$C = \frac{5}{9}(F - 32) \le 30$$
$$\frac{9}{5}\left[\frac{5}{9}(F - 32)\right] \le \frac{9}{5}(30)$$
$$F - 32 \le 54$$
$$F - 32 + 32 \le 54 + 32$$
$$F \le 86$$

The temperature in Houston on a certain summer day is never more than 86 degrees Fahrenheit.

59. Let x = Greg's earnings in October (in dollars).

Average salary	is at least	$250.
↓	↓	↓
$\frac{1}{4}(200 + 300 + 225 + x)$	≥	250

$$\frac{1}{4}(725 + x) \ge 250$$
$$4 \cdot \frac{1}{4}(725 + x) \ge 4(250) \quad \textit{Multiply by 4}$$
$$725 + x \ge 1000$$
$$725 + x - 725 \ge 1000 - 725 \quad \textit{Subtract 725}$$
$$x \ge 275$$

Greg must earn at least $275 during October.

63. $(-2)^6 = (-2)(-2)(-2)(-2)(-2)(-2)$
$= 64$

Chapter 2 Solving Equations and Inequalities

67. $1 - [2 - (-4)] = 1 - [2 + 4]$
$= 1 - 6$
$= 1 + (-6)$
$= -5$

Chapter 2 Review Exercises

In Exercises 1–10, all solutions should be checked by substituting into the original equation. Checks will be shown here for only a few of the exercises.

1. $x - 7 = 2$
$x - 7 + 7 = 2 + 7$
$x = 9$

2. $4r - 6 = 10$
$4r - 6 + 6 = 10 + 6$
$4r = 16$
$\frac{4r}{6} = \frac{16}{4}$
$r = 4$

3. $5x + 8 = 4x + 2$
$5x - 4x + 8 = 4x - 4x + 2$
$x + 8 = 2$
$x + 8 - 8 = 2 - 8$
$x = -6$

4. $8t = 7t + \frac{3}{2}$
$8t - 7t = 7t - 7t + \frac{3}{2}$
$t = \frac{3}{2}$

Check

$8t = 7t + \frac{3}{2}$
$8\left(\frac{3}{2}\right) = 7\left(\frac{3}{2}\right) + \frac{3}{2}$? Let $t = 3/2$
$12 = \frac{21}{2} + \frac{3}{2}$?
$12 = \frac{24}{2}$?
$12 = 12$ True

5. $(4r - 8) - (3r + 12) = 0$
$1(4r - 8) - 1(3r + 12) = 0$
$r - 20 = 0$
$r = 20$

6. $7(2x + 1) = 6(2x - 9)$
$14x + 7 = 12x - 54$
$14x - 12x + 7 = 12x - 12x - 54$
$2x + 7 = -54$
$2x = -61$
$x = -\frac{61}{2}$

Check

$7(2x + 1) = 6(2x - 9)$
$7\left[2\left(-\frac{61}{2}\right) + 1\right] = 6\left[2\left(-\frac{61}{2}\right) - 9\right]$?
                                                              $x = -61/2$
$7(-61 + 1) = 6(-61 - 9)$?
$7(-60) = 6(-70)$?
$-420 = -420$ True

7. $-\frac{6}{5}y = -18$
$\left(-\frac{5}{6}\right)\left(-\frac{6}{5}y\right) = \left(-\frac{5}{6}\right)(-18)$
$y = 15$

8. $\frac{1}{2}r - \frac{1}{6}r + 3 = 2 + \frac{1}{6}r + 1$

The least common denominator is 6.

$6\left(\frac{1}{2}r - \frac{1}{6}r + 3\right) = 6\left(2 + \frac{1}{6}r + 1\right)$

$3r - r + 18 = 12 + r + 6$

$2r + 18 = 18 + r$

$r + 18 = 18$

$r = 0$

9. $3x - (-2x + 6) = 4(x - 4) + x$

$3x - 1(-2x + 6) = 4(x - 4) + x$

$3x + 2x - 6 = 4x - 16 + x$

$5x - 6 = 5x - 16$

$5x - 5x - 6 = 5x - 5x - 16$

$-6 = -16$ *False*

Because the result is a false statement, the given equation has no solution.

10. $.10(x + 80) + .20x = 14$

To clear decimals, multiply both sides by 10.

$1.0(x + 80) + 2.0x = 140$

$x + 80 + 2x = 140$

$3x + 80 = 140$

$3x = 60$

$x = 20$

11. Let x represent the number.

$5x + 7 = 3x$

$2x + 7 = 0$

$2x = -7$

$x = -\frac{7}{2}$

The number is -7/2.

12. Let x represent the number.

$2x - 4 = 36$

$2x = 40$

$x = 20$

The number is 20.

13. Let $\quad$ x = the land area of Rhode Island.

Then x + 5213 = land area of Hawaii.

The areas total 7637 square miles, so

$x + (x + 5213) = 7637.$

Solve this equation.

$2x + 5213 = 7637$

$2x = 2424$

$x = 1212$

The land area of Rhode Island is 1212 square miles and that of Hawaii is 6425 square miles.

14. Let x = the height of Twin Falls;

$\frac{5}{2}x$ = height of Seven Falls.

The sum of the height is 420 feet, so

$x + \frac{5}{2}x = 420.$

To clear fractions, multiply both sides by 2.

$2\left(x + \frac{5}{2}x\right) = 2(420)$

$2x + 5x = 840$

$7x = 840$

$x = 120$

The height of Twin Falls is 120 feet and that of Seven Falls is 300 feet.

15. Let x = the measure of the angle.

Then $90 - x$ = the measure of its complement

and $180 - x$ = the measure of its supplement.

$$180 - x = 10(90 - x)$$
$$180 - x = 900 - 10x$$
$$9x + 180 = 900$$
$$9x = 720$$
$$x = 80$$

The measure of the angle is 80°. Its complement measures 90° − 80° = 10°, and its supplement measures 180° − 80° = 100°. Thus, the measure of the supplement is 10 times the measure of the complement, so the answer checks.

16. Let x = the number of innings pitched by Morris,

$x - 9$ = the number of innings pitched by Saberhagen,

$x + 6$ = the number of innings pitched by Langston.

The total number of innings pitched was 795, so

$$x + (x - 9) + (x + 6) = 795.$$

Solve this equation.

$$3x - 3 = 795$$
$$3x = 798$$
$$x = 266$$

Morris pitched 266 innings, Saberhagen pitched 257 innings, and Langston pitched 272 innings.

$266 + 257 + 272 = 795$, so the answer checks.

In Exercises 17–20, substitute the given values into the given formula and then solve for the remaining variable.

17. $A = \frac{1}{2}bh$; $A = 44$, $b = 8$

$$A = \frac{1}{2}bh$$
$$44 = \frac{1}{2}(8)h$$
$$44 = 4h$$
$$11 = h$$

18. $A = \frac{1}{2}h(b + B)$; $b = 3$, $B = 4$, $h = 8$

$$A = \frac{1}{2}h(b + B)$$
$$A = \frac{1}{2}(8)(3 + 4)$$
$$= \frac{1}{2}(8)(7)$$
$$= (4)(7)$$
$$A = 28$$

19. $C = 2\pi r$; $C = 29.83$, $\pi = 3.14$

$$C = 2\pi r$$
$$29.83 = 2(3.14)r$$
$$29.83 = 6.28r$$
$$\frac{29.83}{6.28} = \frac{6.28r}{6.28}$$
$$4.75 = r$$

20. $V = \frac{4}{3}\pi r^3$; $r = 6$, $\pi = 3.14$

$$V = \frac{4}{3}\pi r^3$$
$$= \frac{4}{3}(3.14)(6)^3$$
$$= \frac{4}{3}(3.14)(216)$$
$$= \frac{4}{3}(678.24)$$
$$V = 904.32$$

21. $A = LW$ for L

$$\frac{A}{W} = \frac{LW}{W} \quad \text{Divide by } W$$

$$\frac{A}{W} = L \quad \text{or} \quad L = \frac{A}{W}$$

22. $A = \frac{1}{2}h(b + B)$ for h

$$2A = 2\left[\frac{1}{2}h(b + B)\right] \quad \text{Multiply by 2}$$

$$2A = h(b + B)$$

$$\frac{2A}{(b + B)} = \frac{h(b + B)}{(b + B)} \quad \text{Divide by } b + B$$

$$\frac{2A}{b + B} = h \quad \text{or} \quad h = \frac{2A}{b + B}$$

23. Because the two angles are supplementary,

$$(8x - 1) + (3x - 6) = 180.$$

Solve this equation.

$$11x - 7 = 180$$
$$11x = 187$$
$$x = 17$$

If $x = 17$,

$$8x - 1 = 8(17) - 1$$
$$= 136 - 1$$
$$= 135,$$

and

$$3x - 6 = 3(17) - 6$$
$$= 51 - 6$$
$$= 45.$$

The measures of the two angles are 135° and 45°.

24. The angles are vertical angles, so their measures are equal.

$$3x + 10 = 4x - 20$$
$$10 = x - 20$$
$$30 = x$$

If $x = 30$,

$$3x + 10 = 3(30) + 10$$
$$= 90 + 10$$
$$= 100,$$

and

$$4x - 20 = 4(30) - 20$$
$$= 120 - 20$$
$$= 100.$$

Each angle has a measure of 100°.

25. The screen is a rectangle with length 92.75 feet and width 70.5 feet, so substitute these values in the formulas for the perimeter and area of a rectangle.

$$P = 2L + 2W$$
$$= 2(92.75) + 2(70.5)$$
$$= 185.5 + 141$$
$$P = 326.5$$

The perimeter is 326.5 feet.

$$A = LW$$
$$= (92.75)(70.5)$$
$$A = 6538.875$$

The area is 6538.875 square feet.

26. First, use the formula for the circumference of a circle to find the value of r.

$$C = 2\pi r$$
$$62.5 = 2(3.14)(r) \quad \text{Let } C = 62.5,$$
$$\pi = 3.14$$
$$62.5 = 6.28r$$
$$\frac{62.5}{6.28} = \frac{6.28r}{6.28}$$
$$9.95 \approx r$$

The radius of the turntable is approximately 9.95 feet.
The diameter is twice the radius, so the diameter is approximately 19.9 feet.

Now use the formula for the area of a circle.

$$A = \pi r^2$$
$$= (3.14)(9.95)^2 \quad \text{Let } \pi = 3.14,$$
$$r = 9.95$$
$$= (3.14)(99.0025)$$
$$A \approx 311$$

The area of the turntable is approximately 311 square feet.

27. 23% of $76 = (23\%)(76)$
$$= (.23)(76)$$
$$= 17.48$$

28. Let x represent the percent as a decimal.
$$x(12) = 21$$
$$12x = 21$$
$$\frac{12x}{12} = \frac{21}{12}$$
$$x = 1.75$$

To change 1.75 to a percent, move the decimal point two places to the right and drop the percent sign.
$$1.75 = 175\%$$
12 is 175% of 21.

29. Let x represent the percent as a decimal.
$$x(18) = 6$$
$$18x = 6$$
$$x = \frac{6}{18} = \frac{1}{3} = .\overline{3}$$

To change $.\overline{3}$ to a percent, move the decimal point two places to the right and drop the decimal point.
$$.\overline{3} = 33.\overline{3}\% \text{ or } 33\tfrac{1}{3}\%$$
6 is $33\tfrac{1}{3}\%$ of 18.

30. Let x represent the number. Change 36% to a decimal.
$$36\% = .36$$
$$.36x = 900$$
$$\frac{.36x}{.36} = \frac{900}{.36}$$
$$x = 2500$$

36% of 2500 is 900.

31. Let x represent the amount of the tax.
$$x = (6.5\%)(17,200)$$
$$x = (.065)(17,200)$$
$$x = 1118$$

Vinh must pay $1118 in sales tax.

32. Let x = the number of miles Alexandia should get per tank with the new tires.

    Alexandia gets 380 miles per tank with her old tires and would get 15% more or .15(380) additional miles with the new tires, so

    $$x = 380 + .15(380)$$
    $$= 380 + 57$$
    $$= 437.$$

    She should get 437 miles per tank with the steel-belted radials.

33. The ratio of 60 centimeters to 40 centimeters is

    $$\frac{60}{40} = \frac{3 \cdot 20}{2 \cdot 20} = \frac{3}{2}.$$

34. To find the ratio of 5 days to 2 weeks, first convert 2 weeks to days.

    $$2 \text{ weeks} = 2 \cdot 7 = 14 \text{ days}$$

    The ratio of 5 days to 2 weeks is thus

    $$\frac{5}{14}.$$

35. To find the ratio of 90 inches to 10 feet, first convert 10 feet to inches.

    $$10 \text{ feet} = 10 \cdot 12 = 120 \text{ inches}$$

    The ratio of 90 inches to 10 feet is thus

    $$\frac{90}{120} = \frac{3 \cdot 30}{4 \cdot 30} = \frac{3}{4}.$$

36. To find the ratio of 3 months to 3 years, first convert 3 years to months.

    $$3 \text{ years} = 3 \cdot 12 = 36 \text{ months}$$

    Thus, the ratio of 3 months to 3 years is

    $$\frac{3}{36} = \frac{1 \cdot 3}{12 \cdot 3} = \frac{1}{12}.$$

37. $\frac{p}{21} = \frac{5}{30}$

    $30p = 105$ *Cross products are equal*

    $\frac{30p}{30} = \frac{105}{30}$ *Divide by 30*

    $p = \frac{105}{30} = \frac{7 \cdot 15}{7 \cdot 2} = \frac{7}{2}$

38. $\frac{5 + x}{3} = \frac{2 - x}{6}$

    $6(5 + x) = 3(2 - x)$ *Cross products are equal*

    $30 + 6x = 6 - 3x$ *Distributive property*

    $30 + 9x = 6$ *Add 3x*

    $9x = -24$ *Subtract 30*

    $x = \frac{-24}{9} = -\frac{8}{3}$

39. $\frac{y}{5} = \frac{6y - 5}{11}$

    $11y = 5(6y - 5)$

    $11y = 30y - 25$

    $-19y = -25$

    $y = \frac{-25}{-19} = \frac{25}{19}$

41. Let x = the number of pounds of fertilizer needed to cover 500 square feet.

$$\frac{x \text{ pounds}}{2 \text{ pounds}} = \frac{150 \text{ square feet}}{500 \text{ square feet}}$$

$$\frac{x}{2} = \frac{150}{500}$$

$$150x = 1000$$

$$x = \frac{1000}{150} = \frac{20 \cdot 50}{3 \cdot 50}$$

$$= \frac{20}{3} = 6\frac{2}{3}$$

6 2/3 pounds of fertilizer will cover 500 square feet.

42. Let x = the number of ounces of medicine to mix with 90 ounces of water.

$$\frac{x \text{ ounces of medicine}}{8 \text{ ounces of medicine}} = \frac{90 \text{ ounces of water}}{20 \text{ ounces of water}}$$

$$\frac{x}{8} = \frac{90}{20}$$

$$20x = 720$$

$$x = 36$$

36 ounces of medicine should be mixed with 90 ounces of water.

43. Let x = the tax on a $36.00 item.

Set up a proportion with one ratio involving sales tax and the other involving the costs of the items.

$$\frac{x \text{ dollars}}{\$2.04} = \frac{\$36}{\$24}$$

$$\frac{x}{2.04} = \frac{36}{24}$$

$$24x = (2.04)(36)$$

$$24x = 73.44$$

$$\frac{24x}{24} = \frac{73.44}{24}$$

$$x = 3.06$$

The sales tax on a $36.00 item is $3.06.

44. Let x = the actual distance in between the second pair of cities (in kilometers).

Set up a proportion with one ratio involving map distances and the other involving actual distances.

$$\frac{x \text{ kilometers}}{150 \text{ kilometers}} = \frac{80 \text{ centimeters}}{32 \text{ centimeters}}$$

$$\frac{x}{150} = \frac{80}{32}$$

$$32x = (150)(80)$$

$$32x = 12,000$$

$$x = \frac{12,000}{32} = 375$$

The cities are 375 kilometers apart.

For Exercises 45–54, see the number line graphs in the answer section of the textbook.

45. $p \geq -4$

Place a solid dot at −4 (to show that −4 is part of the graph) and draw an arrow extending to the right.

46. $x < 7$

Place an open circle at 7 (to show that 7 is not part of the graph) and draw an arrow extending to the left.

47. $-5 \leq y < 6$

Place a solid dot at −5 and an open circle at 6; then draw a line segment between them.

48. $r \geq \frac{1}{2}$

   Place a solid dot at 1/2 and draw an arrow extending to the right.

49. $y + 6 \geq 3$
   $y + 6 - 6 \geq 3 - 6$
   $y \geq -3$

   Place a solid dot at -3 and draw an arrow extending to the right.

50. $5t < 4t + 2$
   $5t - 4t < 4t - 4t + 2$
   $t < 2$

   Place an open circle at 2 and draw an arrow extending to the left.

51. $-6x \leq -18$
   $\frac{-6x}{-6} \geq \frac{-18}{-6}$ *Divide by -6; reverse the symbol*
   $x \geq 3$

   Place a solid dot at 3 and draw arrow extending to the right.

52. $8(k - 5) - (2 + 7k) \geq 4$
   $8(k - 5) - 1(2 + 7k) \geq 4$
   $8k - 40 - 2 - 7k \geq 4$
   $k - 42 \geq 4$
   $k \geq 46$

   Place a solid dot at 46 and draw an arrow extending to the right.

53. $4x - 3x > 10 - 4x + 7x$
   $x > 10 + 3x$
   $-2x > 10$
   $\frac{-2x}{-2} < \frac{10}{-2}$
   $x < -5$

   Place an open circle at -5 and draw an arrow extending to the left.

54. $3(2w + 5) + 4(8 + 3w) < 5(3w + 2) + 2$
   $6w + 15 + 32 + 12w < 15w + 10 + 2w$
   $18w + 47 < 17w + 10$
   $w + 4 < 10$
   $w < -37$

   Place an open circle at -37 and draw an arrow extending to the left.

55. Let x = Carlotta's grade on the third test.

   $\frac{1}{3}(94 + 88 + x) \geq 90$

   $\frac{1}{3}(182 + x) \geq 90$

   $3 \cdot \frac{1}{3}(182 + x) \geq 3(90)$

   $182 + x \geq 270$

   $x \geq 88$

   Carlotta must score 88 or more on her third test.

56. Let x represent the number.

   "At most" means the same thing as "less than or equal to," so we use the inequality symbol $\leq$.

$$6 + 9x \leq 3$$
$$9x \leq -3$$
$$x \leq \frac{-3}{9}$$
$$x \leq -\frac{1}{3}$$

All numbers less than or equal to $-1/3$ satisfy the given condition.

57. $\quad \dfrac{y}{7} = \dfrac{y-5}{2}$

$\quad 2y = 7(y-5) \quad$ *Cross products are equal*

$\quad 2y = 7y - 35$

$\quad -5y = -35$

$\quad y = 7$

58. $I = prt \quad$ for r

$\quad \dfrac{I}{pt} = \dfrac{prt}{pt}$

$\quad \dfrac{I}{pt} = r \text{ or } r = \dfrac{I}{pt}$

59. $\quad -2x > -4$

$\quad \dfrac{-2x}{-2} < \dfrac{-4}{-2} \quad$ *Divide by -2; reverse the symbol*

$\quad x < 2$

60. $\quad 2x - 5 = 4k + 13$

$\quad -2k - 5 = 13 \quad$ *Subtract $2k$*

$\quad -2k = 18 \quad$ *Add 5*

$\quad k = -9 \quad$ *Divide -2*

61. $.05x + .02x = 4.9$

To clear decimals, multiply both sides by 100.

$$100(.05x + .02x) = 100(4.9)$$
$$5x + 2x = 490$$
$$7x = 490$$
$$x = 70$$

62. $\quad 2 - 3(y - 5) = 4 + y$

$\quad 2 - 3y + 15 = 4 + y$

$\quad 17 - 3y = 4 + y$

$\quad 17 - 4y = 4$

$\quad -4y = -13$

$\quad y = \dfrac{-13}{-4} = \dfrac{13}{4}$

63. $\quad 9x - (7x + 2) = 3x + (2 - x)$

$\quad 9x - 1(7x + 2) = 3x + (2 - x)$

$\quad 9x - 7x - 2 = 3x + 2 - x$

$\quad 2x - 2 = 2x + 2$

$\quad 2x - 2x - 2 = 2x - 2x + 2$

$\quad -2 = 2 \quad$ *False*

Because the result is a false statement, the given equation has no solution.

64. $\quad \dfrac{1}{3}s + \dfrac{1}{2}s + 7 = \dfrac{5}{6}s + 5 + 2$

$\quad \dfrac{1}{3}s + \dfrac{1}{2}s + 7 = \dfrac{5}{6}s + 7$

The least common denominator is 6.

$\quad 6\left(\dfrac{1}{3}s + \dfrac{1}{2}s + 7\right) = 6\left(\dfrac{5}{6}s + 7\right)$

$\quad 2s + 3s + 42 = 5s + 42$

$\quad 5s + 42 = 5s + 42$

$\quad 5s - 5s + 42 = 5s - 5s + 42$

$\quad 42 = 42 \quad$ *True*

Because the result is a true statement, all real numbers are solutions.

65. Let x be the number. Then

Two-thirds of a number	added to	the number	is	10.
↓	↓	↓	↓	↓
$\frac{2}{3}x$	+	x	=	10

Solve the equation.

$\frac{2}{3}x + x = 10$

$\frac{5}{3}x = 10$

$\frac{3}{5}(\frac{5}{3}x) = \frac{3}{5}(10)$ *Multiply by 3/5*

$x = 6$

The number is 6.

66. Let x be the number. Then

Twice the number	less	three-fourths of the number	is	15.
↓	↓	↓	↓	↓
2x	−	$\frac{3}{4}x$	=	15

Solve the equation.

$2x - \frac{3}{4}x = 15$

$\frac{5x}{4} = 15$

$\frac{4}{5}(\frac{5x}{4}) = \frac{4}{5}(15)$ *Multiply by 4/5*

$x = 12$

The number is 12.

67. Let x represent the number of votes Bob received.
Then 2x represents the number of votes Buddy received.

Bob's votes		Buddy's votes		total
↓		↓		↓
x	+	2x	=	1800

$x + 2x = 1800$

$3x = 1800$

$x = 600$ *Divide by 3*

Bob received 600 votes and Buddy received 2(600) = 1200 votes.

68. Let x be the number of miles that Gwen commutes.
Then 3x represents the number of miles that John commutes.

Gwen's miles		John's miles		total
↓		↓		↓
x	+	3x	=	112

$x + 3x = 112$

$4x = 112$

$x = 28$ *Divide by 4*

Gwen commutes 28 miles and John commutes 3(28 miles) = 84 miles.

69. Let x be the amount Hoa lost.
Then x + 18 is the amount Duc lost.

The total weight loss		was	42.
↓		↓	↓
x + (x + 18)		=	42

$x + (x + 18) = 42$

$2x + 18 = 42$

$2x + 18 - 18 = 42 - 18$ *Subtract 18*

$2x = 24$

$x = 12$ *Divide by 2*

Hoa lost 12 pounds.

84   Chapter 2   Solving Equations and Inequalities

70. Let x be the distance Steve drove.
Then x + 43 is the distance Rick drove.

The total distance   was   293.
    ↓   ↓   ↓
  x + (x + 43)   =   293

x + (x + 43) = 293
    2x + 43 = 293
2x + 43 − 43 = 293 − 43   *Subtract 43*
      2x = 250
       x = 125   *Divide by 2*

Steve drove 125 miles.

71. Let x be the number of geometry tests.
Then x + 32 is the number of algebra tests.

The total number of tests is 102.
    ↓   ↓   ↓
  x + (x + 32)   = 102

x + (x + 32) = 102
    2x + 32 = 102
2x + 32 − 32 = 102 − 32 *Subtract 32*
      2x = 70
       x = 35   *Divide by 2*

He graded 35 geometry tests.

72. Let x be the number of small cars.
Then x − 11 is the number of large cars.

The total number of cars was 63.
    ↓   ↓   ↓
  x + (x − 11)   = 63

x + (x − 11) = 63
    2x − 11 = 63
2x − 11 + 11 = 63 + 11 *Add 11*
      2x = 74
       x = 37   *Divide by 2*

He parked 37 small cars.

73. Let W be the width of the rectangle.
The formula for the perimeter of a rectangle is P = 2L + 2W.
If L = W + 4, then

the perimeter   is   288.
    ↓   ↓   ↓
2(W + 4) + 2W   =   288

2(W + 4) + 2W = 288
  2W + 8 + 2W = 288   *Distributive property*
    4W + 8 = 288
4W + 8 − 8 = 288 − 8 *Subtract 8*
    4W = 280
     W = 70

The width is 70 feet.

74. Let x = the length of the first side.
Then 2x = the length of the second side.

    x + 2x + 30 = 96
      3x + 30 = 96
        3x = 66
         x = 22

The sides have lengths 22 meters, 44 meters, and 30 meters.
The length of the longest side is 44 meters.

75. Let x = the length of the base.

Use the formula for the area of a triangle.

$$A = \frac{1}{2}bh$$
$$182 = \frac{1}{2}x(14)$$
$$182 = 7x$$
$$26 = x$$

The length of the base is 26 inches.

76. Let x = the length of the rectangle.

Use the formula for the perimeter of a rectangle.

$$P = 2L + 2W$$
$$75 = 2x + 2(17)$$
$$75 = 2x + 34$$
$$41 = 2x$$
$$x = \frac{41}{2} = 20\frac{1}{2}$$

The length of the rectangle is 20 1/2 inches.

77. The angles are vertical angles, so their measurers are equal.

$$3(4x + 1) = 10x + 11$$
$$12x + 3 = 10x + 11$$
$$2x + 3 = 11$$
$$2x = 8$$
$$x = 4$$

If x = 4,

$$3(4x + 1) = 3[4(4) + 1]$$
$$= 3(16 + 1)$$
$$= 3(17) = 51,$$

and

$$10x + 11 = 10(4) + 11$$
$$= 40 + 11 = 51.$$

The measure of each angle is 51°.

78. Let x = Nalima's grade on her third test.

$$\frac{1}{3}(82 + 96 + x) \geq 90$$
$$\frac{1}{3}(178 + x) \geq 90$$
$$3 \cdot \frac{1}{3}(178 + x) \geq 3(90)$$
$$178 + x \geq 270$$
$$x \geq 92$$

Nalima must make 92 or more on the third test.

Chapter 2 Test

For Exercises 1–8, all solutions should be checked by substituting into the original equation. Checks will not be shown here.

1. $$3x - 7 = 11$$
    $$3x - 7 + 7 = 11 + 7$$
    $$3x = 18$$
    $$\frac{3x}{3} = \frac{18}{3}$$
    $$x = 6$$

2. $$5x + 9 = 7x + 21$$
    $$5x - 7x + 9 = 7x - 7x + 21$$
    $$-2x + 9 = 21$$
    $$-2x + 9 - 9 = 21 - 9$$
    $$-2x = 12$$
    $$\frac{-2x}{-2} = \frac{12}{-2}$$
    $$x = -6$$

86 Chapter 2 Solving Equations and Inequalities

3. $2 - 3(y - 5) = 3 + (y - 1)$
$2 - 3y + 15 = 3 + y + 1$
$-3y + 17 = y + 4$
$-3y - y + 17 = y - y + 4$
$-4y + 17 = 4$
$-4y + 17 - 17 = 4 - 17$
$-4y = -13$
$\dfrac{-4y}{-4} = \dfrac{-13}{-4}$
$y = \dfrac{13}{4}$

4. $2.3x + 13.7 = 1.3x + 2.9$
$2.3x - 1.3x + 13.7 = 1.3x - 1.3x + 2.9$
$x + 13.7 = 2.9$
$x + 13.7 - 13.7 = 2.9 - 13.7$
$x = -10.8$

5. $7 - (m - 4) = -3m + 2(m + 1)$
$7 - 1(m - 4) = -3m + 2m + 2$
$7 - m + 4 = -3m + 2m + 2$
$-m + 11 = -m + 2$
$-m + m + 11 = -m + m + 2$
$11 = 2$ *False*

Because the result is a false statement, the equation has no solution.

6. $-\dfrac{4}{7}x = 12$
$\left(-\dfrac{7}{4}\right)\left(-\dfrac{4}{7}x\right) = \left(-\dfrac{7}{4}\right)(-12)$
$x = 21$

7. $.06(x + 20) + .08(x - 10) = 4.6$

To clear decimals, multiply both sides by 100.

$100[.06(x + 20) + .08(x - 10)] = 100(4.6)$
$6(x + 20) + 8(x - 10) = 460$
$6x + 120 + 8x - 80 = 460$
$14x + 40 = 460$
$14x = 420$
$x = 30$

8. $-8(2x + 4) = -4(4x + 8)$
$-16x - 32 = -16x - 32$
$-16x + 16x - 32 = -16x + 16x - 32$
$-32 = -32$ *True*

Because the result is a true statement, all real numbers are solutions.

9. Let x represent the number.

$4x - 3 = 5x - 10$
$4x - 4x - 3 = 5x - 4x - 10$
$-3 = x - 10$
$-3 + 10 = x - 10 + 10$
$7 = x$

The number is 7.

10. Let x = the length of the Brooklyn Bridge.
Then x + 2605 = the length of the Golden Gate Bridge.

$x + (x + 2605) = 5795$
$2x + 2605 = 5795$
$2x = 3190$
$x = 1595$

The length of the Brooklyn Bridge is 1595 feet and that of the Golden Gate Bridge is 4200 feet.

11. Let x = the length of the middle-sized piece;

    3x = the length of the longest piece;

    3x − 23 = the length of the shortest piece.

    $$x + 3x + (3x - 23) = 40$$
    $$7x - 23 = 40$$
    $$7x = 63$$
    $$x = 9$$

    The length of the middle-sized piece is 9 centimeters, of the longest piece is 27 centimeters, and of the shortest piece is 4 centimeters.

12. $P = 2L + 2W$; $P = 116$, $L = 40$

    $$P = 2L + 2W$$
    $$116 = 2(40) + 2W$$
    $$116 = 80 + 2W$$
    $$36 = 2W$$
    $$18 = W$$

    The width is 18.

13. Solve $P = 2L + 2W$ for W.

    $$P - 2L = 2W$$
    $$\frac{P - 2L}{2} = W$$
    $$W = \frac{P - 2L}{2} \text{ or } W = \frac{P}{2} - L$$

14. The angles are supplementary, so

    $$(3x + 55) + (7x - 25) = 180.$$

    Solve this equation.

    $$10x + 30 = 180$$
    $$10x = 150$$
    $$x = 15$$

    If $x = 15$,

    $$3x + 15 = 3(15) + 55$$
    $$= 45 + 55 = 100,$$

    and

    $$7x - 25 = 7(15) - 25$$
    $$= 105 - 25 = 80.$$

    The measures of the angles are 100° and 80°.

15. The angles are vertical angles, so their measures are equal.

    $$3x + 15 = 4x - 5$$
    $$15 = x - 5$$
    $$20 = x$$

    If $x = 20$,

    $$3x + 15 = 3(20) + 15$$
    $$= 60 + 15 = 75,$$

    and

    $$4x - 5 = 4(20) - 5$$
    $$= 80 - 5 = 75.$$

    Both angles have measures of 75°.

16. Let x = the measure of the angle.

    Then 90 − x = the measure of its complement

    and 180 − x = the measure of its supplement.

    $$180 - x = 3(90 - x) + 10$$
    $$180 - x = 270 - 3x + 10$$
    $$180 - x = 280 - 3x$$
    $$180 + 2x = 280$$
    $$2x = 100$$
    $$x = 50$$

    The measure of the angle is 50°.

Chapter 2 Solving Equations and Inequalities

17. $\dfrac{z}{8} = \dfrac{12}{16}$

$16z = 96$

$\dfrac{16z}{16} = \dfrac{96}{16}$

$z = 6$

18. $\dfrac{y + 5}{3} = \dfrac{y - 3}{4}$

$4(y + 5) = 3(y - 3)$

$4y + 20 = 3y - 9$

$y + 20 = -9$

$y = -29$

19.

Size	Unit Cost (dollars per slice)
8 slices	$\dfrac{\$2.19}{8 \text{ slices}} = \$.274$
12 slices	$\dfrac{\$3.30}{12 \text{ slices}} = \$.275$

The better buy is 8 slices for $2.19.

20. Let x = the actual distance between Seattle and Cincinnati.

$\dfrac{x \text{ miles}}{1050 \text{ miles}} = \dfrac{92 \text{ inches}}{42 \text{ inches}}$

$\dfrac{x}{1050} = \dfrac{92}{42}$

$42x = 96{,}600$

$\dfrac{42x}{42} = \dfrac{96{,}600}{42}$

$x = 2300$

The actual distance between Seattle and Cincinnati is 2300 miles.

21. 25% of 56 = (.25)(56) = 14

Fourteen of the signers were judges.

For Exercises 22 and 23, see the number line graphs in the answer section of the textbook.

22. $-3x > -33$

$\dfrac{-3x}{-3} < \dfrac{-33}{-3}$

$x < 11$

Place an open circle at 11 and draw an arrow extending to the left.

23. $-4x + 2(x - 3) \geq 4x - (3 + 5x) - 7$

$-4x + 2(x - 3) \geq 4x - 1(3 + 5x) - 7$

$-4x + 2x - 6 \geq 4x - 3 - 5x - 7$

$-2x - 6 \geq -x - 10$

$-x - 6 \geq -10$

$-x \geq -4$

$\dfrac{-1x}{-1} \leq \dfrac{-4}{-1}$

$x \leq 4$

Place a solid dot at 4 and draw an arrow extending to the left.

24. Let x = Julie's score on her third test.

$\dfrac{1}{3}(84 + 100 + x) \geq 90$

$3 \cdot \dfrac{1}{3}(184 + x) \geq 3(90)$

$184 + x \geq 270$

$x \geq 86$

She must score 86 or more on her third test.

25. When both sides of an inequality are multiplied or divided by a negative number, the inequality symbol must be reversed.

Cumulative Review: Chapters R–2

1. $\dfrac{15}{40} = \dfrac{3 \cdot 5}{8 \cdot 5} = \dfrac{3}{8}$

2. $\dfrac{108}{144} = \dfrac{3 \cdot 36}{4 \cdot 36} = \dfrac{3}{4}$

3. $\dfrac{5}{6} + \dfrac{1}{4} + \dfrac{7}{15} = \dfrac{50}{60} + \dfrac{15}{60} + \dfrac{28}{60}$

   $= \dfrac{93}{60}$

   $= \dfrac{31}{20}$

4. $16\dfrac{7}{8} - 3\dfrac{1}{10} = \dfrac{135}{8} - \dfrac{31}{10}$

   $= \dfrac{675}{40} - \dfrac{124}{40}$

   $= \dfrac{551}{40}$ or $13\dfrac{31}{40}$

5. $\dfrac{9}{8} \cdot \dfrac{16}{3} = \dfrac{3 \cdot 3 \cdot 2 \cdot 8}{8 \cdot 3} = 6$

6. $\dfrac{3}{4} \div \dfrac{5}{8} = \dfrac{3}{4} \cdot \dfrac{8}{5} = \dfrac{3 \cdot 2 \cdot 4}{4 \cdot 5} = \dfrac{6}{5}$

7. $4.8 + 12.5 + 16.73$

   ```
       4.80
      12.50
   +  16.73
   ─────────
      34.03
   ```

8. $56.3 - 28.99$

   ```
      56.30
   -  28.99
   ─────────
      27.31
   ```

9. $(67.8)(.45)$

   ```
        6 7.8
     ×    .4 5
     ─────────
        3 3 9 0
      27 1 2
     ─────────
      30.5 1 0
   ```

   $67.8(.45) = 30.51$

10. $236.46 \div 4.2$

    ```
              56.3
       4.2.)236.4.6
             210
            ─────
             26 4
             25 2
             ─────
              1 2 6
              1 2 6
              ─────
                  0
    ```

    $236.46 \div 4.2 = 56.3$

11. $56\left(\dfrac{5}{8}\right) = \dfrac{56}{1} \cdot \dfrac{5}{8} = \dfrac{280}{8} = 35$

    To make 56 dresses, 35 yards of trim would be used.

12. If the recipe for 6 people requires 1 1/4 cups of cheese, the amount needed for 1 person is

    $1\dfrac{1}{4} \div 6 = \dfrac{5}{4} \div 6 = \dfrac{5}{4} \cdot \dfrac{1}{6}$

    $= \dfrac{5}{24}$ cup.

    For 20 people, the amount of cheese needed is

    $20\left(\dfrac{5}{24}\right) = \dfrac{5 \cdot 4 \cdot 5}{4 \cdot 6} = \dfrac{25}{6} = 4\dfrac{1}{6}$ cups.

Chapter 2 Solving Equations and Inequalities

13. First dog's weight $\rightarrow$ and second dog's weight $\rightarrow$

   $8\frac{1}{3}$ + $12\frac{5}{8}$ is total weight of both dogs.

   $= \frac{25}{3} + \frac{101}{8}$

   $= \frac{200}{24} + \frac{303}{24}$

   $= \frac{503}{24}$ or $20\frac{23}{24}$

   The total weight is $20\frac{23}{24}$ pounds.

14. 3 desks at $211.40 each and 3 chairs

   $3(211.40) + 195 + 189.95 + 168.50$
   $= 1187.65$

   The total cost is $1187.65.

15. $\frac{8(7) - 5(6 + 2)}{3 \cdot 5 + 1} \geq 1$

   $\frac{8(7) - 5(8)}{3 \cdot 5 + 1} \geq 1$

   $\frac{56 - 40}{15 + 1} \geq 1$

   $\frac{16}{16} \geq 1$

   $1 \geq 1$

   The statement is true.

16. $\frac{4(9 + 3) - 8(4)}{2 + 3 - 3} \geq 2$

   $\frac{4(12) - 8(4)}{2 + [3 + (-3)]} \geq 2$

   $\frac{48 - 32}{2 + 0} \geq 2$

   $\frac{16}{2} \geq 2$

   $8 \geq 2$

   The statement is true.

17. $-11 + 20 + (-2) = 9 + (-2) = 7$

18. $13 + (-19) + 7 = -6 + 7 = 1$

19. $9 - (-4) = 9 + 4 = 13$

20. $-2(-5)(-4) = 10(-4) = -40$

21. $\frac{4 \cdot 9}{-3} = \frac{36}{-3} = -12$

22. $\frac{8}{7 - 7} = \frac{8}{0}$ undefined

23. $(-5 + 8) + (-2 - 7) = 3 + (-9) = -6$

24. $(-7 - 1)(-4) + (-4) = (-8)(-4) + (-4)$
    $= 32 + (-4)$
    $= 28$

25. $\frac{-3 - (-5)}{1 - (-1)} = \frac{-3 + 5}{1 + 1} = \frac{2}{2} = 1$

26. $\frac{6(-4) - (-2)(12)}{3^2 + 7^2} = \frac{(-24) - (-24)}{9 + 49}$
    $= \frac{(-24) + 24}{58}$
    $= \frac{0}{58} = 0$

27. $\dfrac{(-3)^2 - (-4)(2^4)}{5 \cdot 2 - (-2)^3}$

 $= \dfrac{9 - (-4)(16)}{5 \cdot 2 - (-2)(-2)(-2)}$ *Do exponent first*

 $= \dfrac{9 - (-64)}{10 - (-8)}$ *Multiply*

 $= \dfrac{9 + 64}{10 + 8}$

 $= \dfrac{73}{18}$

28. $\dfrac{-2(5^3) - 6}{4^2 + 2(-5) + (-2)}$

 $= \dfrac{-2(125) - 6}{16 + (-10) + (-2)}$

 $= \dfrac{-250 - 6}{4}$

 $= \dfrac{-256}{4}$

 $= -64$

In Exercises 29 and 30, replace x with -2, y with -4, and z with 3.

29. $xz^3 - 5y^2 = -2(3)^3 - 5(-4)^2$

 $= -2(27) + (-5)(16)$

 $= -54 + (-80)$

 $= -134$

30. $\dfrac{3x - y^3}{-4z} = \dfrac{3(-2) - (-4)^3}{-4(3)}$

 $= \dfrac{-6 - (-64)}{-12}$

 $= \dfrac{-6 + 64}{-12}$

 $= \dfrac{58}{-12}$

 $= -\dfrac{29}{6}$

31. $7(k + m) = 7k + 7m$

 The multiplication of 7 is distributed over the sum, which illustrates the distributive property.

32. $3 + (5 + 2) = 3 + (2 + 5)$

 The order of the numbers added in the parentheses is changed, which illustrates the commutative property.

33. $7 + (-7) = 0$

 Inverses are added to give 0, which illustrates the inverse property.

34. $3.5(1) = 3.5$

 The value of the number 3.5 is unchanged by multiplying by one. This illustrates the identity property.

35. $4p - 6 + 3p - 8 = 4p + 3p - 6 - 8$
 $= 7p - 14$

36. $-4(k + 2) + 3(2k - 1)$
 $= (-4)(k) + (-4)(2) + (3)(2k) + (3)(-1)$
 $= -4k - 8 + 6k - 3$
 $= -4k + 6k - 8 - 3$
 $= 2k - 11$

For Exercises 37–44, all solutions should be checked by substituting into the original equation. The checks will not be shown here.

37.
$$2r - 6 = 8$$
$$2r - 6 + 6 = 8 + 6$$
$$2r = 14$$
$$\frac{2r}{2} = \frac{14}{2}$$
$$r = 7$$

38.
$$2(p - 1) = 3p + 2$$
$$2p - 2 = 3p + 2$$
$$2p - 2 - 2p = 3p + 2 - 2p$$
$$-2 = p + 2$$
$$-2 - 2 = p + 2 - 2$$
$$-4 = p$$

39.
$$4 - 5(a + 2) = 3(a + 1) - 1$$
$$4 - 5a - 10 = 3a + 3 - 1$$
$$-5a - 6 = 3a + 2$$
$$-5a - 6 - 3a = 3a + 2 - 3a$$
$$-8a - 6 = 2$$
$$-8a - 6 + 6 = 2 + 6$$
$$-8a = 8$$
$$\frac{-8a}{-8} = \frac{8}{-8}$$
$$a = -1$$

40.
$$2 - 6(z + 1) = 4(z - 2) + 10$$
$$2 - 6z - 6 = 4z - 8 + 10$$
$$-6z - 4 = 4z + 2$$
$$-6z - 4 - 4z = 4z + 2 - 4z$$
$$-10z - 4 = 2$$
$$-10z - 4 + 4 = 2 + 4$$
$$-10z = 6$$
$$\frac{-10z}{-10} = \frac{6}{-10}$$
$$z = -\frac{3}{5}$$

41.
$$-(m - 1) = 3 - 2m$$
$$-1(m - 1) = 3 - 2m$$
$$-m + 1 = 3 - 2m$$
$$-m + 1 + 2m = 3 - 2m + 2m$$
$$m + 1 = 3$$
$$m + 1 - 1 = 3 - 1$$
$$m = 2$$

42.
$$\frac{y - 2}{3} = \frac{2y + 1}{5}$$
$$(y - 2)(5) = (3)(2y + 1) \quad \text{Cross products}$$
$$(y)(5) - (2)(5) = (3)(2y) + (3)(1)$$
$$5y - 10 = 6y + 3$$
$$5y - 10 - 5y = 6y + 3 - 5y$$
$$-10 = y + 3$$
$$-10 - 3 = y + 3 - 3$$
$$-13 = y$$

43.
$$\frac{2x + 3}{5} = \frac{x - 4}{2}$$
$$(2x + 3)(2) = (5)(x - 4) \quad \text{Cross products}$$
$$(2x)(2) + (3)(2) = (5)(x) + (5)(-4)$$
$$4x + 6 = 5x - 20$$
$$4x + 6 - 4x = 5x - 20 - 4x$$
$$6 = x - 20$$
$$6 + 20 = x - 20 + 20$$
$$26 = x$$

44. $\frac{2}{3}y + \frac{3}{4}y = -17$

The least common denominator is 12.

$$12\left(\frac{2}{3}y + \frac{3}{4}y\right) = 12(-17)$$
$$8y + 9y = -204$$
$$17y = -204$$
$$y = -12$$

45. $P = a + b + c$ for c

$$P = (a + b) + c$$
$$P - (a + b) = (a + b) - (a + b) + c$$
$$P - a - b = c \text{ or } c = P - a - b$$

46. $P = 4s$ for s

$$\frac{P}{4} = \frac{4s}{4}$$
$$\frac{P}{4} = s \text{ or } s = \frac{P}{4}$$

For Exercises 47 and 48, see the number line graphs in the answer section of the textbook.

47. $-5z \geq 4z - 18$
$-5z - 4z \geq 4z - 18 - 4z$
$-9z \geq -18$
$\frac{-9z}{-9} \leq \frac{-18}{-9}$
$z \leq 2$

48. $6(r - 1) + 2(3r - 5) \leq -4$
$6r - 6 + 6r - 10 \leq -4$
$6r + 6r - 6 - 10 \leq -4$
$12r - 16 \leq -4$
$12r - 16 + 16 \leq -4 + 16$
$12r \leq 12$
$\frac{12r}{12} \leq \frac{12}{12}$
$r \leq 1$

49. $6\frac{1}{4}\%$ of $1187.65 is

$.0625(1187.65) = 74.228$ (rounded).
The sales tax is $74.23, so the final bill is $1261.88.

50. 25% off $5000 is
$(.25)(5000) = 1250.$

The price less the trade-in is the cost.
   5000   −   1250   =   3750

The price with the trade-in is $3750.

51. Let x be the price of the books. Then

price of books	and	sales tax	is	total cost.
↓	↓	↓	↓	↓
x	+	.06x	=	52.47

$x(1 + .06) = 52.47$
$1.06x = 52.47$
$\frac{1.06x}{1.06} = \frac{52.47}{1.06}$
$x = 49.50$

The cost of the textbooks is $49.50.

52. Let p be the amount of her purchases. Use the formula $I = prt$ with $r = 1\frac{1}{2}\%$ and $t = 1$ to find the interest.

Total bill	is	purchases	and	interest
↓	↓	↓	↓	↓
104.93	=	p	+	p(.015)1

and late charge.
  +   5

$$104.93 = p + .015p + 5$$
$$104.93 = p(1 + .015) + 5$$
$$104.93 = p(1.015) + 5$$
$$104.93 - 5 = 1.015p + 5 - 5$$
$$99.93 = 1.015p$$
$$\frac{99.93}{1.015} = \frac{1.015p}{1.015}$$
$$98.453 = P$$

Her purchases amounted to $98.45.

53. To find the length, substitute $P = 98$ and $W = 19$ in the formula for the perimeter of a rectangle.

$$P = 2L + 2W$$
$$98 = 2L + 2(19)$$
$$98 = 2L + 38$$
$$98 - 38 = 2L + 38 - 38$$
$$60 = 2L$$
$$\frac{60}{2} = \frac{2L}{2}$$
$$30 = L$$

The length is 30 centimeters.

54. To find the height, substitute $A = 104$ and $b = 13$ in the formula for the area of a triangle.

$$A = \tfrac{1}{2}bh$$
$$104 = \tfrac{1}{2}(13)h$$
$$104 = \left(\tfrac{13}{2}\right)h$$
$$\left(\tfrac{2}{13}\right)(104) = \left(\tfrac{2}{13}\right)\left(\tfrac{13}{2}\right)h$$
$$16 = h$$

The height is 16 inches.

55. Let x be the number of miles Wally and Joanna drive on the third day. Take 1/3 of the sum of the miles traveled in three days to find the average.

Average	is at least	450.
↓	↓	↓
$\tfrac{1}{3}(430 + 470 + x)$	≥	450

Solve the inequality.

$$\tfrac{1}{3}(430 + 470 + x) \geq 450$$
$$3 \cdot \tfrac{1}{3}(430 + 470 + x) \geq 3 \cdot 450 \quad \textit{Multiply by 3}$$
$$900 + x \geq 1350$$
$$900 + x - 900 \geq 1350 - 900 \quad \textit{Subtract 900}$$
$$x \geq 450$$

They must drive 450 miles or more on the third day.

56. Let x be the amount Vern paid to tune up the Oldsmobile. Then $x + 57$ is the amount he paid to tune up the Bronco.

The amount for the Bronco	and	the amount for the Oldsmobile	is	the total.
↓	↓	↓	↓	↓
(x + 57)	+	x	=	257

$(x + 57) + x = 257$

$2x + 57 = 257$

$2x + 57 - 57 = 257 - 57$ *Subtract 57*

$2x = 200$

$\dfrac{2x}{2} = \dfrac{200}{2}$ *Divide by 2*

$x = 100$

He paid $100 to tune up the Oldsmobile.

CHAPTER 3 EXPONENTS AND POLYNOMIALS

Section 3.1 The Product Rule and Power Rules for Exponents

3.1 Margin Exercises

1. $2 \cdot 2 \cdot 2 \cdot 2$

   Since 2 is a factor four times, the base is 2 and the exponent is 4. The exponential expression is 2^4. The value may be found by multiplying from left to right as follows.

   $$2^4 = 2 \cdot 2 \cdot 2 \cdot 2 = 4 \cdot 2 \cdot 2$$
   $$= 8 \cdot 2 = 16$$

2. (a) $(-2)^5 = (-2)(-2)(-2)(-2)(-2)$
   $$= -32$$

   The base is -2 and the exponent is 5.

   (b) $-2^5 = -1 \cdot (2 \cdot 2 \cdot 2 \cdot 2 \cdot 2)$
   $$= -1 \cdot 32$$
   $$= -32$$

   The base is 2 and the exponent is 5.

   (c) $-4^2 = -1 \cdot (4 \cdot 4)$
   $$= -1 \cdot 16$$
   $$= -16$$

   The base is 4 and the exponent is 2.

   (d) $(-4)^2 = (-4)(-4) = 16$

   The base is -4 and the exponent is 2.

3. (a) $8^2 \cdot 8^5 = 8^{2+5} = 8^7$

   Since each factor in the product $8^2 \cdot 8^5$ has the same base, 8, the product rule applies.

   (b) $(-7)^5 \cdot (-7)^3 = (-7)^{5+3} = (-7)^8$

   The base is -7 in each factor, so the product rule applies.

   (c) $y^3 \cdot y = y^3 \cdot y^1 = y^{3+1} = y^4$

   The base is y in each factor, so the product rule applies.

   (d) $4^2 \cdot 3^5$

   The product rule does not apply because the bases are not the same.

   (e) $6^4 + 6^2$

   The product rule does not apply because the expression is a sum, not a product.

4. (a) $5m^2 \cdot 2m^6 = 5 \cdot 2 \cdot m^2 \cdot m^6$
   $$= 10 \cdot m^{2+6} \quad \textit{Product rule}$$
   $$= 10m^8$$

   (b) $3p^5 \cdot 9p^4 = 3 \cdot 9 \cdot p^5 \cdot p^4$
   $$= 27p^{5+4} \quad \textit{Product rule}$$
   $$= 27p^9$$

   (c) $-7p^5 \cdot (3p^8) = (-7 \cdot 3)p^5 \cdot p^8$
   $$= -21p^{5+8} \quad \textit{Product rule}$$
   $$= -21p^{13}$$

5. (a) $(5^3)^4 = 5^{3 \cdot 4} = 5^{12}$ *Power rule (a)*

   (b) $(6^2)^5 = 6^{2 \cdot 5} = 6^{10}$ *Power rule (a)*

   (c) $(3^2)^4 = 3^{2 \cdot 4} = 3^8$ *Power rule (a)*

   (d) $(a^6)^5 = a^{6 \cdot 5} = a^{30}$ *Power rule (a)*

6. **(a)** $5(mn)^3 = 5m^3n^3$ *Power rule (b)*

   (b) $(3a^2b^4)^5 = 3^5 \cdot (a^2)^5 \cdot (b^4)^5$
   Power rule (b)
   $= 3^5 \cdot a^{2 \cdot 5} \cdot b^{4 \cdot 5}$
   Power rule (a)
   $= 3^5 a^{10} b^{20}$

   (c) $(5m^2)^3 = 5^3 \cdot (m^2)^3$
   Power rule (b)
   $= 5^3 \cdot m^{2 \cdot 3}$
   Power rule (a)
   $= 5^3 m^6$

7. **(a)** $\left(\dfrac{5}{2}\right)^4 = \dfrac{5^4}{2^4}$ *Power rule (c)*

   (b) $\left(\dfrac{p}{q}\right)^2 = \dfrac{p^2}{q^2}$ *Power rule (c)*

   (c) $\left(\dfrac{r}{t}\right)^3 = \dfrac{r^3}{t^3}$ *Power rule (c)*

8. **(a)** $(2m)^3(2m)^4 = (2m)^{3+4}$ *Product rule*
   $= (2m)^7$
   $= 2^7 m^7$ *Power rule (b)*

   (b) $\left(\dfrac{5k^3}{3}\right)^2 = \dfrac{(5k^3)^2}{3^2}$ *Power rule (c)*
   $= \dfrac{5^2(k^3)^2}{3^2}$ *Power rule (b)*
   $= \dfrac{5^2 k^{3 \cdot 2}}{3^2}$ *Power rule (a)*
   $= \dfrac{5^2 k^6}{3^2}$

   (c) $\left(\dfrac{1}{5}\right)^4 (2x)^2 = \dfrac{1^4}{5^4}(2x)^2$ *Power rule (c)*
   $= \dfrac{1}{5^4}(2^2 x^2)$ *Power rule (b)*
   $= \dfrac{2^2 x^2}{5^4}$

(d) $(3xy^2)^3 (x^2 y)^4$
$= 3^3 x^3 (y^2)^3 (x^2)^4 y^4$
Power rule (b)
$= 3^3 x^3 y^{2 \cdot 3} x^{2 \cdot 4} y^4$
Power rule (a)
$= 3^3 x^3 y^6 x^8 y^4$
$= 3^3 x^{11} y^{10}$ *Product rule*

3.1 Section Exercises

3. $(-6)(-6)(-6)(-6)$

   Since -6 occurs as a factor 4 times, the base is -6 and the exponent is 4. The exponential expression is $(-6)^4$.

7. $\dfrac{1}{4 \cdot 4 \cdot 4 \cdot 4}$

   Since 4 occurs as a factor 4 times, the base is 4 and the exponent is 4. The exponential expression is $\dfrac{1}{4^4}$.

11. $\left(\dfrac{1}{2}\right)\left(\dfrac{1}{2}\right)\left(\dfrac{1}{2}\right)\left(\dfrac{1}{2}\right)\left(\dfrac{1}{2}\right)\left(\dfrac{1}{2}\right)$

    Since 1/2 occurs as a factor 6 times, the base is 1/2 and the exponent is 6. The exponential expression is $\left(\dfrac{1}{2}\right)^6$.

15. In the expression 3^5, the base is 3 and the exponent is 5.

    $3^5 = 3 \cdot 3 \cdot 3 \cdot 3 \cdot 3 = 243$

19. In the expression $(-6x)^4$, the base is $-6x$ and the exponents is 4.

$$(-6x)^4 = (-6x)(-6x)(-6x)(-6x)$$

23. $5^2 + 5^3$ cannot be evaluated by the product rule because you are adding powers of 5, not multiplying them.

$$5^2 + 5^3 = 5 \cdot 5 + 5 \cdot 5 \cdot 5$$
$$= 25 + 125 = 150$$

Note that the value of this expression, 150, is not a power of 5.

27. $4^2 \cdot 4^7 \cdot 4^3 = 4^{2+7+3}$ Product rule
$$= 4^{12}$$

31. $t^3 \cdot t^8 \cdot t^{13} = t^{3+8+13} = t^{24}$

35. $(-6p^5)(-7p^5) = (-6)(-7)p^5 \cdot p^5$
$$= (-6)(-7)p^{5+5}$$
$$= 42p^{10}$$

39. $5x^4 + 9x^4 = (5+9)x^4 = 14x^4$
$5x^4 \cdot 9x^4 = 5 \cdot 9 \cdot x^4 \cdot x^4$
$$= 5 \cdot 9 \cdot x^{4+4}$$
$$= 45x^8$$

43. $(4^3)^2 = 4^{3 \cdot 2}$ Power rule (a)
$$= 4^6$$

47. $(7r)^3 = 7^3 r^3$ Power rule (b)

51. $(-5^2)^6 = (-1 \cdot 5^2)^6$
$$= (-1)^6 \cdot 5^{2 \cdot 6}$$
$$= 1 \cdot 5^{12} = 5^{12}$$

55. $8(qr)^3 = 8q^3 r^3$

59. $\left(\dfrac{a}{b}\right)^3$ $(b \neq 0) = \dfrac{a^3}{b^3}$ Power rule (c)

63. $\left(\dfrac{5}{2}\right)^3 \cdot \left(\dfrac{5}{2}\right)^2 = \left(\dfrac{5}{2}\right)^{3+2}$ Power rule (a)
$$= \left(\dfrac{5}{2}\right)^5$$
$$= \dfrac{5^5}{2^5} \quad \text{Power rule (c)}$$

67. $(2x)^9 (2x)^3 = (2x)^{9+3}$
$$= (2x)^{12}$$
$$= 2^{12} x^{12}$$

71. $(6x^2 y^3)^5 = 6^5 (x^2)^5 (y^3)^5$
$$= 6^5 x^{2 \cdot 5} y^{3 \cdot 5}$$
$$= 6^5 x^{10} y^{15}$$

75. $(2w^2 x^3 y)^2 (x^4 y)^5$
$$= [2^2 (w^2)^2 (x^3)^2 y^2][(x^4)^5 y^5]$$
$$\hspace{4em} \text{Power rule (b)}$$
$$= (2^2 w^4 x^6 y^2)(x^{20} y^5) \quad \text{Power rule (a)}$$
$$= 2^2 w^4 (x^6 x^{20})(y^2 y^5)$$
$$\hspace{4em} \text{Commutative and associative properties}$$
$$= 2^2 w^4 x^{26} y^7$$

79. $\left(\dfrac{5a^2 b^5}{c^6}\right)^3$ $(c \neq 0)$
$$= \dfrac{(5a^2 b^5)^3}{(c^6)^3} \quad \text{Power rule (c)}$$
$$= \dfrac{5^3 (a^2)^3 (b^5)^3}{(c^6)^3} \quad \text{Power rule (b)}$$
$$= \dfrac{5^3 a^6 b^{15}}{c^{18}} \quad \text{Power rule (a)}$$

83. $8 - (-2) = 8 + 2 = 10$

Section 3.2 Integer Exponents and the Quotient Rule

3.2 Margin Exercises

1. **(a)** 28^0

   The base is 28, and 28 is not zero, so $28^0 = 1$.

   (b) $(-16)^0$

   The base is -16, and -16 is not zero, so $(-16)^0 = 1$.

   (c) -7^0

   The base is 7, and 7 is not zero, so $-7^0 = -(7^0) = -(1) = -1$.

   (d) m^0, $m \neq 0$

   The base is m and $m \neq 0$, so $m^0 = 1$.

   (e) $-p^0$, $p \neq 0$

   The base is p and $p \neq 0$, so $-p^0 = -(p^0) = -(1) = -1$.

2. **(a)** $4^{-3} = \dfrac{1}{4^3}$ Definition of negative exponent

   (b) $6^{-2} = \dfrac{1}{6^2}$

   (c) $\left(\dfrac{2}{3}\right)^{-2} = \left(\dfrac{3}{2}\right)^2$

   (d) $2^{-1} + 5^{-1} = \dfrac{1}{2} + \dfrac{1}{5} = \dfrac{5}{10} + \dfrac{2}{10}$
   $= \dfrac{7}{10}$

   (e) m^{-5} ($m \neq 0$) $= \dfrac{1}{m^5}$

   (f) $\dfrac{1}{z^{-4}}$ ($z \neq 0$) $= \dfrac{1^{-4}}{z^{-4}} = \left(\dfrac{1}{z}\right)^{-4} = z^4$

3. **(a)** $\dfrac{7^{-1}}{5^{-4}} = \dfrac{5^4}{7^1} = \dfrac{5^4}{7}$

   (b) $\dfrac{x^{-3}}{y^{-2}} = \dfrac{y^2}{x^3}$

   (c) $\dfrac{4h^{-5}}{m^{-2}k} = \dfrac{4}{k} \cdot \dfrac{h^{-5}}{m^{-2}} = \dfrac{4m^2}{h^5k}$

   (d) $p^2 q^{-5} = p^2\left(\dfrac{1}{q^5}\right) = \dfrac{p^2}{q^5}$

4. **(a)** $\dfrac{5^{11}}{5^8} = 5^{11-8} = 5^3$

   (b) $\dfrac{4^7}{4^{10}} = 4^{7-10} = 4^{-3} = \dfrac{1}{4^3}$

   (c) $\dfrac{6^{-5}}{6^{-2}} = 6^{-5-(-2)} = 6^{-3} = \dfrac{1}{6^3}$

   (d) $\dfrac{8^4 \cdot m^9}{8^5 \cdot m^{10}} = \dfrac{8^4}{8^5} \cdot \dfrac{m^9}{m^{10}}$
   $= 8^{4-5} \cdot m^{9-10}$
   $= 8^{-1} \cdot m^{-1}$
   $= \dfrac{1}{8} \cdot \dfrac{1}{m} = \dfrac{1}{8m}$

5. **(a)** $12^5 \cdot 12^{-7} \cdot 12^6$
   $= 12^{5+(-7)+6}$ Product rule
   $= 12^{-2+6}$
   $= 12^4$

   (b) $y^{-2} \cdot y^5 \cdot y^{-8} = y^{-2+5+(-8)}$ Product rule
   $= y^{3+(-8)}$
   $= y^{-5}$
   $= \dfrac{1}{y^5}$

   (c) $\dfrac{(6x)^{-1}}{(3x^2)^{-2}} = \dfrac{(3x^2)^2}{6x}$
   $= \dfrac{9x^4}{6x}$
   $= \dfrac{3}{2}(x^{4-1})$
   $= \dfrac{3x^3}{2}$

Chapter 3 Exponents and Polynomials

(d) $\dfrac{3^9 \cdot (x^2y)^{-2}}{3^3 \cdot x^{-4}y} = \dfrac{3^9}{3^3} \cdot \dfrac{(x^2y)^{-2}}{x^{-4}} \cdot \dfrac{1}{y}$

$= 3^{9-3} \cdot \dfrac{x^4}{(x^2y)^2} \cdot \dfrac{1}{y}$

$= 3^6 \cdot \dfrac{x^4}{x^4 y^2} \cdot \dfrac{1}{y}$

$= 3^6 x^{4-4} \left(\dfrac{1}{y^{2+1}}\right)$

$= \dfrac{3^6 x^0}{y^3}$

$= \dfrac{3^6}{y^3}$

3.2 Section Exercises

3. $(-4)^0 = 1$ Definition of zero exponent

7. $(-2)^0 - 2^0 = 1 - 1 = 0$

11. $x^0 = 1$ if $x \neq 0$

15. $4^{-3} = \dfrac{1}{4^3}$ Definition of negative exponent

$= \dfrac{1}{64}$

19. $\left(\dfrac{6}{7}\right)^{-2} = \left(\dfrac{7}{6}\right)^2$ 6/7 and 7/6 are reciprocals

$= \dfrac{7^2}{6^2}$ Power rule (c)

$= \dfrac{49}{36}$

23. $5^{-1} + 3^{-1} = \dfrac{1}{5} + \dfrac{1}{3}$

$= \dfrac{3}{15} + \dfrac{5}{15} = \dfrac{8}{15}$

27. $-2^4 = -(2^4) = -16$, which is negative.

31. $1 - 5^0 = 1 - 1 = 0$, so the value of the expression is zero.

35. $\dfrac{9^4}{9^5} = 9^{4-5}$ Quotient rule

$= 9^{-1}$

$= \dfrac{1}{9}$ Definition of negative exponent

39. $\dfrac{x^{12}}{x^{-3}} = x^{12-(-3)} = x^{12+3} = x^{15}$

43. $\dfrac{2}{r^{-4}} = 2r^4$

47. $p^5 q^{-8} = \dfrac{p^5}{q^8}$

51. $\dfrac{6^4 x^8}{6^5 x^3} = 6^{4-5} x^{8-3} = 6^{-1} x^5 = \dfrac{x^5}{6}$

55. $x^{-3} \cdot x^5 \cdot x^{-4}$

$= x^{-3+5+(-4)}$ Product rule

$= x^{-2}$

$= \dfrac{1}{x^2}$ Definition of negative exponent

59. $\left(\dfrac{x^{-1}y}{z^2}\right)^{-2} = \dfrac{(x^{-1}y)^{-2}}{(z^2)^{-2}}$ Power rule (c)

$= \dfrac{(x^{-1})^{-2} y^{-2}}{(z^2)^{-2}}$ Power rule (b)

$= \dfrac{x^2 y^{-2}}{z^{-4}}$ Power rule (a)

$= \dfrac{x^2 z^4}{y^2}$ Definition of negative exponent

63. $\dfrac{(m^7n)^{-2}}{m^{-4}n^3} = \dfrac{(m^7)^{-2}n^{-2}}{m^{-4}n^3} = \dfrac{m^{7(-2)}n^{-2}}{m^{-4}n^3}$

$= \dfrac{m^{-14}n^{-2}}{m^{-4}n^3} = m^{-14-(-4)}n^{-2-3}$

$= m^{-10}n^{-5} = \dfrac{1}{m^{10}n^5}$

67. $\dfrac{12}{10,000}$

When dividing by 10,000, move the decimal point 4 places to the left.

$\dfrac{12}{10,000} = .0012$

Section 3.3 An Application of Exponents: Scientific Notation

3.3 Margin Exercises

1. **(a)** $63,000 = 6.3 \times 10^4$

   The decimal point has been moved 4 places to put it after the first nonzero digit. Since 6.3 is smaller than the original number, it must be multiplied by a number larger than 1 to get 63,000. Thus, the exponent on 10 must be positive.

   (b) $5,870,000 = 5.87 \times 10^6$

   (c) $.0571 = 5.71 \times 10^{-2}$

   The decimal point has been moved 2 places to put it after the first nonzero digit. Since 5.71 is larger than the original number, it must be multiplied by a number smaller than 1 to get .0571. Thus, the exponent on 10 must be negative.

   (d) $.000062 = 6.2 \times 10^{-5}$

2. **(a)** $4.2 \times 10^3 = 4200$

   Multiplying by a positive power of 10 gives a larger number, so move the decimal point 3 places to the right.

   (b) $8.7 \times 10^5 = 870,000$

   Move the decimal point 5 places to the right.

   (c) $6.42 \times 10^{-3} = .00642$

   Multiplying by a negative power of 10 gives a smaller number, so move the decimal point 3 places to the left.

3. **(a)** $(2.6 \times 10^4)(2 \times 10^{-6})$

   $= (2.6 \times 2) \times (10^4 \times 10^{-6})$

   $= 5.2 \times 10^{4+(-6)}$

   $= 5.2 \times 10^{-2}$

   $= .052$

   (b) $\dfrac{4.8 \times 10^2}{2.4 \times 10^{-3}} = \dfrac{4.8}{2.4} \times \dfrac{10^2}{10^{-3}}$

   $= 2 \times 10^{2-(-3)}$

   $= 2 \times 10^{2+3}$

   $= 2 \times 10^5$

   $= 200,000$

3.3 Section Exercises

3. 5,600,000 is not in scientific notation. It can be written in scientific notation as

   $5.6 \times 10^6.$

7. .004 is not in scientific notation because .004 is not between 1 and 10. It can be written in scientific notation as

   $4 \times 10^{-3}.$

11. 5,876,000,000

   Move the decimal point to the right of the first nonzero digit and count the number of places the decimal point was moved.

   5.876,000,000   *9 places*

   Because moving the decimal point to the *left* made the number *smaller*, we must multiply by a *positive* power of 10 so that the product 5.876×10^n will be equal to the larger number. Thus, n = 9, and

   $5,876,000,000 = 5.876 \times 10^9$.

15. .000007

   Move the decimal point to the right of the first nonzero digit.

   .000007.   *6 places*

   Since moving the decimal point to the *right* made the number *larger*, we must multiply by a *negative* power of 10 so that the product 7×10^n will equal the smaller number. Thus, n = −6, and

   $.000007 = 7 \times 10^{-6}$.

19. 7.5×10^5

   Because the exponent is positive, make 7.5 larger by moving the decimal point 5 places to the right.

   $7.5 \times 10^5 = 750,000$

23. -6.21×10^0

   Because the exponent is 0, the decimal point should not be moved.

   $-6.21 \times 10^0 = -6.21$

   We know this result is correct because $10^0 = 1$.

27. 5.134×10^{-9}

   Because the exponent is negative, make 5·134 smaller by moving the decimal point 9 places to the left.

   $5.134 \times 10^{-9} = .000000005134$

31. $(5 \times 10^4) \times (3 \times 10^2)$
   $= (5 \times 3) \times (10^4 \times 10^2)$
   $= 15 \times 10^6$
   $= 15,000,000$

35. $\dfrac{9 \times 10^{-5}}{3 \times 10^{-1}} = \dfrac{9}{3} \times \dfrac{10^{-5}}{10^{-1}} = 3 \times 10^{-4}$

39. $\dfrac{2.6 \times 10^{-3}}{2 \times 10^2} = \dfrac{2.6}{2} \times \dfrac{10^{-3}}{10^2} = 1.3 \times 10^{-5}$

43. $1,150,000 = 1.15 \times 10^6$

47. $8x + 4x = 12x$

Section 3.4 Addition and Subtraction of Polynomials

3.4 Margin Exercises

1. **(a)** In $3m^2$, the coefficient is 3.

   (b) In $2x^3 - x$, the term $2x^3$ has coefficient 2. The coefficient of x is -1 since $2x^3 - x$ can be written $2x^3 + (-1x)$.

   (c) In $x + 8$, the coefficient of x is 1 because $x = 1 \cdot x$. The coefficient of 8 is 8 because $8 = 8x^0$.

2. **(a)** $5x^4 + 7x^4 = (5 + 7)x^4 = 12x^4$

   (b) $9pq + 3pq - 2pq = (9 + 3 - 2)pq$
   $= (12 - 2)pq$
   $= 10pq$

   (c) $r^2 + 3r + 5r^2 = r^2 + 5r^2 + 3r$
   $= (1 + 5)r^2 + 3r$
   $= 6r^2 + 3r$

   (d) $8t + 6w$ has unlike terms, which cannot be added.

3. **(a)** $3m^3 + 5m^2 - 2m + 1$ is (1) a polynomial and (2) a polynomial written in descending order.

   (b) $2p^4 + p^6$ is (1) a polynomial.

   (c) $\frac{1}{x} + 2x^2 + 3$ is (3) not a polynomial because of the term $1/x$, which is x^{-1}; the exponent on x is not a whole number.

   (d) $x - 3$ is (1) a polynomial and (2) a polynomial written in descending order.

4. **(a)** $3x^2 + 2x - 4$ is already simplified; its degree is 2, and it is a trinomial because it has three terms.

   (b) $x^3 + 4x^3 = (1 + 4)x^3 = 5x^3$ is the simplified form; it has degree 3, and it is a monomial because it has one term when simplified.

   (c) $x^8 - x^7 + 2x^8 = x^8 + 2x^8 - x^7$
   $= (1 + 2)x^8 - x^7$
   $= 3x^8 - x^7$

   It has degree 8 and is a binomial because it has two terms when simplified.

5. **(a)** $(2x^4 - 6x^2 + 7) + (-3x^4 + 5x^2 + 2)$
   $= 2x^4 + (-6x^2) + 7 + (-3x^4) + 5x^2 + 2$
   Remove parentheses
   $= 2x^4 + (-3x^4) + (-6x^2) + 5x^2 + 7 + 2$
   Place like terms together
   $= [2 + (-3)]x^4 + (-6 + 5)x^2 + (7 + 2)$
   Combine like terms
   $= (-1)x^4 + (-1)x^2 + 9$
   $= -x^4 - x^2 + 9$

   (b) $(3x^2 + 4x + 2) + (6x^3 - 5x - 7)$
   $= 3x^2 + 4x + 2 + 6x^3 + (-5x) + (-7)$
   Remove parentheses
   $= 6x^3 + 3x^2 + 4x + (-5x) + 2 + (-7)$
   Place like terms together
   $= 6x^3 + 3x^2 + [4 + (-5)]x + [2 + (-7)]$
   Combine like terms
   $= 6x^3 + 3x^2 - x - 5$

6. **(a)** $\quad 4x^3 - 3x^2 + 2x$
   $\quad\underline{6x^3 + 2x^2 - 3x}$
   $\quad 10x^3 - x^2 - x$

   Add the coefficients of like terms. Note that, in this problem, like powers are "lined up" beneath each other. This makes it easy to see what numbers to add.

(b) $\quad x^2 - 2x + 5$
$4x^2 + 3x - 2$
$\overline{5x^2 + x + 3}$

7. (a) $(14y^3 - 6y^2 + 2y - 5)$
$\quad - (2y^3 - 7y^2 - 4y + 6)$
$= (14y^3 - 6y^2 + 2y - 5)$
$\quad + [-(2y^3 - 7y^2 - 4y + 6)]$
$\quad$ Definition of subtraction
$= (14y^3 - 6y^2 + 2y - 5)$
$\quad + (-2y^3 + 7y^2 + 4y - 6)$
$= [14y^3 + (-2y^3)] + (-6y^2 + 7y^2)$
$\quad + (2y + 4y) + [-5 + (-6)]$
$\quad$ Add like terms
$= 12y^3 + y^2 + 6y - 11$

Check

$(2y^3 - 7y^2 - 4y + 6)$
$\quad + (12y^3 + y^2 + 6y - 11)$
$= (2y^3 + 12y^3) + (-7y^2 + y^2)$
$\quad + (-4y + 6y) + (6 - 11)$
$= 14y^3 - 6y^2 + 2y - 5$

The answer is correct.

(b) $(\frac{7}{2}y^2 - \frac{11}{3}y + 8) - (-\frac{3}{2}y^2 + \frac{4}{3}y + 6)$
$= (\frac{7}{2}y^2 - \frac{11}{3}y + 8)$
$\quad + [-(-\frac{3}{2}y^2 + \frac{4}{3}y + 6)]$
$\quad$ Definition of subtraction
$= (\frac{7}{2}y^2 - \frac{11}{3}y + 8) + (\frac{3}{2}y^2 - \frac{4}{3}y - 6)$
$= (\frac{7}{2}y^2 + \frac{3}{2}y^2) + [(-\frac{11}{3}y) + (-\frac{4}{3}y)]$
$\quad + [8 + (-6)] \quad$ Group like terms
$= \frac{10}{2}y^2 - \frac{15}{3}y + 2 \quad$ Add like terms
$= 5y^2 - 5y + 2$

Check

$(-\frac{3}{2}y^2 + \frac{4}{3}y + 6) + (5y^2 - 5y + 2)$
$= (-\frac{3}{2}y^2 + 5y^2) + (\frac{4}{3}y - 5y) + (6 + 2)$
$= (-\frac{3}{2}y^2 + \frac{10}{2}y^2) + (\frac{4}{3}y - \frac{15}{3}y) + (6 + 2)$
$= \frac{7}{2}y^2 - \frac{11}{3}y + 8$

8. (a) $(14y^3 - 6y^2 + 2y) - (2y^3 - 7y^2 + 6)$

$\; 14y^3 - 6y^2 + 2y$
Step 1: $-(2y^3 - 7y^2 + 6)$

Step 2: $14y^3 - 6y^2 + 2y$
$\; -2y^3 + 7y^2 - 6$
Step 3: $\overline{12y^3 + y^2 + 2y - 6}$ Add

(b) $(6p^4 - 8p^3 + 2p - 1) - (-7p^4 + 6p^2 - 12)$

$\; 6p^4 - 8p^3 + 2p - 1$
Step 1: $-(-7p^4 + 6p^2 - 12)$

Step 2: $6p^4 - 8p^3 + 2p - 1$
$\; 7p^4 - 6p^2 + 12$
Step 3: $\overline{13p^4 - 8p^3 - 6p^2 + 2p + 11}$
 Add

9. (a) $(3mn + 2m - 4n) + (-mn + 4m + n)$
$= [3mn + (-mn)] + (2m + 4m)$
$\quad + [(-4n) + n]$
$= 2mn + 6m - 3n$

This polynomial has degree 2.

(b) $(5p^2q^2 - 4p^2 + 2q) - (2p^2q^2 - p^2 - 3q)$
$= (5p^2q^2 - 4p^2 + 2q)$
$\quad + [-(2p^2q^2 - p^2 - 3q)]$
$= (5p^2q^2 - 4p^2 + 2q)$
$\quad + (-2p^2q^2 + p^2 + 3q)$
$= [5p^2q^2 + (-2p^2q^2)] + [(-4p^2) + p^2]$
$\quad + (2q + 3q)$
$= 3p^2q^2 - 3p^2 + 5q$

This polynomial has degree 4.

3.4 Section Exercises

3. t^4 has 1 term. Since $t^4 = 1 \cdot t^4$, the coefficient is 1.

7. $x + 8x^2$ has 2 terms. The coefficient of x is 1 and the coefficient of x^2 is 8.

11. $2r^5 + (-3r^5) = [2 + (-3)]r^5$
    $= -1r^5 = -r^5$

15. $(-3x^5 + 2x^5 - 4x^5) = (-3 + 2 - 4)x^5$
    $= -5x^5$

19. $-4y^2 + 3y^2 - 2y^2 + y^2$
    $= (-4 + 3 - 2 + 1)y^2$
    $= -2y^2$

23. A trinomial is *always* a polynomial, by definition.

27. $6x^4 - 9x$

    This polynomial has no like terms, so it is already simplified. It is already written in descending powers of the variable x. The highest degree of any nonzero term is 4, so the degree of the polynomial is 4. There are two terms, so this is a binomial.

31. $\frac{5}{3}x^4 - \frac{2}{3}x^4 + \frac{1}{3}x^2 - 4$
    $= \left(\frac{5}{3} - \frac{2}{3}\right)x^4 + \frac{1}{3}x^2 - 4$
    $= \frac{3}{3}x^4 + \frac{1}{3}x^2 - 4$
    $= x^4 + \frac{1}{3}x^2 - 4$

    This polynomial has degree 4. The simplified polynomial has three terms, so it is a trinomial.

35. Add.

    $3m^2 + 5m$
    $\underline{2m^2 - 2m}$
    $5m^2 + 3m$

39. Add.

    $\frac{2}{3}x^2 + \frac{1}{5}x + \frac{1}{6}$
    $\underline{\frac{1}{2}x^2 - \frac{1}{3}x + \frac{2}{3}}$

    Rewrite the fractions so that the fractions in each column have a common denominator; then add column by column.

    $\frac{4}{6}x^2 + \frac{3}{15}x + \frac{1}{6}$
    $\underline{\frac{3}{6}x^2 - \frac{5}{15}x + \frac{4}{6}}$
    $\frac{7}{6}x^2 - \frac{2}{15}x + \frac{5}{6}$

47. $(8m^2 - 7m) - (3m^2 + 7m - 6)$
    $= (8m^2 - 7m) + (-3m^2 - 7m + 6)$
    $= 8m^2 + (-7m) + (-3m^2) + (-7m) + 6$
    $= [8m^2 + (-3m^2)] + [(-7m) + (-7m)]$
    $\quad + 6$
    $= 5m^2 - 14m + 6$

51. $(7y^4 + 3y^2 + 2y) - (18y^5 - 5y^3 + y)$
$= (7y^4 + 3y^2 + 2y) + (-18y^5 + 5y^3 - y)$
$= 7y^4 + 3y^2 + 2y + (-18y^5) + 5y^3 + (-y)$
$= (-18y^5) + 7y^4 + 5y^3 + 3y^2$
$\quad + [2y + (-y)]$
$= -18y^5 + 7y^4 + 5y^3 + 3y^2 + y$

55. The perimeter of a rectangle of length $4x^2 + 3x + 1$ and width $x + 2$ is
$P = 2L + 2W$
$= 2(4x^2 + 3x + 1) + 2(x + 2)$
$= 8x^2 + 6x + 2 + 2x + 4$
$= 8x^2 + 6x + 2x + 2 + 4$
$= 8x^2 + 8x + 6.$

59. Subtract $9x^2 - 3x + 7$ from $-2x^2 - 6x + 4$.

$(-2x^2 - 6x + 4) - (9x^2 - 3x + 7)$
$= (-2x^2 - 6x + 4) + (-9x^2 + 3x - 7)$
$= -2x^2 + (-6x) + 4 + (-9x^2)$
$\quad + 3x + (-7)$
$= [-2x^2 + (-9x^2)] + [(-6x) + 3x]$
$\quad + 4 + (-7)$
$= -11x^2 - 3x - 3$

63. $(9a^2b - 3a^2 + 2b) + (4a^2b - 4a^2 - 3b)$
$= 9a^2b + (-3a^2) + 2b + 4a^2b$
$\quad + (-4a^2) + (-3b)$
$= (9a^2b + 4a^2b) + [(-3a^2) + (-4a^2)]$
$\quad + [2b + (-3b)]$
$= 13a^2b + (-7a^2) + (-b)$ or
$= 13a^2b - 7a^2 - b$

For the term $13a^2b = 13a^2b^1$, the sum of the exponents on the variables is $2 + 1 = 3$, so this term has degree 3, and the polynomial $13a^2b - 7a^2 - b$ has degree 3.

67. Subtract.

$9m^3n - 5m^2n^2 + 4mn^2$
$-3m^3n + 6m^2n^2 + 8mn^2$

Change all signs in the second row; then add.

$9m^3n - 5m^2n^2 + 4mn^2$
$\underline{3m^3n - 6m^2n^2 - 8mn^2}$
$12m^3n - 11m^2n^2 - 4mn^2$

In the answer, the degree of the first term is $3 + 1 = 4$, the degree of the second term is $2 + 2 = 4$, and the degree of the third term is $1 + 2 = 3$. Therefore, the degree of the answer is 4.

71. $(-4y^3)(-8y) = -4(-8)(y^3)(y)$
$= 32y^{3+1}$
$= 32y^4$

Section 3.5 Multiplication of Polynomials

3.5 Margin Exercises

1. (a) $5m^3(2m + 7)$
$= (5m^3)(2m) + (5m^3)(7)$
Distributive property
$= 10m^4 + 35m^3$
Multiply monomials

(b) $2x^4(3x^2 + 2x - 5)$
$= (2x^4)(3x^2) + (2x^4)(2x)$
$\quad + (2x^4)(-5)$
Distributive property
$= 6x^6 + 4x^5 + (-10x^4)$
Multiply monomials
$= 6x^6 + 4x^5 - 10x^4$

3.5 Margin Exercises

(c) $-4y^2(3y^3 + 2y^2 - 4y + 8)$

$= (-4y^2)(3y^3) + (-4y^2)(2y^2)$
$+ (-4y^2)(-4y) + (-4y^2)(8)$
Distributive property

$= -12y^5 - 8y^4 + 16y^3 - 32y^2$
Multiply monomials

2. (a) $(4x + 3)(2x + 1)$

$= (4x + 3)(2x) + (4x + 3)(1)$
Distributive property

$= 4x(2x) + 3(2x) + 4x(1) + 3(1)$
Distributive property again

$= 8x^2 + 6x + 4x + 3$
Multiply

$= 8x^2 + 10x + 3$
Add like terms

(b) $(3k - 2)(2k + 1)$

$= (3k - 2)(2k) + (3k - 2)(1)$
Distributive property

$= 3k(2k) + (-2)(2k) + 3k(1)$
$+ (-2)(1)$
Distributive property

$= 6k^2 + (-4k) + 3k + (-2)$
Multiply

$= 6k^2 - k - 2$
Add like terms

(c) $(m + 5)(3m - 4)$

$= (m + 5)(3m) + (m + 5)(-4)$

$= m(3m) + 5(3m) + m(-4) + 5(-4)$

$= 3m^2 + 15m + (-4m) + (-20)$

$= 3m^2 + 11m - 20$

3. (a) $(m^3 - 2m + 1)(2m^2 + 4m + 3)$

$= m^3(2m^2) + m^3(4m) + m^3(3)$
$- 2m(2m^2) - 2m(4m) - 2m(3)$
$+ 1(2m^2) + 1(4m) + 1(3)$

$= 2m^5 + 4m^4 + 3m^3 - 4m^3 - 8m^2$
$- 6m + 2m^2 + 4m + 3$

$= 2m^5 + 4m^4 - m^3 - 6m^2 - 2m + 3$

(b) $(6p^2 + 2p - 4)(3p^2 - 5)$

$= 6p^2(3p^2) + 6p^2(-5) + 2p(3p^2)$
$+ 2p(-5) - 4(3p^2) - 4(-5)$

$= 18p^4 - 30p^2 + 6p^3 - 10p$
$- 12p^2 + 20$

$= 18p^4 + 6p^3 - 42p^2 - 10p + 20$

4. (a)

```
      4k -  6
      2k +  5
     ─────────
     20k - 30        ← Multiply top row by 5
 8k² - 12k           ← Multiply top row by 2k,
─────────────           lining up like terms
 8k² +  8k - 30     ← Add
```

(b)

```
      3x² + 4x -  5
             x +  4
     ────────────────
          12x² + 16x - 20   ← Multiply top
                               row by 4
 3x³ + 4x² -  5x            ← Multiply top row
─────────────────────          by x, lining up
 3x³ + 16x² + 11x - 20      ← Add                like terms
```

5. (a)

```
       2m +  3p
       5m -  4p
      ───────────
      -8mp - 12p²    ← Multiply top row
                        by -4p
                        Multiply top row by
 10m² + 15mp         ← 5m, lining up like
──────────────────      terms
 10m² +  7mp - 12p²  ← Add
```

(b)

```
  k³ - k² + k + 1
              k + 1
 ──────────────────
  k³ - k² + k + 1      ← Multiply by 1
 k⁴ - k³ + k² + k      ← Multiply by k
 ──────────────────
 k⁴         + 2k + 1   ← Add
```

Product: $k^4 + 2k + 1$

(c) Follow this procedure.

(1) Multiply top row by 5, leaving room for terms that have missing powers.

(2) Multiply by 6a, leaving room again.

(3) Multiply by $2a^2$, leaving room.
(4) Add like terms in columns.

$$
\begin{array}{r}
a^3 + 3a - 4 \\
2a^2 + 6a + 5 \\
\hline
5a^3 + 15a - 20 \quad (1) \\
6a^4 + 18a^2 - 24a \quad (2) \\
2a^5 + 6a^3 - 8a^2 \quad (3) \\
\hline
2a^5 + 6a^4 + 11a^3 + 10a^2 - 9a - 20 \quad (4)
\end{array}
$$

6. (a) $(3k - 2)^2$

$$
\begin{array}{r}
3k - 2 \\
3k - 2 \\
\hline
-6k + 4 \quad \leftarrow \text{Multiply by } -2 \\
9k^2 - 6k \leftarrow \text{Multiply by } 3k \\
\hline
9k^2 - 12k + 4 \quad \leftarrow \text{Add}
\end{array}
$$

(b) $(2x^2 - 3x + 4)^2$

$$
\begin{array}{r}
2x^2 - 3x + 4 \\
2x^2 - 3x + 4 \\
\hline
8x^2 - 12x + 16 \quad \leftarrow \text{Multiply by } 4 \\
-6x^3 + 9x^2 - 12x \leftarrow \text{Multiply by } -3x \\
4x^4 - 6x^3 + 8x^2 \leftarrow \text{Multiply by } 2x^2 \\
\hline
4x^4 - 12x^3 + 25x^2 - 24x + 16 \quad \leftarrow \text{Add}
\end{array}
$$

(c) $(m + 1)^3$

Since $(m + 1)^3 = (m + 1)(m + 1)(m + 1)$, the first step is to find the product $(m + 1)(m + 1)$.

$$(m + 1)(m + 1) = m^2 + m + m + 1$$
$$= m^2 + 2m + 1$$

Now multiply this result by $m + 1$.

$$(m + 1)^3 = (m + 1)(m^2 + 2m + 1)$$
$$= m^3 + 2m^2 + m + m^2 + 2m + 1$$
$$= m^3 + 3m^2 + 3m + 1$$

3.5 Section Exercises

3. $(-5a^9)(-8a^5) = (-5)(-8)a^9 \cdot a^5$
$$= 40a^{9+5}$$
$$= 40a^{14}$$

7. $3p(8 - 6p + 12p^3)$
$$= (3p)(8) + (3p)(-6p) + (3p)(12p^3)$$
$$= 24p + (-18p^2) + 36p^4$$
$$= 24p - 18p^2 + 36p^4$$

11. $(n - 2)(n + 3)$
$$= (n - 2)(n) + (n - 2)(3)$$
$$= (n)(n) + (-2)(n) + (n)(3)$$
$$ + (-2)(3)$$
$$= n^2 + (-2n) + 3n + (-6)$$
$$= n^2 + n + (-6)$$
$$= n^2 + n - 6$$

15. $(3x + 2)(3x - 2)$
$$= (3x + 2)(3x) + (3x + 2)(-2)$$
$$= (3x)(3x) + (2)(3x) + (3x)(-2)$$
$$ + (2)(-2)$$
$$= 9x^2 + 6x + (-6x) + (-4)$$
$$= 9x^2 + (-4)$$
$$= 9x^2 - 4$$

19. $(3t + 4s)(2t + 5s)$
$$= (3t + 4s)(2t) + (3t + 4s)(5s)$$
$$= (3t)(2t) + (4s)(2t) + (3t)(5s)$$
$$ + (4s)(5s)$$
$$= 6t^2 + 8st + 15ts + 20s^2$$
$$= 6t^2 + 8st + 15st + 20s^2$$
$$= 6t^2 + 23st + 20s^2$$

23. The area of a square is found by multiplying the length of a side by itself, so the area is

$(6x + 2)(6x + 2)$
$= (6x + 2)(6x) + (6x + 2)(2)$
$= (6x)(6x) + (2)(6x) + (6x)(2)$
    $+ (2)(2)$
$= 36x^2 + 12x + 12x + 4$
$= 36x^2 + 24x + 4.$

In Exercises 27-43, we can multiply the polynomials horizontally or vertically. The following solutions will illustrate these two methods.

27. $(6x + 1)(2x^2 + 4x + 1)$
$= (6x)(2x^2) + (6x)(4x) + (6x)(1)$
    $+ (1)(2x^2) + (1)(4x) + (1)(1)$
$= 12x^3 + 24x^2 + 6x + 2x^2 + 4x + 1$
$= 12x^3 + 26x^2 + 10x + 1$

31. $(5x^2 + 2x + 1)(x^2 - 3x + 5)$

Multiply vertically.

$$\begin{array}{r} 5x^2 + 2x + 1 \\ x^2 - 3x + 5 \\ \hline 25x^2 + 10x + 5 \\ -15x^3 - 6x^2 - 3x \\ 5x^4 + 2x^3 + x^2 \\ \hline 5x^4 - 13x^3 + 20x^2 + 7x + 5 \end{array}$$

35. $(3t + 1)^2$
$= (3t + 1)(3t + 1)$
$= (3t)(3t) + (3t)(1) + (1)(3t)$
    $+ (1)(1)$
$= 9t^2 + 3t + 3t + 1$
$= 9t^2 + 6t + 1$

39. $(h - 5)^3 = (h - 5)(h - 5)(h - 5)$

The first step is to find the product $(h - 5)(h - 5)$.

$(h - 5)(h - 5)$
$= (h - 5)(h) + (h - 5)(-5)$
$= h^2 - 5h - 5h + 25$
$= h^2 - 10h + 25$

Now multiply this result by $h - 5$.

$(h - 5)^3$
$= (h - 5)(h^2 - 10h + 25)$
$= (h - 5)(h^2) + (h - 5)(-10h)$
    $+ (h - 5)(25)$
$= h^3 - 5h^2 - 10h^2 + 50h + 25h - 125$
$= h^3 - 15h^2 + 75h - 125$

43. $(3x^2 + x - 4)^2$
$= (3x^2 + x - 4)(3x^2 + x - 4)$

Multiply vertically.

$$\begin{array}{r} 3x^2 + x - 4 \\ 3x^2 + x - 4 \\ \hline -12x^2 - 4x + 16 \\ 3x^3 + x^2 - 4x \\ 9x^4 + 3x^3 - 12x^2 \\ \hline 9x^4 + 6x^3 - 23x^2 - 8x + 16 \end{array}$$

47. Find two numbers having a product of -56 and a sum of -1.

Since the product is negative, one number must be positive, the other negative. The only possibilities are

56 and -1
-56 and 1
28 and -2
-28 and 2

14 and −4
−14 and 4
7 and −8
−7 and 8.

Of all these possibilities, only 7 and −8 have a sum of −1.

Section 3.6 Products of Binomials

3.6 Margin Exercises

1. (a) $(2p - 5)(3p + 7)$
   $2p(3p) = 6p^2$

   (b) $(2p - 5)(3p + 7)$
   $2p(7) = 14p$

   (c) $(2p - 5)(3p + 7)$
   $-5(3p) = -15p$

   (d) $(2p - 5)(3p + 7)$
   $-5(7) = -35$

   (e) $(2p - 5)(3p + 7)$
   $= 6p^2 + 14p + (-15p) + (-35)$
   $= 6p^2 - p - 35$

2. (a) $(m + 4)(m - 3)$
   F: $m(m) = m^2$
   O: $m(-3) = -3m$
   I: $4(m) = 4m$
   L: $4(-3) = -12$
   $(m + 4)(m - 3) = m^2 - 3m + 4m - 12$
   $= m^2 + m - 12$

   (b) $(y + 7)(y + 2)$
   F: $y(y) = y^2$
   O: $y(2) = 2y$
   I: $7(y) = 7y$
   L: $7(2) = 14$
   $(y + 7)(y + 2) = y^2 + 2y + 7y + 14$
   $= y^2 + 9y + 14$

   (c) $(r - 8)(r - 5)$
   F: $r(r) = r^2$
   O: $r(-5) = -5r$
   I: $-8(r) = -8r$
   L: $-8(-5) = 40$
   $(r - 8)(r - 5) = r^2 - 5r - 8r + 40$
   $= r^2 - 13r + 40$

3. (a) $(4k - 1)(2k + 3)$
   F: $4k(2k) = 8k^2$
   O: $4k(3) = 12k$
   I: $-1(2k) = -2k$
   L: $-1(3) = -3$
   $(4k - 1)(2k + 3) = 8k^2 + 12k - 2k - 3$
   $= 8k^2 + 10k - 3$

   (b) $(6m + 5)(m - 4)$
   F: $6m(m) = 6m^2$
   O: $6m(-4) = -24m$
   I: $5(m) = 5m$
   L: $5(-4) = -20$
   $(6m + 5)(m - 4) = 6m^2 - 24m + 5m - 20$
   $= 6m^2 - 19m - 20$

   (c) $(8y + 3)(2y + 1)$
   F: $8y(2y) = 16y^2$
   O: $8y(1) = 8y$
   I: $3(2y) = 6y$
   L: $3(1) = 3$
   $(8y + 3)(2y + 1) = 16y^2 + 8y + 6y + 3$
   $= 16y^2 + 14y + 3$

(d) $(3r + 2t)(3r + 4t)$

F: $3r(3r) = 9r^2$
O: $3r(4t) = 12rt$
I: $2t(3r) = 6rt$
L: $2t(4t) = 8t^2$

$(3r + 2t)(3r + 4t)$
$= 9r^2 + 12rt + 6rt + 8t^2$
$= 9r^2 + 18rt + 8t^2$

4. **(a)** $(t + u)^2 = t^2 + 2(t)(u) + u^2$
$= t^2 + 2tu + u^2$

(b) $(2m - p)^2 = (2m)^2 - 2(2m)(p) + p^2$
$= 4m^2 - 4mp + p^2$

(c) $(4p + 3q)^2$
$= (4p)^2 + 2(4p)(3q) + (3q)^2$
$= 16p^2 + 24pq + 9q^2$

(d) $(5r - 6s)^2$
$= (5r)^2 - 2(5r)(6s) + (6s)^2$
$= 25r^2 - 60rs + 36s^2$

(e) $\left(3k - \frac{1}{2}\right)^2$
$= (3k)^2 - 2(3k)\left(\frac{1}{2}\right) + \left(\frac{1}{2}\right)^2$
$= 9k^2 - 3k + \frac{1}{4}$

5. **(a)** $(6a + 3)(6a - 3) = (6a)^2 - 3^2$
$= 36a^2 - 9$

(b) $(10m + 7)(10m - 7) = (10m)^2 - 7^2$
$= 100m^2 - 49$

(c) $(7p + 2q)(7p - 2q)$
$= (7p)^2 - (2q)^2$
$= 49p^2 - 4q^2$

(d) $\left(3r - \frac{1}{2}\right)\left(3r + \frac{1}{2}\right) = (3r)^2 - \left(\frac{1}{2}\right)^2$
$= 9r^2 - \frac{1}{4}$

3.6 Section Exercises

3. $(r + 1)(r + 3) \overset{\text{F O I L}}{=} r^2 + 3r + r + 3$
$= r^2 + 4r + 3$

7. $(s - 12)(s + 4) \overset{\text{F O I L}}{=} s^2 + 4s - 12s - 48$
$= s^2 - 8s - 48$

11. $(9x + 2)(3x + 7)$
$\overset{\text{F O I L}}{=} 27x^2 + 63x + 6x + 14$
$= 27x^2 + 69x + 14$

15. $(3 - 2x)(5 - 3x) \overset{\text{F O I L}}{=} 15 - 9x - 10x + 6x$
$= 15 - 19x + 6x^2$

19. $(-8 + 3k)(-2 - k) \overset{\text{F O I L}}{=} 16 + 8k - 6k - 3k$
$= 16 + 2k - 3k^2$

23. $(-8p + 3s)(2p + s)$
$\overset{\text{F O I L}}{=} -16p^2 - 8ps + 6ps + 3s^2$
$= -16p^2 - 2ps + 3s^2$

27. $\left(x - \frac{2}{3}\right)\left(x + \frac{1}{4}\right) \overset{\text{F O I L}}{=} x^2 + \frac{1}{4}x - \frac{2}{3}x - \frac{1}{6}$
$= x^2 - \frac{5}{12}x - \frac{1}{6}$

Chapter 3   Exponents and Polynomials

In Exercises 31-39, use one of the following formulas for the square of a binomial:

$$(a + b)^2 = a^2 + 2ab + b^2$$
$$(a - b)^2 = a^2 - 2ab + b^2.$$

31. $(p + 2)^2 = (p)^2 + 2(p)(2) + (2)^2$
    $= p^2 + 4p + 4$

35. $(4x - 3)^2 = (4x)^2 - 2(4x)(3) + (3)^2$
    $= 16x^2 - 24x + 9$

39. $\left(5x + \frac{2}{5}y\right)^2 = (5x)^2 + 2(5x)\left(\frac{2}{5}y\right) + \left(\frac{2}{5}y\right)^2$
    $= 25x^2 + 4xy + \frac{4}{25}y^2$

In Exercises 43-51, use the formula for the product of the sum and difference of two terms:

$$(a + b)(a - b) = a^2 - b^2.$$

43. $(q + 2)(q - 2) = (q)^2 - (2)^2$
    $= q^2 - 4$

47. $(10x + 3y)(10x - 3y)$
    $= (10x)^2 - (3y)^2$
    $= 100x^2 - 9y^2$

51. $\left(7x + \frac{3}{7}\right)\left(7x - \frac{3}{7}\right) = (7x)^2 - \left(\frac{3}{7}\right)^2$
    $= 49x^2 - \frac{9}{49}$

55. $\dfrac{64m^9}{8m^2} = 8m^{9-2} = 8m^7$

Section 3.7   Division of a Polynomial by a Monomial

3.7 Margin Exercises

1. (a) $\dfrac{6p^4 + 18p^7}{3p^2} = \dfrac{6p^4}{3p^2} + \dfrac{18p^7}{3p^2}$
   $= 2p^2 + 6p^5$

   (b) $\dfrac{12m^6 + 18m^5 + 30m^4}{6m^2}$
   $= \dfrac{12m^6}{6m^2} + \dfrac{18m^5}{6m^2} + \dfrac{30m^4}{6m^2}$
   $= 2m^4 + 3m^3 + 5m^2$

   (c) $(18r^7 - 9r^2) \div (3r)$
   $= \dfrac{18r^7 - 9r^2}{3r}$
   $= \dfrac{18r^7}{3r} - \dfrac{9r^2}{3r}$
   $= 6r^6 - 3r$

2. (a) $\dfrac{20x^4 - 25x^3 + 5x}{5x^2}$
   $= \dfrac{20x^4}{5x^2} - \dfrac{25x^3}{5x^2} + \dfrac{5x}{5x^2}$
   $= 4x^2 - 5x + \dfrac{1}{x}$

   (b) $\dfrac{50m^4 - 30m^3 + 20m}{10m^3}$
   $= \dfrac{50m^4}{10m^3} - \dfrac{30m^3}{10m^3} + \dfrac{20m}{10m^3}$
   $= 5m - 3 + \dfrac{2}{m^2}$

3. (a) $\dfrac{8y^7 - 9y^6 - 11y - 4}{y^2}$
   $= \dfrac{8y^7}{y^2} - \dfrac{9y^6}{y^2} - \dfrac{11y}{y^2} - \dfrac{4}{y^2}$
   $= 8y^5 - 9y^4 - \dfrac{11}{y} - \dfrac{4}{y^2}$

(b) $\dfrac{12p^5 + 8p^4 + 3p^3 - 5p^2}{3p^3}$

$= \dfrac{12p^5}{3p^3} + \dfrac{8p^4}{3p^3} + \dfrac{3p^3}{3p^3} - \dfrac{5p^2}{3p^3}$

$= 4p^2 + \dfrac{8p}{3} + 1 - \dfrac{5}{3p}$

(c) $\dfrac{45x^4 + 30x^3 - 60x^2}{-15x^2}$

$= \dfrac{45x^4}{-15x^2} + \dfrac{30x^3}{-15x^2} - \dfrac{60x^2}{-15x^2}$

$= -3x^2 - 2x + 4$

3.7 Section Exercises

3. $\dfrac{15x^9 y^4}{3x^2 y^3} = 5x^{9-2} y^{4-3} = 5x^7 y$

7. $\dfrac{60m^4 - 20m^2 + 10m}{2m}$

$= \dfrac{60m^4}{2m} - \dfrac{20m^2}{2m} + \dfrac{10m}{2m}$

$= 30m^3 - 10m + 5$

11. $\dfrac{8m^5 - 4m^3 + 4m^2}{2m} = \dfrac{8m^5}{2m} - \dfrac{4m^3}{2m} + \dfrac{4m^2}{2m}$

$= 4m^4 - 2m^2 + 2m$

15. $\dfrac{12x^5 - 9x^4 + 6x^3}{3x^2} = \dfrac{12x^5}{3x^2} - \dfrac{9x^4}{3x^2} + \dfrac{6x^3}{3x^2}$

$= 4x^3 - 3x^2 + 2x$

19. $\dfrac{36x + 24x^2 + 6x^3}{3x^2}$

$= \dfrac{36x}{3x^2} + \dfrac{24x^2}{3x^2} + \dfrac{6x^3}{3x^2}$

$= \dfrac{12}{x} + 8 + 2x$

23. $\dfrac{27r^4 - 36r^3 - 6r^2 + 26r - 2}{3r}$

$= \dfrac{27r^4}{3r} - \dfrac{36r^3}{3r} - \dfrac{6r^2}{3r} + \dfrac{26r}{3r} - \dfrac{2}{3r}$

$= 9r^3 - 12r^2 - 2r + \dfrac{26}{3} - \dfrac{2}{3r}$

27. $(20a^4 - 15a^5 + 25a^3) \div (5a^4)$

$= \dfrac{20a^4}{5a^4} - \dfrac{15a^5}{5a^4} + \dfrac{25a^3}{5a^4}$

$= 4 - 3a + \dfrac{5}{a}$

31. $\dfrac{2}{3x}$ is not the same as $\dfrac{2}{3}x$.

$\dfrac{2}{3}x = \dfrac{2}{3} \cdot \dfrac{x}{1} = \dfrac{2x}{3}$

$\dfrac{4}{3}x^2 = \dfrac{4}{3} \cdot \dfrac{x^2}{1} = \dfrac{4x^2}{3}$

Therefore, $\dfrac{4x^2}{3}$ is the same as $\dfrac{4}{3}x^2$.

35. $-3k(8k^2 - 12k + 2)$

$= (-3k)(8k^2) + (-3k)(-12k)$
$\quad + (-3k)(2)$

$= -24k^3 + 36k^2 - 6k$

39. Subtract.

$\begin{array}{r} -4x^3 + 2x^2 - 3x + 7 \\ -4x^3 - 8x^2 + x - 4 \\ \hline \end{array}$

Change all signs in the second row; then add.

$\begin{array}{r} -4x^3 + 2x^2 - 3x + 7 \\ 4x^3 + 8x^2 - x + 4 \\ \hline 10x^2 - 4x + 11 \end{array}$

Chapter 3 Exponents and Polynomials

Section 3.8 The Quotient of Two Polynomials

3.8 Margin Exercises

1. **(a)** $(2y^2 - y - 21) \div (y + 3)$

   y divides into $2y^2$ **2y** times.

   $$\begin{array}{r} 2y \\ y + 3 \overline{\smash{\big)} 2y^2 - y - 21} \\ \underline{2y^2 + 6y} \quad \leftarrow 2y(y+3) \\ -7y - 21 \quad \text{Subtract and} \\ \text{bring down} \end{array}$$

   y divides into $-7y$ **-7** times.

   $$\begin{array}{r} 2y - 7 \\ y + 3 \overline{\smash{\big)} 2y^2 - y - 21} \\ \underline{2y^2 + 6y} \\ -7y - 21 \\ \underline{-7y - 21} \leftarrow -7(y+3) \\ 0 \quad \text{Subtract} \end{array}$$

   $(2y^2 - y - 21) \div (y + 3) = 2y - 7$

   (b) $(x^3 + x^2 + 4x - 6) \div (x - 1)$

   $$\begin{array}{r} x^2 + 2x + 6 \\ x - 1 \overline{\smash{\big)} x^3 + x^2 + 4x - 6} \\ \underline{x^3 - x^2} \\ 2x^2 + 4x \\ \underline{2x^2 - 2x} \\ 6x - 6 \\ \underline{6x - 6} \\ 0 \end{array}$$

   $(x^3 + x^2 + 4x - 6) \div (x - 1)$
   $= x^2 + 2x + 6$

   (c) $\dfrac{p^3 - 2p^2 - 5p + 9}{p + 2}$

   $$\begin{array}{r} p^2 - 4p + 3 \\ p + 2 \overline{\smash{\big)} p^3 - 2p^2 - 5p + 9} \\ \underline{p^3 + 2p^2} \\ -4p^2 - 5p \\ \underline{-4p^2 - 8p} \\ 3p + 9 \\ \underline{3p + 6} \\ 3 \\ \uparrow \\ \text{Remainder} \end{array}$$

   Write the remainder in the numerator of a fraction with the divisor as the denominator. Add this fraction to the quotient to get the answer.

   $$\dfrac{p^3 - 2p^2 - 5p + 9}{p + 2}$$
   $$= p^2 - 4p + 3 + \dfrac{3}{p + 2}$$

2. **(a)** $\dfrac{r^2 - 5}{r + 4}$

   $$\begin{array}{r} r - 4 \\ r + 4 \overline{\smash{\big)} r^2 + 0r - 5} \leftarrow \text{Insert 0r for} \\ \underline{r^2 + 4r} \quad \text{missing term} \\ -4r - 5 \\ \underline{-4r - 16} \\ 11 \leftarrow \text{Remainder} \end{array}$$

   $\dfrac{r^2 - 5}{r + 4} = r - 4 + \dfrac{11}{r + 4}$

   (b) $(x^3 - 8) \div (x - 2)$

   $$\begin{array}{r} x^2 + 2x + 4 \\ x - 2 \overline{\smash{\big)} x^3 + 0x^2 + 0x - 8} \leftarrow \text{Insert} \\ \underline{x^3 - 2x^2} \quad \text{missing} \\ 2x^2 + 0x \quad \text{terms} \\ \underline{2x^2 - 4x} \\ 4x - 8 \\ \underline{4x - 8} \\ 0 \end{array}$$

   $(x^3 - 8) \div (x - 2) = x^2 + 2x + 4$

3. **(a)**

   $(2x^4 + 3x^3 - x^2 + 6x + 5) \div (x^2 - 1)$

   Insert 0x for the missing term in the divisor
   ↓

   $$\begin{array}{r} 2x^2 + 3x + 1 \\ x^2 + 0x - 1 \overline{\smash{\big)} 2x^4 + 3x^3 - x^2 + 6x + 5} \\ \underline{2x^4 + 0x^3 - 2x^2} \\ 3x^3 + x^2 + 6x \\ \underline{3x^3 + 0x^2 - 3x} \\ x^2 + 9x + 5 \\ \underline{x^2 + 0x - 1} \\ 9x + 6 \\ \uparrow \\ \text{Remainder} \end{array}$$

   $(2x^4 + 3x^2 - x^2 + 6x + 5) \div (x^2 - 1)$
   $= 2x^2 + 3x + 1 + \dfrac{9x + 6}{x^2 - 1}$

3.8 Section Exercises

(b) $\dfrac{2m^5 + m^4 + 6m^3 - 3m^2 - 18}{m^2 + 3}$

$$
\begin{array}{r}
2m^3 + m^2 - 6 \\
m^2 + 0m + 3 \overline{\smash{\big)}\, 2m^5 + m^4 + 6m^3 - 3m^2 + 0m - 18} \\
\underline{2m^5 + 0m^4 + 6m^3 } \\
m^4 + 0m^3 - 3m^2 \\
\underline{m^4 + 0m^3 + 3m^2 } \\
-6m^2 + 0m - 18 \\
\underline{-6m^2 + 0m - 18} \\
0
\end{array}
$$

$\dfrac{2m^5 + m^4 + 6m^3 - 3m^2 - 18}{m^2 + 3}$
$= 2m^3 + m^2 - 6$

3.8 Section Exercises

3. $\dfrac{2y^2 + 9y - 35}{y + 7}$

$$
\begin{array}{r}
2y - 5 \\
y + 7 \overline{\smash{\big)}\, 2y^2 + 9y - 35} \\
\underline{2y^2 + 14y } \\
-5y - 35 \\
\underline{-5y - 35} \\
0
\end{array}
$$

The remainder is 0. The answer is the quotient, $2y - 5$.

7. $(r^2 - 8r + 15) \div (r - 3)$

$$
\begin{array}{r}
r - 5 \\
r - 3 \overline{\smash{\big)}\, r^2 - 8r + 15} \\
\underline{r^2 - 3r } \\
-5r + 15 \\
\underline{-5r + 15} \\
0
\end{array}
$$

The remainder is 0. The answer is the quotient, $r - 5$.

11. $\dfrac{4a^2 - 22a + 32}{2a + 3}$

$$
\begin{array}{r}
2a - 14 \\
2a + 3 \overline{\smash{\big)}\, 4a^2 - 22a + 32} \\
\underline{4a^2 + 6a } \\
-28a + 32 \\
\underline{-28a - 42} \\
74
\end{array}
$$

The remainder is 74. The answer is
$2a - 14 + \dfrac{74}{2a + 3}.$

15. In the problem
$(4x^4 + 2x^3 - 14x^2 + 19x + 10) \div (2x + 5)$
$= 2x^3 - 4x^2 + 3x + 2,$
the divisor is $2x + 5$ and the quotient is $2x^3 - 4x^2 + 3x + 2.$

19. $\dfrac{3k^3 - 4k^2 - 6k + 10}{k^2 - 2}$

$$
\begin{array}{r}
3k - 4 \\
k^2 + 0k + 2 \overline{\smash{\big)}\, 3k^3 - 4k^2 - 6k + 10} \\
\underline{3k^3 + 0k^2 - 6k } \\
-4k^2 + 0k + 10 \\
\underline{-4k^2 + 0k + 8} \\
2
\end{array}
$$

The remainder is 2. The answer is
$3k - 4 + \dfrac{2}{k^2 - 2}.$

23. $\dfrac{6p^4 - 15p^3 + 14p^2 - 5p + 10}{3p^2 + 1}$

$$
\begin{array}{r}
2p^2 - 5p + 4 \\
3p^2 + 0p + 1 \overline{\smash{\big)}\, 6p^4 - 15p^3 + 14p^2 - 5p + 10} \\
\underline{6p^4 + 0p^3 + 2p^2 } \\
-15p^3 + 12p^2 - 5p \\
\underline{-15p^3 + 0p^2 - 5p } \\
12p^2 + 0p + 10 \\
\underline{12p^2 + 0p + 4} \\
6
\end{array}
$$

The remainder is 6. The answer is
$2p^2 - 5p + 4 + \dfrac{6}{3p^2 + 1}.$

116   Chapter 3   Exponents and Polynomials

27. $\dfrac{x^4 - 1}{x^2 - 1}$

$$\begin{array}{r} x^2 + 1 \\ x^2 + 0x - 1 \overline{\smash{)} x^4 + 0x^3 + 0x^2 + 0x - 1} \\ \underline{x^4 + 0x^3 - x^2 } \\ x^2 + 0x - 1 \\ \underline{x^2 + 0x - 1} \\ 0 \end{array}$$

The remainder is 0. The answer is the quotient, $x^2 + 1$.

31. $48 = 1 \cdot 48$
    $= 2 \cdot 24$
    $= 3 \cdot 16$
    $= 4 \cdot 12$
    $= 6 \cdot 8$

The positive integer factors are 1, 2, 3, 4, 6, 8, 12, 16, 24, and 48.

Chapter 3   Review Exercises

1. $4^3 \cdot 4^8 = 4^{3+8} = 4^{11}$

2. $(-5)^6(-5)^5 = (-5)^{6+5} = (-5)^{11}$

3. $(-8x^4)(9x^3) = (-8)(9)(x^4)(x^3)$
   $= -72x^{4+3} = -72x^7$

4. $(2x^2)(5x^3)(x^9) = (2)(x^2)(5)(x^3)(x^9)$
   $= 10x^{2+3+9} = 10x^{14}$

5. $(19x)^5 = 19^5 x^5$

6. $(-4y)^7 = (-4)^7 y^7$

7. $5(pt)^4 = 5p^4 t^4$

8. $\left(\dfrac{7}{5}\right)^6 = \dfrac{7^6}{5^6}$

9. $(3x^2 y^3)^3 = 3^3 (x^2)^3 (y^3)^3$
   $= 3^3 x^{2 \cdot 3} y^{3 \cdot 3} = 3^3 x^6 y^9$

10. $(t^4)^8 (t^2)^5 = t^{4 \cdot 8} \cdot t^{2 \cdot 5} = t^{32} \cdot t^{10}$
    $= t^{32+10} = t^{42}$

11. $(6x^2 z^4)^2 (x^3 y z^2)^4$
    $= 6^2 (x^2)^2 (z^4)^2 (x^3)^4 (y)^4 (z^2)^4$
    $= 6^2 x^4 z^8 x^{12} y^4 z^8$
    $= 6^2 x^{4+12} y^4 z^{8+8}$
    $= 6^2 x^{16} y^4 z^{16}$

12. The product rule does not apply to $7^2 + 7^4$ because you are adding powers of 7, not multiplying them.

13. $5^0 + 8^0 = 1 + 1 = 2$

14. $2^{-5} = \dfrac{1}{2^5} = \dfrac{1}{2 \cdot 2 \cdot 2 \cdot 2 \cdot 2} = \dfrac{1}{32}$

15. $\left(\dfrac{6}{5}\right)^{-2} = \left(\dfrac{5}{6}\right)^2$
    $= \dfrac{5^2}{6^2}$ or $\dfrac{25}{36}$

16. $4^{-2} - 4^{-1} = \dfrac{1}{4^2} - \dfrac{1}{4} = \dfrac{1}{16} - \dfrac{1}{4}$
    $= \dfrac{1}{16} - \dfrac{4}{16} = -\dfrac{3}{16}$

17. $\dfrac{6^{-3}}{6^{-5}} = 6^{(-3)-(-5)} = 6^2$

18. $\dfrac{x^{-7}}{x^{-9}} = x^{(-7)-(-9)} = x^2$

19. $\dfrac{p^{-8}}{p^4} = p^{-8-4} = p^{-12} = \dfrac{1}{p^{12}}$

20. $\dfrac{r^{-2}}{r^{-6}} = r^{(-2)(-6)} = r^4$

21. $(2^4)^2 = 2^{4 \cdot 2} = 2^8$

22. $(9^3)^{-2} = 9^{(3)(-2)} = 9^{-6} = \dfrac{1}{9^6}$

23. $(5^{-2})^{-4} = 5^{(-2)(-4)} = 5^8$

24. $(8^{-3})^4 = 8^{(-3)(4)} = 8^{-12} = \dfrac{1}{8^{12}}$

25. $\dfrac{(m^2)^3}{(m^4)^2} = \dfrac{m^6}{m^8} = m^{6-8} = m^{-2} = \dfrac{1}{m^2}$

26. $\dfrac{y^4 \cdot y^{-2}}{y^{-5}} = \dfrac{y^{4+(-2)}}{y^{-5}} = \dfrac{y^2}{y^{-5}}$
    $= y^{2-(-5)} = y^7$

27. $\dfrac{r^9 \cdot r^{-5}}{r^{-2} \cdot r^{-7}} = \dfrac{r^{9+(-5)}}{r^{(-2)+(-7)}} = \dfrac{r^4}{r^{-9}}$
    $= r^{4-(-9)} = r^{13}$

28. $(-5m^3)^2 = (-5)^2(m^3)^2 = (-5)^2 m^{3 \cdot 2}$
    $= (-5)^2 m^6$

29. $(2y^{-4})^{-3} = 2^{-3}(y^{-4})^{-3} = 2^{-3}y^{(-4)(-3)}$
    $= 2^{-3}y^{12} = \dfrac{1}{2^3} \cdot y^{12} = \dfrac{y^{12}}{2^3}$

30. $\dfrac{ab^{-3}}{a^4 b^2} = a^{1-4}b^{(-3)-2} = a^{-3}b^{-5} = \dfrac{1}{a^3 b^5}$

31. $\dfrac{(6r^{-1})^2 \cdot (2r^{-4})}{r^{-5}(r^2)^{-3}} = \dfrac{6^2(r^{-1})^2(2r^{-4})}{r^{-5}r^{2(-3)}}$
    $= \dfrac{6^2 r^{(-1)(2)}(2r^{-4})}{r^{-5}r^{-6}}$
    $= \dfrac{6^2 r^{-2}(2r^{-4})}{r^{-5}r^{-6}}$
    $= \dfrac{2 \cdot 6^2 \cdot r^{(-2)+(-4)}}{r^{(-5)+(-6)}}$
    $= \dfrac{2 \cdot 6^2 \cdot r^{-6}}{r^{-11}}$
    $= 2 \cdot 6^2 \cdot r^{-6-(-11)}$
    $= 2 \cdot 6^2 \cdot r^5$

32. $\dfrac{(2m^{-5}n^2)^3(3m^2)^{-1}}{m^{-2}n^{-4}(m^{-1})^2}$
    $= \dfrac{2^3(m^{-5})^3(n^2)^3 3^{-1}(m^2)^{-1}}{m^{-2}n^{-4}(m^{-1})^2}$
    $= \dfrac{2^3 m^{(-5)(3)} n^{2 \cdot 3} 3^{-1} m^{2(-1)}}{m^{-2}n^{-4}m^{(-1)(2)}}$
    $= \dfrac{2^3 m^{-15} n^6 \cdot 3^{-1} m^{-2}}{m^{-2}n^{-4}m^{-2}}$
    $= \dfrac{2^3 \cdot 3^{-1} m^{(-15)+(-2)} n^6}{m^{(-2)+(-2)} n^{-4}}$
    $= \dfrac{2^3 \cdot 3^{-1} m^{-17} n^6}{m^{-4} n^{-4}}$
    $= 2^3 \cdot 3^{-1} m^{-17-(-4)} n^{6-(-4)}$
    $= 2^3 \cdot 3^{-1} m^{-13} n^{10} = \dfrac{2^3 n^{10}}{3 m^{13}}$

33. $48{,}000{,}000 = 4.8 \times 10^7$

34. $28{,}988{,}000{,}000 = 2.8988 \times 10^{10}$

35. $.000065 = 6.5 \times 10^{-5}$

118 Chapter 3 Exponents and Polynomials

36. $.0000000824 = 8.24 \times 10^{-8}$

37. $2.4 \times 10^4 = 24{,}000$

    Move the decimal point 4 places to the right.

38. $7.83 \times 10^7 = 78{,}300{,}000$

    Move the decimal point 7 places to the right.

39. $8.97 \times 10^{-7} = .000000897$

    Move the decimal point 7 places to the left.

40. $9.95 \times 10^{-12} = .00000000000995$

    Move the decimal point 12 places to the left.

41. $(2 \times 10^{-3}) \times (4 \times 10^5)$
    $= (2 \times 4) \times (10^{-3} \times 10^5)$
    $= 8 \times 10^{-3+5} = 8 \times 10^2$
    $= 800$

42. $\dfrac{8 \times 10^4}{2 \times 10^{-2}} = \dfrac{8}{2} \times \dfrac{10^4}{10^{-2}} = 4 \times 10^{4-(-2)}$
    $= 4 \times 10^6 = 4{,}000{,}000$

43. $\dfrac{12 \times 10^{-5} \times 5 \times 10^4}{4 \times 10^3 \times 6 \times 10^{-2}}$
    $= \dfrac{12 \times 5 \times 10^{-5} \times 10^4}{4 \times 6 \times 10^3 \times 10^{-2}}$
    $= \dfrac{60 \times 10^{-5+4}}{24 \times 10^{3+(-2)}}$
    $= \dfrac{60 \times 10^{-1}}{24 \times 10^1} = \dfrac{60}{24} \times \dfrac{10^{-1}}{10^1}$
    $= 2.5 \times 10^{-1-1} = 2.5 \times 10^{-2}$
    $= .025$

44. $\dfrac{2.5 \times 10^5 \times 4.8 \times 10^{-4}}{7.5 \times 10^8 \times 1.6 \times 10^{-5}}$
    $= \dfrac{2.5 \times 4.8 \times 10^5 \times 10^{-4}}{7.5 \times 1.6 \times 10^8 \times 10^{-5}}$
    $= \dfrac{2.5}{7.5} \times \dfrac{4.8}{1.6} \times \dfrac{10^{5+(-4)}}{10^{8+(-5)}}$
    $= \dfrac{1}{3} \times \dfrac{3}{1} \times \dfrac{10^1}{10^3} = 1 \times 10^{1-3}$
    $= 1 \times 10^{-2} = .01$

45. $9m^2 + 11m^2 + 2m^2 = 22m^2$

    degree 2; monomial (1 term)

46. $-4p + p^3 - p^2 + 8p + 2$
    $= p^3 - p^2 - 4p + 8p + 2$
    $= p^3 - p^2 + 4p + 2$

    degree 3; none of these (4 terms)

47. $2r^4 - r^3 + 8r^4 + r^3 - 6r^4 + 8r^5$
    $= 8r^5 + 2r^4 + 8r^4 - 6r^4 - r^3 + r^3$
    $= 8r^5 + 4r^4$

    degree 5; binomial (2 terms)

48. $12a^5 + 19a^4 + 8a^3 + 2a^3 - 9a^4 + 3a^5$
    $= 12a^5 + 3a^5 + 19a^4 - 9a^4$
    $\quad + 8a^3 + 2a^3$
    $= 15a^5 + 10a^4 + 10a^3$

    degree 5; trinomial (3 terms)

49. Add.

    $\begin{array}{r} -2a^3 + 5a^2 \\ -3a^3 - a^2 \\ \hline -5a^3 + 4a^2 \end{array}$

50. Add.

$$\begin{array}{r} 4r^3 - 8r^2 + 6r \\ -2r^3 + 5r^2 + 3r \\ \hline 2r^3 - 3r^2 + 9r \end{array}$$

51. Subtract.

$$\begin{array}{r} 6y^2 - 8y + 2 \\ -5y^2 + 2y - 7 \end{array}$$

Change all signs in the second row; then add.

$$\begin{array}{r} 6y^2 - 8y + 2 \\ 5y^2 - 2y + 7 \\ \hline 11y^2 - 10y + 9 \end{array}$$

52. Subtract.

$$\begin{array}{r} -12k^4 - 8k^2 + 7k - 5 \\ k^4 + 7k^2 + 11k + 1 \end{array}$$

Change all signs in the second row; then add.

$$\begin{array}{r} -12k^4 - 8k^2 + 7k - 5 \\ -k^4 - 7k^2 - 11k - 1 \\ \hline -13k^4 - 15k^2 - 4k - 6 \end{array}$$

53. $(2m^3 - 8m^2 + 4) + (8m^3 + 2m^2 - 7)$
    $= 2m^3 - 8m^2 + 4 + 8m^3 + 2m^2 - 7$
    $= 2m^3 + 8m^3 - 8m^2 + 2m^2 + 4 - 7$
    $= 10m^3 - 6m^2 - 3$

54. $(-5y^2 + 3y + 11) + (4y^2 - 7y + 15)$
    $= -5y^2 + 3y + 11 + 4y^2 - 7y + 15$
    $= -5y^2 + 4y^2 + 3y - 7y + 11 + 15$
    $= -y^2 - 4y + 26$

55. $(6p^2 - p - 8) - (-4p^2 + 2p + 3)$
    $= (6p^2 - p - 8) + (4p^2 - 2p - 3)$
    $= 6p^2 - p - 8 + 4p^2 - 2p - 3$
    $= 6p^2 + 4p^2 - p - 2p - 8 - 3$
    $= 10p^2 - 3p - 11$

56. $(12r^4 - 7r^3 + 2r^2)$
    $\quad - (5r^4 - 3r^3 + 2r^2 + 1)$
    $= (12r^4 - 7r^3 + 2r^2)$
    $\quad + (-5r^4 + 3r^3 - 2r^2 - 1)$
    $= 12r^4 - 7r^3 + 2r^2 - 5r^4$
    $\quad + 3r^3 - 2r^2 - 1$
    $= 12r^4 - 5r^4 - 7r^3 + 3r^3 + 2r^2$
    $\quad - 2r^2 - 1$
    $= 7r^4 - 4r^3 - 1$

57. $5x(2x + 14) = (5x)(2x) + (5x)(14)$
    $\qquad\qquad\quad = 10x^2 + 70x$

58. $-3p^3(2p^2 - 5p)$
    $= (-3p^2)(2p^3) + (-3p^3)(-5p)$
    $= -6p^5 + 15p^4$

59. $(m - 9)(m + 2)$
    $= (m)(m) + (m)(2) + (-9)(m)$
    $\quad + (-9)(2)$
    $= m^2 + 2m - 9m - 18$
    $= m^2 - 7m - 18$

60. $(3k - 6)(2k + 1)$
    $= (3k)(2k) + (3k)(1) + (-6)(2k)$
    $\quad + (-6)(1)$
    $= 6k^2 + 3k - 12k - 6$
    $= 6k^2 - 9k - 6$

61. $(3r - 2)(2r^2 + 4r - 3)$
    $= (3r)(2r^2) + (3r)(4r) + (3r)(-3)$
    $\quad + (-2)(2r^2) + (-2)(4r) + (-2)(-3)$
    $= 6r^3 + 12r^2 - 9r - 4r^2 - 8r + 6$
    $= 6r^3 + 8r^2 - 17r + 6$

62. $(2y + 3)(4y^2 - 6y + 9)$
    $= (2y)(4y^2) + (2y)(-6y) + (2y)(9)$
    $\quad + (3)(4y^2) + (3)(-6y) + (3)(9)$
    $= 8y^3 - 12y^2 + 18y + 12y^2 - 18y + 27$
    $= 8y^3 + 27$

63. $(r + 2)^3$
    $= (r + 2)(r + 2)(r + 2)$
    $= (r + 2)(r^2 + 4r + 4)$
    $= r(r^2 + 4r + 4) + 2(r^2 + 4r + 4)$
    $= r^3 + 4r^2 + 4r + 2r^2 + 8r + 8$
    $= r^3 + 6r^2 + 12r + 8$

64. $(a + b)^2 \neq a^2 + b^2$ because
    $(a + b)^2 = (a + b)(a + b)$
    $= (a)(a) + (a)(b) + (b)(a) + (b)(b)$
    $= a^2 + ab + ab + b^2$
    $= a^2 + 2ab + b^2.$

65. $(3k + 1)(2k + 3)$
    $= (3k)(2k + 3) + (1)(2k + 3)$
    $= (3k)(2k) + (3k)(3) + (1)(2k)$
    $\quad + (1)(3)$
    $= 6k^2 + 9k + 2k + 3$
    $= 6k^2 + 11k + 3$

66. $(a + 3b)(2a - b)$
    $= a(2a - b) + 3b(2a - b)$
    $= (a)(2a) + a(-b) + (3b)(2a) + (3b)(-b)$
    $= 2a^2 - ab + 6ab - 3b^2$
    $= 2a^2 + 5ab - 3b^2$

67. $(6k - 3q)(2k - 7q)$
     F $\quad$ O $\quad$ I $\quad$ L
    $= (6k)(2k)+(6k)(-7q)+(-3q)(2k)+(-3q)(-7q)$
    $= 12k^2 - 42kq - 6kq + 21q^2$
    $= 12k^2 - 48kq + 21q^2$

68. $(a + 4)^2 = (a)^2 + 2(a)(4) + (4)^2$
    $= a^2 + 8a + 16$

69. $(3p - 2)^2 = (3p)^2 - 2(3p)(2) + (2)^2$
    $= 9p^2 - 12p + 4$

70. $(2r + 5s)^2 = (2r)^2 + 2(2r)(5s) + (5s)^2$
    $= 4r^2 + 20rs + 25s^2$

71. $(6m - 5)(6m + 5) = (6m)^2 - (5)^2$
    $= 36m^2 - 25$

72. $(2z + 7)(2z - 7) = (2z)^2 - 7^2$
    $= 4z^2 - 49$

73. $(5a + 6b)(5a - 6b) = (5a)^2 - (6b)^2$
    $= 25a^2 - 36b^2$

74. $(2x^2 + 5)(2x^2 - 5) = (2x^2)^2 - (5)^2$
    $= 4x^4 - 25$

75. $\dfrac{-15y^4}{-9y^2} = \dfrac{-15}{-9} \cdot \dfrac{y^4}{y^2} = \dfrac{5}{3} \cdot y^{4-2} = \dfrac{5y^2}{3}$

76. $\dfrac{-12x^3y^2}{6xy} = \dfrac{-12}{6} \cdot \dfrac{x^3}{x} \cdot \dfrac{y^2}{y} = -2x^{3-1}y^{2-1}$
    $= -2x^2y$

77. $\dfrac{6y^4 - 12y^2 + 18y}{-6y} = \dfrac{6y^4}{-6y} - \dfrac{12y^2}{-6y} + \dfrac{18y}{-6y}$
    $= -y^3 + 2y - 3$

78. $\dfrac{2p^3 - 6p^2 + 5p}{2p^2} = \dfrac{2p^3}{2p^2} - \dfrac{6p^2}{2p^2} + \dfrac{5p}{2p^2}$

$= p - 3 + \dfrac{5}{2p}$

79. $(5x^{13} - 10x^{12} + 20x^7 - 35x^5) \div (-5x^4)$

$= \dfrac{5x^{13}}{-5x^4} - \dfrac{10x^{12}}{-5x^4} + \dfrac{20x^7}{-5x^4} - \dfrac{35x^5}{-5x^4}$

$= -x^9 + 2x^8 - 4x^3 + 7x$

80. $(-10m^4n^2 + 5m^3n^3 + 6m^2n^4) \div (5m^2n)$

$= \dfrac{-10m^4n^2}{5m^2n} + \dfrac{5m^3n^3}{5m^2n} + \dfrac{6m^2n^4}{5m^2n}$

$= -2m^2n + mn^2 + \dfrac{6n^3}{5}$

81. $(2r^2 + 3r - 14) \div (r - 2)$

$$\begin{array}{r}
2r + 7 \\
r - 2 \overline{\smash{)}2r^2 + 3r - 14} \\
\underline{2r^2 - 4r } \\
7r - 14 \\
\underline{7r - 14} \\
0
\end{array}$$

Answer: $2r + 7$

82. $\dfrac{12m^2 - 11m - 10}{3m - 5}$

$$\begin{array}{r}
4m + 3 \\
3m - 5 \overline{\smash{)}12m^2 - 11m - 10} \\
\underline{12m^2 - 20m } \\
9m - 10 \\
\underline{9m - 15} \\
5
\end{array}$$

Answer: $4m + 3 + \dfrac{5}{3m - 5}$

83. $\dfrac{10a^3 + 5a^2 - 14a + 9}{5a^2 - 3}$

$$\begin{array}{r}
2a + 1 \\
5a^2 + 0a - 3 \overline{\smash{)}10a^3 + 5a^2 - 14a + 9} \\
\underline{10a^3 + 0a^2 - 6a } \\
5a^2 - 8a + 9 \\
\underline{5a^2 + 0a - 3} \\
-8a + 12
\end{array}$$

Answer: $2a + 1 + \dfrac{-8a + 12}{5a^2 - 3}$

84. $\dfrac{2k^4 + 4k^3 + 9k^2 - 8}{2k^2 + 1}$

$$\begin{array}{r}
k^2 + 2k + 4 \\
2k^2 + 0k + 1 \overline{\smash{)}2k^4 + 4k^3 + 9k^2 + 0k - 8} \\
\underline{2k^4 + 0k^3 + k^2 } \\
4k^3 + 8k^2 + 0k \\
\underline{4k^3 + 0k^2 + 2k} \\
8k^2 - 2k - 8 \\
\underline{8k^2 + 0k + 4} \\
-2k - 12
\end{array}$$

Answer: $k^2 + 2k + 4 + \dfrac{-2k - 12}{2k^2 + 1}$

85. $19^0 - 3^0 = 1 - 1 = 0$

86. $(3p)^4(3p^{-7}) = (3^4)(p^4)(3p^{-7})$

$= (3^4)(3)(p^4)(p^{-7})$

$= 3^{4+1}p^{4+(-7)}$

$= 3^5 p^{-3} = \dfrac{3^5}{p^3}$

87. $7^{-2} = \dfrac{1}{7^2}$

88. $(-7 + 2k)^2 = (-7)^2 + 2(-7)(2k) + (2k)^2$

$= 49 - 28k + 4k^2$

89. $\dfrac{2y^3 + 17y^2 + 37y + 7}{2y + 7}$

$$\begin{array}{r}
y^2 + 5y + 1 \\
2y + 7 \overline{\smash{)}2y^3 + 17y^2 + 37y + 7} \\
\underline{2y^3 + 7y^2 } \\
10y^2 + 37y \\
\underline{10y^2 + 35y} \\
2y + 7 \\
\underline{2y + 7} \\
0
\end{array}$$

Answer: $y^2 + 5y + 1$

90. $\left(\dfrac{6r^2s}{5}\right)^4 = \dfrac{(6r^2s)^4}{5^4} = \dfrac{6^4(r^2)^4 s^4}{5^4} = \dfrac{6^4 r^8 s^4}{5^4}$

122 Chapter 3 Exponents and Polynomials

91. $-m^5(8m^2 + 10m + 6)$
$= (-m^5)(8m^2) + (-m^5)(10m) + (-m^5)(6)$
$= -8m^{5+2} - 10m^{5+1} - 6m^5$
$= -8m^7 - 10m^6 - 6m^5$

92. $\left(\frac{1}{2}\right)^{-5} = \left(\frac{2}{1}\right)^5 = 2^5$

93. $(25x^2y^3 - 8xy^2 + 15x^3y) \div (5x)$
$= \frac{25x^2y^3}{5x} - \frac{8xy^2}{5x} + \frac{15x^3y}{5x}$
$= 5xy^3 - \frac{8y^2}{5} + 3x^2y$

94. $(6r^{-2})^{-1} = 6^{-1}(r^{-2})^{-1} = 6^{-1}r^2 = \frac{r^2}{6}$

95. $(2x + y)^3 = (2x + y)(2x + y)(2x + y)$

First find $(2x + y)(2x + y) = (2x + y)^2$.

$(2x + y)^2 = (2x)^2 + 2(2xy) + y^2$
$= 4x^2 + 4xy + y^2$

Now multiply this result by $2x + y$.

$$\begin{array}{r} 4x^2 + 4xy + y^2 \\ 2x + y \\ \hline 4x^2y + 4xy^2 + y^3 \\ 8x^3 + 8x^2y + 2xy^2 \\ \hline 8x^3 + 12x^2y + 6xy^2 + y^3 \end{array}$$

Thus,

$(2x + y)^3 = 8x^3 + 12x^2y + 6xy^2 + y^3$.

96. $2^{-1} + 4^{-1} = \frac{1}{2} + \frac{1}{4} = \frac{2}{4} + \frac{1}{4} = \frac{3}{4}$

97. $(a + 2)(a^2 - 4a + 1)$
$= a(a^2 - 4a + 1) + 2(a^2 - 4a + 1)$
$= a^3 - 4a^2 + a + 2a^2 - 8a + 2$
$= a^3 - 2a^2 - 7a + 2$

98. $(5y^3 - 8y^2 + 7) - (-3y^3 + y^2 + 2)$
$= (5y^3 - 8y^2 + 7) + (3y^3 - y^2 - 2)$
$= 5y^3 - 8y^2 + 7 + 3y^3 - y^2 - 2$
$= 5y^3 + 3y^3 - 8y^2 - y^2 + 7 - 2$
$= 8y^3 - 9y^2 + 5$

99. $(2r + 5)(5r - 2)$

$\ \ \ \ \ \ \ \ \text{F} \ \ \ \ \ \ \ \ \ \ \ \ \text{O} \ \ \ \ \ \ \ \ \ \ \ \ \text{I} \ \ \ \ \ \ \ \ \ \ \ \ \text{L}$
$= (2r)(5r) + (2r)(-2) + (5)(5r) + (5)(-2)$
$= 10r^2 - 4r + 25r - 10$
$= 10r^2 + 21r - 10$

100. $(12a + 1)(12a - 1) = (12a)^2 - (1)^2$
$ = 144a^2 - 1$

Chapter 3 Test

1. $5^{-4} = \frac{1}{5^4} = \frac{1}{625}$

2. $(-3)^0 + 4^0 = 1 + 1 = 2$

3. $4^{-1} + 3^{-1} = \frac{1}{4} + \frac{1}{3} = \frac{3}{12} + \frac{4}{12} = \frac{7}{12}$

4. $\frac{8^{-1} \cdot 8^4}{8^{-2}} = \frac{8^{(-1)+4}}{8^{-2}} = \frac{8^3}{8^{-2}} = 8^{3-(-2)} = 8^5$

5. $\frac{(x^{-3})^{-2}(x^{-1}y)^2}{(xy^{-2})^2} = \frac{(x^{-3})^{-2}(x^{-1})^2(y)^2}{(x)^2(y^{-2})^2}$
$= \frac{x^6 x^{-2} y^2}{x^2 y^{-4}}$
$= \frac{x^4 y^2}{x^2 y^{-4}}$
$= x^{4-2} y^{2-(-4)}$
$= x^2 y^6$

6. (a) $344,000,000,000 = 3.44 \times 10^{11}$

   (b) $.00000557 = 5.57 \times 10^{-6}$

7. (a) $2.96 \times 10^7 = 29,600,000$

   Move the decimal point 7 places to the right.

   (b) $6.07 \times 10^{-8} = .0000000607$

   Move the decimal point 8 places to the left.

8. $5x^2 + 8x - 12x^2 = 5x^2 - 12x^2 + 8x$
   $= -7x^2 + 8x$

   degree 2; binomial (2 terms)

9. $13n^3 - n^2 + n^4 + 3n^4 - 9n^2$
   $= n^4 + 3n^4 + 13n^3 - n^2 - 9n^2$
   $= 4n^4 + 13n^3 - 10n^2$

   degree 4; trinomial (3 terms)

10. $(5t^4 - 3t^2 + 7t + 3)$
    $- (t^4 - t^3 + 3t^2 + 8t + 3)$
    $= (5t^4 - 3t^2 + 7t + 3)$
    $+ (-t^4 + t^3 - 3t^2 - 8t - 3)$
    $= 5t^4 - 3t^2 + 7t + 3 - t^4$
    $+ t^3 - 3t^2 - 8t - 3$
    $= 5t^4 - t^4 + t^3 - 3t^2 - 3t^2$
    $+ 7t - 8t + 3 - 3$
    $= 4t^4 + t^3 - 6t^2 - t$

11. $(2y^2 - 8y + 8) + (-3y^2 + 2y + 3)$
    $- (y^2 + 3y - 6)$
    $= (2y^2 - 8y + 8) + (-3y^2 + 2y + 3)$
    $+ (-y^2 - 3y + 6)$
    $= 2y^2 - 8y + 8 - 3y^2 + 2y + 3 - y^2 - 3y + 6$
    $= 2y^2 - 3y^2 - y^2 - 8y + 2y - 3y + 8 + 3 + 6$
    $= -2y^2 - 9y + 17$

12. Subtract.

    $$\begin{array}{r} 9t^3 - 4t^2 + 2t + 2 \\ 9t^3 + 8t^2 - 3t - 6 \end{array}$$

    Change all signs in the second row; then add.

    $$\begin{array}{r} 9t^3 - 4t^2 + 2t + 2 \\ -9t^3 - 8t^2 + 3t + 6 \\ \hline -12t^2 + 5t + 8 \end{array}$$

13. $3x^2(-9x^3 + 6x^2 - 2x + 1)$
    $= (3x^2)(-9x^3) + (3x^2)(6x^2)$
    $+ (3x^2)(-2x) + (3x^2)(1)$
    $= -27x^5 + 18x^4 - 6x^3 + 3x^2$

14. $\overset{\,\text{F}\text{O}\text{I}\text{L}}{(t - 8)(t + 3)} = t^2 + 3t - 8t - 24$
    $= t^2 - 5t - 24$

15. $(4x + 3y)(2x - y)$

    $\,\text{F}\text{O}\text{I}\text{L}$
    $= 8x^2 - 4xy + 6xy - 3y^2$
    $= 8x^2 + 2xy - 3y^2$

16. $(5x - 2y)^2 = (5x)^2 - 2(5x)(2y) + (2y)^2$
    $= 25x^2 - 20xy + 4y^2$

17. $(10v + 3w)(10v - 3w)$
    $= (10v)^2 - (3w)^2$
    $= 100v^2 - 9w^2$

18. $(2r - 3)(r^2 + 2r - 5)$

    Multiply vertically.

    $$\begin{array}{r} r^2 + 2r - 5 \\ 2r - 3 \\ \hline -3r^2 - 6r + 15 \\ 2r^3 + 4r^2 - 10r \\ \hline 2r^3 + r^2 - 16r + 15 \end{array}$$

Chapter 3 Exponents and Polynomials

19. $(x + 1)^3$

   $= (x + 1)(x + 1)(x + 1)$
   $= (x + 1)(x^2 + 2x + 1)$
   $= x(x^2 + 2x + 1) + 1(x^2 + 2x + 1)$
   $= x^3 + 2x^2 + x + x^2 + 2x + 1$
   $= x^3 + 2x^2 + x^2 + x + 2x + 1$
   $= x^3 + 3x^2 + 3x + 1$

20. $\dfrac{8y^3 - 6y^2 + 4y + 10}{2y}$

   $= \dfrac{8y^3}{2y} - \dfrac{6y^2}{2y} + \dfrac{4y}{2y} + \dfrac{10}{2y}$

   $= 4y^2 - 3y + 2 + \dfrac{5}{y}$

21. $(-9x^2y^3 + 6x^4y^3 + 12xy^3) \div (3xy)$

   $= \dfrac{-9x^2y^3}{3xy} + \dfrac{6x^4y^3}{3xy} + \dfrac{12xy^3}{3xy}$

   $= -3xy^2 + 2x^3y^2 + 4y^2$

22. $\dfrac{2x^2 + x - 36}{x - 4}$

   ```
              2x + 9
   x - 4 ) 2x² +  x - 36
           2x² - 8x
                9x - 36
                9x - 36
                      0
   ```

   Answer: $2x + 9$

23. $(3x^3 - x + 4) \div (x - 2)$

   ```
              3x² + 6x + 11
   x - 2 ) 3x³ + 0x² -  x +  4
           3x³ - 6x²
                 6x² -   x
                 6x² - 12x
                        11x +  4
                        11x - 22
                              26
   ```

   Answer: $3x^2 + 6x + 11 + \dfrac{26}{x - 2}$

24. Use the formula for the area of a square, $A = s^2$, with $s = 3x + 9$.

   $A = s^2$
   $A = (3x + 9)^2$
   $ = (3x)^2 + 2(3x)(9) + 9^2$
   $ = 9x^2 + 54x + 81$

25. For the sum of two fourth degree polynomials in x to be a third degree polynomial in x, the degree 4 terms would have to have a sum of 0. Therefore, they must be opposites of each other. For example,

   $(-4x^4 + 3x^3 + 2x + 1) + (4x^4 - 8x^3 + 2x + 7)$
   $= -5x^3 + 4x + 8.$

   Notice that the degree 4 terms, $-4x^4$ and $4x^4$, are opposites of each other.

Cumulative Review: Chapters R-3

1. $\dfrac{28}{16} = \dfrac{7 \cdot 4}{4 \cdot 4} = \dfrac{7}{4}$

2. $\dfrac{55}{11} = \dfrac{5 \cdot 11}{1 \cdot 11} = 5$

3. $\dfrac{2}{3} + \dfrac{1}{8} = \dfrac{16}{24} + \dfrac{3}{24} = \dfrac{19}{24}$

4. $\dfrac{7}{4} - \dfrac{9}{5} = \dfrac{35}{20} - \dfrac{36}{20} = -\dfrac{1}{20}$

5. $8.32 - 4.6$

   ```
     8.32
    -4.60
     3.72
   ```

Cumulative Review: Chapters R–3   125

6. 7.21×8.6

$$\begin{array}{r} 7.21 \\ \times\ 8.6 \\ \hline 4\ 32\ 6 \\ 57\ 68 \\ \hline 62.00\ 6 \end{array}$$

7. Each shed requires 1 1/2 cubic yards of concrete, so the total amount of concrete needed for 25 sheds would be

$$25 \times 1\tfrac{1}{4} = 25 \times \tfrac{5}{4}$$
$$= \tfrac{125}{4}$$
$$= 31\tfrac{1}{4} \text{ cubic yards.}$$

8. Use the formula for simple interest, $I = Prt$, with $P = \$34{,}000$, $r = 5.4\%$, and $t = 1$.

    $I = Prt$
    $= (34{,}000)(.054)(1)$
    $= 1836$

    She earned $1836 in interest.

9. $\dfrac{4x - 2y}{x + y} = \dfrac{4(-2) - 2(4)}{(-2) + 4}$ Let $x = -2$, $y = 4$
    $= \dfrac{-8 - 8}{2} = \dfrac{-16}{2} = -8$

10. $x^3 - 4xy = (-2)^3 - 4(-2)(4)$
    Let $x = -2$, $y = 4$
    $= -8 + 32 = 24$

11. $\dfrac{(-13 + 15) - (3 + 2)}{6 - 12} = \dfrac{2 - 5}{-6} = \dfrac{-3}{-6} = \dfrac{1}{2}$

12. $-7 - 3[2 + (5 - 8)] = -7 - 3[2 + (-3)]$
    $= -7 - 3[-1]$
    $= -7 + 3 = -4$

13. $(9 + 2) + 3 = 9 + (2 + 3)$

    The numbers are in the same order but grouped differently, so this is an example of the associative property of addition.

14. $-7 + 7 = 0$

    The sum of the two numbers is 0, so they are additive inverses (or opposites) of each other. This is an example of the additive inverse property.

15. $6(4 + 2) = 6(4) + 6(2)$

    The number 6 outside the parentheses is "distributed" over the 4 and the 2. This is an example of the distributive property.

16. $2x - 7x + 8x = 30$
    $3x = 30$
    $x = 10$

17. $2 - 3(t - 5) = 4 + t$
    $2 - 3t + 15 = 4 + t$
    $-3t + 17 = 4 + t$
    $-4t + 17 = 4$
    $-4t = -13$
    $t = \dfrac{13}{4}$

18. $2(5h + 1) = 10h + 4$

      $10h + 2 = 10h + 4$

           $2 = 4$ *False*

The false statement indicates that the equation has no solution.

19. $d = rt$ for r

$$\frac{d}{t} = \frac{rt}{t}$$

$$\frac{d}{t} = r$$

20. $\dfrac{x}{5} = \dfrac{x - 2}{7}$

    $7x = 5(x - 2)$ *Cross products are equal*

    $7x = 5x - 10$

    $2x = -10$

     $x = -5$

21. $\dfrac{1}{3}p - \dfrac{1}{6}p = -2$

To clear fractions, multiply both sides of the equation by the least common denominator, which is 6.

$$6\left(\tfrac{1}{3}p - \tfrac{1}{6}p\right) = (6)(-2)$$

$$6\left(\tfrac{1}{3}p\right) - 6\left(\tfrac{1}{6}p\right) = -12$$

          $2p - p = -12$

              $p = -12$

22. $.05x + .15(50 - x) = 5.50$

To clear decimals, multiply both sides of the equation by 100.

$100[.05x + .15(50 - x)] = 100(5.50)$

$100(.05x) + 100[.15(50 - x)] = 100(5.50)$

      $5x + 15(50 - x) = 550$

        $5x + 750 - 15x = 550$

            $-10x + 750 = 550$

                  $-10x = -200$

                      $x = 20$

23. $4 - (3x + 12) = (2x - 9) - (5x - 1)$

    $4 - 1(3x + 12) = (2x - 9) - 1(5x - 1)$

    $4 - 3x - 12 = 2x - 9 - 5x + 1$

        $-3x - 8 = -3x - 8$ *True*

The true statement indicates that all real numbers are solutions of the equation.

24. Let x = Louis's allowance;
    $6x$ = Janet's allowance.

      $x + 6x = 56$

          $7x = 56$

           $x = 8$

Louis's allowance is $8. Janet's allowance is $6(8) = \$48$.

25. Let x = the unknown number.

    $3(x - 8) = -3x$

    $3x - 24 = -3x$

        $-24 = -6x$

          $4 = x$

The unknown number is 4.

26. Let x = the number of roosters;
$x + 28$ = the number of hens.

$$x + (x + 28) = 190$$
$$x + x + 28 = 190$$
$$2x + 28 = 190$$
$$2x = 162$$
$$x = 81$$

There are 81 roosters and $81 + 28 = 109$ hens.

27. Let x = one side of the triangle;
$2x$ = the second side of the triangle.

The perimeter of the triangle cannot be more than 50 feet. This is equivalent to stating that the sum of the lengths of the side must be less than or equal to 50 feet. Write this statement as an inequality and solve.

$$x + 2x + 17 \leq 50$$
$$3x + 17 \leq 50$$
$$3x \leq 33$$
$$x \leq 11$$

One side cannot be more than 11 feet. The other side cannot be more than $2 \cdot 11 = 22$ feet.

28. $-8x \leq -80$

$\dfrac{-8x}{-8} \geq \dfrac{-80}{-8}$ *Divide by -8; reverse the symbol*

$x \geq 10$

29. $-2(x + 4) > 3x + 6$

$-2x - 8 > 3x + 6$ *Distributive property*

$-2x - 8 + 8 > 3x + 6 + 8$ *Add 8*

$-2x > 3x + 14$

$-2x - 3x > 3x + 14 - 3x$ *Subtract 3x*

$-5x > 14$

$\dfrac{-5x}{-5} < \dfrac{14}{-5}$ *Divide by -5; reverse the symbol*

$x < -\dfrac{14}{5}$

30. $-3 \leq 2x + 5 < 9$

$-3 - 5 \leq 2x + 5 - 5 < 9 - 5$ *Subtract 5*

$-8 \leq 2x < 4$

$\dfrac{-8}{2} \leq \dfrac{2x}{2} < \dfrac{4}{2}$ *Divide by 2*

$-4 \leq x < 2$

31. $4^{-1} + 3^0 = \dfrac{1}{4} + 1 = 1\dfrac{1}{4}$ or $\dfrac{5}{4}$

32. $2^{-4} \cdot 2^5 = 2^{-4+5} = 2^1 = 2$

33. $\dfrac{8^{-5} \cdot 8^7}{8^2} = \dfrac{8^{-5+7}}{8^2} = \dfrac{8^2}{8^2} = 8^{2-2} = 8^0 = 1$

34. $\dfrac{(a^{-3}b^2)^2}{(2a^{-4}b^{-3})^{-1}} = \dfrac{(a^{-3})^2(b^2)^2}{2^{-1}(a^{-4})^{-1}(b^{-3})^{-1}}$

$= \dfrac{a^{-6}b^4}{2^{-1}a^4b^3} = \dfrac{2a^{-6}b^4}{a^4b^3}$

$= 2a^{-6-4}b^{4-3} = 2a^{-10}$

$= \dfrac{2b}{a^{10}}$

128   Chapter 3   Exponents and Polynomials

35. $34,500 = 3.45 \times 10^4$

36. $(7x^3 - 12x^2 - 3x + 8) + (6x^2 + 4)$
$\quad - (-4x^3 + 8x^2 - 2x - 2)$
$\quad = (7x^3 - 12x^2 - 3x + 8) + (6x^2 + 4)$
$\quad\quad + (4x^3 - 8x^2 + 2x + 2)$
$\quad = 7x^3 - 12x^2 - 3x + 8 + 6x^2 + 4$
$\quad\quad + 4x^3 - 8x^2 + 2x + 2$
$\quad = 7x^3 + 4x^3 - 12x^2 + 6x^2 - 8x^2$
$\quad\quad - 3x + 2x + 8 + 4 + 2$
$\quad = 11x^3 - 14x^2 - x + 14$

37. $6x^5(3x^2 - 9x + 10)$
$\quad = (6x^5)(3x^2) + (6x^5)(-9x)$
$\quad\quad + (6x^5)(10)$
$\quad = 18x^7 - 54x^6 + 60x^5$

38. $(7x + 4)(9x + 3)$
$\quad = 63x^2 + 21x + 36x + 12 \quad FOIL$
$\quad = 63x^2 + 57x + 12$

39. $(5x + 8)^2 = (5x)^2 + 2(5x)(8) + (8)^2$
$\quad\quad\quad = 25x^2 + 80x + 64$

40. $\dfrac{y^3 - 3y^2 + 8y - 6}{y - 1}$

$$\begin{array}{r}
y^2 - 2y + 6 \\
y - 1 \overline{\smash{)}y^3 - 3y^2 + 8y - 6} \\
\underline{y^3 - y^2} \\
-2y^2 + 8y \\
\underline{-2y^2 + 2y} \\
6y - 6 \\
\underline{6y - 6} \\
0
\end{array}$$

Answer: $y^2 - 2y + 6$

CHAPTER 4 FACTORING

Section 4.1 Factors; The Greatest Common Factor

4.1 Margin Exercises

1. **(a)** 30, 20, 15

   Write each number in prime factored form.

   $30 = 2 \cdot 3 \cdot 5$, $20 = 2^2 \cdot 5$, $15 = 3 \cdot 5$

   Take each prime the *least* number of times it appears in *all* the factored forms. The greatest common factor is 5.

   (b) 42, 28, 35

   Write each number in prime factored form.

   $42 = 2 \cdot 3 \cdot 7$, $28 = 2^2 \cdot 7$, $35 = 5 \cdot 7$

   Take each prime the least number of times it appears. The greatest common factor is 7.

   (c) 12, 18, 26, 32

   Write each number in prime factored form.

   $12 = 2^2 \cdot 3$, $18 = 2 \cdot 3^2$,
   $26 = 2 \cdot 13$, $32 = 2^5$

   Take each prime the least number of times it appears. The greatest common factor is 2.

   (d) 10, 15, 21

   Write each number in prime factored form.

   $10 = 2 \cdot 5$, $15 = 3 \cdot 5$, $21 = 3 \cdot 7$

   Take each prime the least number of times it appears. Since no prime appears in all the factored forms, the greatest common factor is 1.

2. **(a)** $6m^4$, $9m^2$, $12m^5$

   The greatest common factor of the coefficients 6, 9, and 12 is 3. The greatest common factor of the terms m^4, m^2, and m^5 is m^2, since 2 is the smallest exponent on m. Thus, the greatest common factor of these terms is $3m^2$.

   (b) $-12p^5$, $-18q^4$

   The greatest common factor of -12 and -18 is 6. Since these terms do not contain the same variables, no variable is common to them. The greatest common factor is 6.

   (c) y^4z^2, y^6z^8, z^9

   The smallest exponent on z is 2, and no y occurs in z^9; thus, the greatest common factor is z^2.

   (d) $12p^{11}$, $17q^5$

   Since $12p^{11}$ and $17q^5$ do not contain the same variables and 1 is the greatest common factor of 12 and 17, the greatest common factor is 1.

3. **(a)** $32p^2 + 16p + 48$

   Since 16 is the greatest common factor of 32, 16, and 48, and p does not appear in the third term, 16 is all that can be factored out.

$32p^2 + 16p + 48$
$= 16(2p^2) + 16(p) + 16(3)$
$= 16(2p^2 + p + 3)$

(b) $10y^5 - 8y^4 + 6y^2$

2 is the greatest common factor of 10, 8, and 6, so 2 can be factored out. Since y occurs in every term and its lowest exponent is 2, y^2 can be factored out. Hence, $2y^2$ is the greatest common factor.

$10y^5 - 8y^4 + 6y^2$
$= (2y^2)(5y^3) - (2y^2)(4y^2)$
$\quad + (2y^2)(3)$
$= 2y^2(5y^3 - 4y^2 + 3)$

(c) $m^7 + m^9$

m occurs in every term and 7 is the lowest exponent on m, so m^7 is the greatest common factor.

$m^7 + m^9 = m^7 \cdot 1 + m^7 \cdot m^2$
$= m^7(1 + m^2)$

(d) $8p^5q^2 + 16p^6q^3 - 12p^4q^7$

4 is the greatest common factor of 8, 16, and 12. The lowest power of p is p^4; the lowest power of q is q^2. Hence, $4p^4q^2$ is the greatest common factor.

$8p^5q^2 + 16p^6q^3 - 12p^4q^7$
$= (4p^4q^2)(2p) + (4p^4q^2)(4p^2q)$
$\quad - (4p^4q^2)(3q^5)$
$= 4p^4q^2(2p + 4p^2q - 3q^5)$

(e) $13x^2 - 27$

The greatest common factor of 13 and 27 is 1, and x does not appear in the second term. Thus, there is no common factor (except 1).

(f) $r(t - 4) + 5(t - 4)$

The binomial $t - 4$ is the greatest common factor here.

$r(t - 4) + 5(t - 4)$
$= (t - 4)(r) + (t - 4)(5)$
$= (t - 4)(r + 5)$

4. (a) $pq + 5q + 2p + 10$
$= (pq + 5q) + (2p + 10)$
$= q(p + 5) + 2(p + 5)$
$= (p + 5)(q + 2)$

(b) $2mn - 8n + 3m - 12$
$= (2mn - 8n) + (3m - 12)$
$= 2n(m - 4) + 3(m - 4)$
$= (m - 4)(2n + 3)$

(c) $6x - yx + 2y - 3x^2$
$= (6x + 2y) + (-3x^2 - yx)$
   Rearrange and group terms
$= 2(3x + y) - x(3x + y)$
$= (3x + y)(2 - x)$

(d) $2a + 3ax - 2b^2 - 3b^2x$
$= (2a + 3ax) + (-2b^2 - 3b^2x)$
$= a(2 + 3x) - b^2(2 + 3x)$
$= (2 + 3x)(a - b^2)$

4.1 Section Exercises

3. 40, 20, 4

Find the prime factored form of each number.

$40 = 2^3 \cdot 5$
$20 = 2^2 \cdot 5$
$4 = 2^2$

The greatest common factor is $2^2 = 4$.

4.1 Section Exercises

7. 4, 9, 12

   Write each number in prime factored form.

   $$4 = 2^2$$
   $$9 = 3^3$$
   $$12 = 2^2 \cdot 3$$

   There are no primes common to all three numbers, so the greatest common factor is 1.

11. 16y, 24

    $$16y = 2^4 \cdot y$$
    $$24 = 2^3 \cdot 3$$

    The greatest common factor is $2^3 = 8$.

15. $12m^3n^2$, $18m^5n^4$, $36m^8n^3$

    $$12m^3n^2 = 2^2 \cdot 3 \cdot m^3 \cdot n^2$$
    $$18m^5n^4 = 2 \cdot 3^2 \cdot m^5 \cdot n^4$$
    $$36m^8n^3 = 2^2 \cdot 3^2 \cdot m^8 \cdot n^3$$

    The greatest common factor is
    $2 \cdot 3 \cdot m^3 \cdot n^2 = 6m^3n^2$.

19. $42ab^3$, $-36a$, $90b$, $-48ab$

    $$42ab^3 = 2 \cdot 3 \cdot 7 \cdot a \cdot b^3$$
    $$-36a = -2^2 \cdot 3^2 \cdot a$$
    $$90b = 2 \cdot 3^2 \cdot 5 \cdot b$$
    $$-48ab = -2^4 \cdot 3 \cdot a \cdot b$$

    The greatest common factor is $2 \cdot 3 = 6$.

23. $12 = 6()$

    Since $12/6 = 2$, $12 = 6(2)$.

27. $9m^4 = 3m^2()$

    Since $9m^4/3m^2 = 3m^2$,
    $$9m^4 = 3m^2(3m^2).$$

31. $6m^4n^5 = 3m^3n()$

    Since $6m^4n^5/3m^3n = 2mn^4$,
    $$6m^4n^5 = 3m^3n(2mn^4).$$

35. $12y - 24$

    The greatest common factor is 12.

    $$12y - 24 = 12 \cdot y - 12 \cdot 2$$
    $$= 12(y - 2)$$

39. $65y^{10} + 35y^6$

    The greatest common factor is $5y^6$.

    $$65y^{10} + 35y^6$$
    $$= (5y^6)(13y^4) + (5y^6)(7)$$
    $$= 5y^6(13y^4 + 7)$$

43. $8m^2n^3 + 24m^2n^2$

    The greatest common factor is $8m^2n^2$.

    $$8m^2n^3 + 24m^2n^2$$
    $$= (8m^2n^2)(n) + (8m^2n^2)(3)$$
    $$= 8m^2n^2(n + 3)$$

47. $45q^4p^5 + 36qp^6 + 81q^2p^3$

    The greatest common factor is $9qp^3$.

    $$45q^4p^5 + 36qp^6 + 81q^2p^3$$
    $$= (9qp^3)(5q^3p^2) + (9qp^3)(4p^3)$$
    $$ + (9qp^3)(9q)$$
    $$= 9qp^3(5q^3p^2 + 4p^3 + 9q)$$

51. $p^2 + 4p + 3p + 12$

The first two terms have a common factor of p, and the last two terms have a common factor of 3. Thus,

$p^2 + 4p + 3p + 12$
$= (p^2 + 4p) + (3p + 12)$
$= p(p + 4) + 3(p + 4)$.

Now we have two terms which have a common binomial factor of $p + 4$. Thus,

$p^2 + 4p + 3p + 12$
$= p(p + 4) + 3(p + 4)$
$= (p + 4)(p + 3)$.

55. $7z^2 + 14z - az - 2a$

$= (7z^2 + 14z) + (-az - 2a)$
    Group the terms

$= 7z(z + 2) - a(z + 2)$
    Factor each group

$= (z + 2)(7z - a)$
    Factor out the common binomial factor

59. $3a^3 + 3ab^2 + 2a^2b + 2b^3$

$= (3a^3 + 3ab^2) + (2a^2b + 2b^3)$
    Group the terms

$= 3a(a^2 + b^2) + 2b(a^2 + b^2)$
    Factor each group

$= (a^2 + b^2)(3a + 2b)$
    Factor out the common binomial factor

63. $16m^3 - 4m^2p^2 - 4mp + p^3$

$= (16m^3 - 4m^2p^2) + (-4mp + p^3)$
$= 4m^2(4m - p^2) - p(4m - p^2)$
$= (4m - p^2)(4m^2 - p)$

67. $(x + 6)(x - 9)$

          F   O   I   L
$= (x)(x) + (x)(-9) + (6)(x) + (6)(-9)$
$= x^2 - 9x + 6x - 54$
$= x^2 - 3x - 54$

Section 4.2 Factoring Trinomials

4.2 Margin Exercises

1. **(a)** All pairs of positive integers whose product is 6 are

    1, 6 and 2, 3.

(b) The pair 2, 3 has a sum of 5.

2. $y^2 + 12y + 20$

Factors of 20	Sum of factors
20, 1	$20 + 1 = 21$
10, 2	$10 + 2 = 12$
5, 4	$5 + 4 = 9$

From the list, 10 and 2 are the required integers. Thus,

$y^2 + 12y + 20 = (y + 10)(y + 2)$.

3. **(a)** $a^2 - 9a - 22$

Find two integers whose product is -22 and whose sum is -9. Because the last term is negative, the pair must include one positive and one negative integer.

Factors of -22	Sum of factors
22, -1	$22 + (-1) = 21$
11, -2	$11 + (-2) = 9$
-22, 1	$-22 + 1 = -21$
-11, 2	$-11 + 2 = -9$

The required integers are -11 and 2. Thus,
$$a^2 - 9a - 22 = (a - 11)(a + 2).$$

(b) $r^2 - 6r - 16$

Find two integers whose product is -16 and whose sum is -6.

Factors of -16	Sum of factors
$16, -1$	$16 + (-1) = 15$
$8, -2$	$8 + (-2) = 6$
$4, -4$	$4 + (-4) = 0$
$-16, 1$	$-16 + 1 = -15$
$-8, 2$	$-8 + 2 = -6$

We can stop here since we have found the required integers, -8 and 2. Thus,
$$r^2 - 6r - 16 = (r - 8)(r + 2).$$

4. (a) $r^2 - 3r - 4$

Find two integers whose product is -4 and whose sum is -3.

Factors of -4	Sum of factors
$2, 2$	$2 + (-2) = 0$
$4, -1$	$4 + (-1) = 3$
$-4, 1$	$-4 + 1 = -3$

The required integers are -4 and 1. Thus,
$$r^2 - 3r - 4 = (r - 4)(r + 1).$$

(b) $m^2 - 2m + 5$

There is no pair of integers whose product is 5 and whose sum is -2, so $m^2 - 2m + 5$ is a prime polynomial.

5. (a) $b^2 - 3ab - 4a^2$

Two expressions whose product is $-4a^2$ and whose sum is $-3a$ are $-4a$ and a, so
$$b^2 - 3ab - 4a^2 = (b - 4a)(b + a).$$

(b) $r^2 - 6rs + 8s^2$

Two expressions whose product is $8s^2$ and whose sum is $-6s$ are $-4s$ and $-2s$, so
$$r^2 - 6rs + 8s^2 = (r - 4s)(r - 2s).$$

6. (a) $2p^3 + 6p^2 - 8p$

First, factor out the greatest common factor, $2p$.
$$2p^3 + 6p^2 - 8p = 2p(p^2 + 3p - 4)$$

Now factor $p^2 + 3p - 4$.

The integers 4 and -1 have a product of -4 and a sum of 3, so
$$p^2 + 3p - 4 = (p + 4)(p - 1).$$

The complete factored form is
$$2p^3 + 6p^2 - 8p = 2p(p + 4)(p - 1).$$

(b) $3x^4 - 15x^3 + 18x^2$
$$= 3x^2(x^2 - 5x + 6)$$

Factor $x^2 - 5x + 6$.

The integers -3 and -2 have a product of 6 and a sum of -5, so
$$x^2 - 5x + 6 = (x - 3)(x - 2).$$

The complete factored form is
$$3x^4 - 15x^3 + 18x^2$$
$$= 3x^2(x - 3)(x - 2).$$

4.2 Section Exercises

3. Product: −24 Sum: −5

List all pairs of integers whose product is −24, and then find the sum of each pair.

Factors of −24	Sum of factors
1, −24	1 + (−24) = −23
−1, 24	−1 + 24 = 23
2, −12	2 + (−12) = −10
−2, 12	−2 + 12 = 10
3, −8	3 + (−8) = −5 *Sum is −5*
−3, 8	−3 + 8 = 5
4, −6	4 + (−6) = −2
−4, 6	−4 + 6 = 2

The pair of integers whose product is −24 and whose sum is −5 is 3 and −8.

7. Product: −48 Sum: 2

Factors of −48	Sum of factors
1, −48	1 + (−48) = −47
−1, 48	−1 + 48 = 47
2, −24	2 + (−24) = −22
−2, 24	−2 + 24 = 22
3, −16	3 + (−16) = −13
−3, 16	−3 + 16 = 13
4, −12	4 + (−12) = −8
−4, 12	−4 + 12 = 8
6, −8	6 + (−8) = −2
−6, 8	−6 + 8 = 2 *Sum is 2*

The pair of integers whose product is −48 and whose sum is 2 is −6 and 8.

11. $p^2 + 11p + 30 = (p + 5)()$

Since we need to find two integers whose product is 30, and 5 is already given as one of these integers, the other is 30/5 = 6. Notice that 5 + 6 = 11. Hence, the other factor is $p + 6$.

15. $x^2 - 9x + 8 = (x - 1)()$

Since we need to find two integers whose product is 8, and −1 is already given as one of these integers, the other is 8/−1 = −8. Notice that $(-1) + (-8) = -9$. Hence, the other factor is $x - 8$.

19. $x^2 + 9x - 22 = (x - 2)()$

Since we need to find two integers whose product is −22, and −2 is already given as one of these integers, the other is −22/−2 = 11. Notice that $-2 + 11 = 9$. Hence, the other factor is $x + 11$.

23. $y^2 + 9y + 8$

Look for two integers whose product is 8 and whose sum is 9. Both integers must be positive because b and c are both positive.

Factors of 8	Sum of factors
8, 1	8 + 1 = 9 *Sum is 9*
4, 2	4 + 2 = 6

Thus,

$$y^2 + 9y + 8 = (y + 8)(y + 1).$$

27. $m^2 + m - 20$

Look for two integers whose product is -20 and whose sum is 1. Because c is negative, one integer must be positive and one must be negative.

Factors of -20	Sum of factors
$-1, 20$	$-1 + 20 = 19$
$1, -20$	$1 + (-20) = -19$
$-2, 10$	$-2 + 10 = 8$
$2, -10$	$2 + (-10) = -8$
$-4, 5$	$-4 + 5 = 1$
$4, -5$	$4 + (-5) = -1$

Thus,

$$m^2 + m - 20 = (m - 4)(m + 5).$$

31. $r^2 - r - 30$

Look for two integers whose product is -30 and whose sum is -1. Because c is negative, one integer must be positive and the other must be negative.

Factors of -30	Sum of factors
$-1, 30$	$-1 + 30 = 29$
$1, -30$	$1 + (-30) = -29$
$-2, 15$	$-2 + 15 = 13$
$2, -15$	$2 + (-15) = -13$
$-3, 10$	$-3 + 10 = 7$
$3, -10$	$3 + (-10) = -7$
$-5, 6$	$-5 + 6 = 1$
$5, -6$	$5 + (-6) = -1$
	Sum is -1

Thus,

$$r^2 - r - 30 = (r + 5)(r - 6).$$

35. $r^2 + 3ra + 2a^2$

Find two expressions whose product is $2a^2$ and whose sum is $3a$. They are $2a$ and a, so

$$r^2 + 3ra + 2a^2 = (r + 2a)(r + a).$$

39. $x^2 + 4xy + 3y^2$

Find two expressions whose product is $3y^2$ and whose sum is $4y$. They are $3y$ and y, so

$$x^2 + 4xy + 3y^2 = (x + 3y)(x + y).$$

43. $4x^2 + 12x - 40$

First, factor out the greatest common factor, 4.

$$4x^2 + 12x - 40 = 4(x^2 + 3x - 10)$$

Now factor $x^2 + 3x - 10$.

Factors of -10	Sum of factors
$-1, 10$	$-1 + 10 = 9$
$1, -10$	$1 + (-10) = -9$
$2, -5$	$2 + (-5) = -3$
$-2, 5$	$-2 + 5 = 3$
	Sum is 3

Thus,

$$x^2 + 3x - 10 = (x - 2)(x + 5).$$

The complete factored form is

$$4x^2 + 12x - 40 = 4(x - 2)(x + 5).$$

47. $2x^6 + 8x^5 - 42x^4$

First, factor out the greatest common factor, $2x^4$.

$$2x^6 + 8x^5 - 42x^4$$
$$= 2x^4(x^2 + 4x - 21).$$

Now factor $x^2 + 4x - 21$.

Factors of -21	Sum of factors
1, -21	$1 + (-21) = -20$
-1, 21	$-1 + 21 = 20$
3, -7	$3 + (-7) = -4$
-3, 7	$-3 + 7 = 4$
	Sum is 4

Thus,

$$x^2 + 4x - 21 = (x - 3)(x + 7).$$

The complete factored form is

$$2x^6 + 8x^5 - 42x^4$$
$$= 2x^4(x - 3)(x + 7).$$

51. $(2x + 4)(x - 3)$

$$\ \ \ \mathrm{F}\ \ \ \ \ \ \ \ \mathrm{O}\ \ \ \ \ \ \ \ \ \mathrm{I}\ \ \ \ \ \ \ \ \ \mathrm{L}$$
$$= (2x)(x) + (2x)(-3) + (4)(x) + (4)(-3)$$
$$= 2x^2 - 6x + 4x - 12$$
$$= 2x^2 - 2x - 12$$

It is incorrect to completely factor $2x^2 - 2x - 12$ as $(2x + 4)(x - 3)$ because $2x + 4$ can be factored further as $2(x + 2)$. The first step should be to factor out the greatest common factor, 2. The correct factorization is

$$2x^2 - 2x - 12 = 2(x^2 - x - 6)$$
$$= 2(x + 2)(x - 3).$$

55. $(2y - 7)(y + 4) = 2y^2 + 8y - 7y - 28$
$$= 2y^2 + y - 28$$

59. $(4p + 1)(2p - 3) = 8p^2 - 12p + 2p - 3$
$$= 8p^2 - 10p - 3$$

Section 4.3 More on Factoring Trinomials

4.3 Margin Exercises

1. (a) By FOIL,

$$(2x + 1)(x + 6) = 2x^2 + 12x + x + 6$$
$$= 2x^2 + 13x + 6.$$

$(2x + 1)(x + 6)$ is an incorrect factoring of $2x^2 + 7x + 6$.

(b) By FOIL,

$$(2x + 6)(x + 1) = 2x^2 + 2x + 6x + 6$$
$$= 2x^2 + 8x + 6.$$

$(2x + 6)(x + 1)$ is an incorrect factoring of $2x^2 + 7x + 6$.

(c) By FOIL,

$$(2x + 2)(x + 3) = 2x^2 + 6x + 2x + 6$$
$$= 2x^2 + 8x + 6.$$

$(2x + 2)(x + 3)$ is an incorrect factoring of $2x^2 + 7x + 6$.

2. (a) $2p^2 + 9p + 9$

The only factors of $2p^2$ are $2p$ and p. The only factors of 9 are 3 and 3, or 9 and 1. Try various possibilities.

$$(2p + 9)(p + 1) = 2p^2 + 11p + 9$$
$$\text{Incorrect}$$
$$(2p + 3)(p + 3) = 2p^2 + 9p + 9$$
$$\text{Correct}$$

(b) $6p^2 + 19p + 10$

The factors of $6p^2$ are $2p$ and $3p$, or $6p$ and p. The factors of 10 are 10 and 1, or 5 and 2. Try various possibilities.

$$(3p + 5)(2p + 2)\quad 2p + 2 = 2(p + 1)$$

Since 2 can be factored out of 2p + 2 but it cannot be factored out of $6p^2 + 19p + 10$, this factorization cannot be correct.

$$(3p + 2)(2p + 5) = 6p^2 + 19p + 10$$
Correct

(c) $8x^2 + 14x + 3$

The factors of $8x^2$ are 8x and x, or 4x and 2x. The factors of 3 are 3 and 1. Try various possibilities.

$$(8x + 3)(x + 1) = 8x^2 + 11x + 3$$
Incorrect

$$(8x + 1)(x + 3) = 8x^2 + 25x + 3$$
Incorrect

$$(4x + 1)(2x + 3) = 8x^2 + 14x + 3$$
Correct

3. (a) $6x^2 + 5x - 4$

6 and -4 each have several factors. Since the middle coefficient, 5, is not large, we try 3x and 2x as factors of 6, rather than 6x and x.

$$(3x \quad)(2x \quad)$$

-4 has factors -4 and 1, 4 and -1, and -2 and 2. Try -4 and 1.

$$(3x - 4)(2x + 1) = 6x^2 - 5x - 4$$
Incorrect

Try 4 and -1.

$$(3x + 4)(2x - 1) = 6x^2 + 5x - 4$$
Correct

Thus,

$$6x^2 + 5x - 4 = (3x + 4)(2x - 1).$$

(b) $6m^2 - 11m - 10$

The factors of $6m^2$ are 6m and m, or 3m and 2m. Since -11, the middle coefficient, is not very large, we start with 3m and 2m.

$$(3m \quad)(2m \quad)$$

Some factors of -10 are -10 and 1, or 2 and -5. Try various possibilities.

$$(3m - 10)(2m + 1) = 6m^2 - 17m - 10$$
Incorrect

$$(3m + 2)(2m - 5) = 6m^2 - 11m - 10$$
Correct

Thus,

$$6m^2 - 11m - 10 = (3m + 2)(2m - 5).$$

(c) $4x^2 - 3x - 7$

The factors of $4x^2$ are 4x and x, or 2x and 2x. Since -3, the middle coefficient, is not very large, we start with 2x and 2x.

$$(2x \quad)(2x \quad)$$

The factors of -7 are -7 and 1. Try various possibilities.

$$(2x - 7)(2x + 1) = 4x^2 - 12x - 7$$
Incorrect

It is obvious that 2x and 2x won't work, so we will try 4x and x.

$$(4x - 7)(x + 1) = 4x^2 - 3x - 7$$
Correct

Thus,

$$4x^2 - 3x - 7 = (4x - 7)(x + 1).$$

4. **(a)** $2x^2 - 5xy - 3y^2$

Try various possibilities.

$(2x - y)(x + 3y) = 2x^2 + 5xy - 3y^2$
   Incorrect

The middle terms differ only in sign, so reverse the signs of the two factors.

$(2x + y)(x - 3y) = 2x^2 - 5xy - 3y^2$
   Correct

Thus,

$2x^2 - 5xy - 3y^2 = (2x + y)(x - 3y).$

(b) $8a^2 + 2ab - 3b^2$

Try various possibilities.

$(8a + 3b)(a - b) = 8a^2 - 5ab - 3b^2$
   Incorrect

$(4a + 3b)(2a - b) = 8a^2 + 2ab - 3b^2$
   Correct

Thus,

$8a^2 + 2ab - 3b^2 = (4a + 3b)(2a - b).$

(c) $12x^2 - 16xy - 3y^2$

Try various possibilities.

$(4x - 3y)(3x + y) = 12x^2 - 5xy - 3y^2$
   Incorrect

$(6x + y)(2x - 3y) = 12x^2 - 16xy - 3y^2$
   Correct

Thus,

$12x^2 - 16xy - 3y^2 = (6x + y)(2x - 3y).$

5. **(a)** $4x^2 - 2x - 30$

First factor out the greatest common factor, 2.

$4x^2 - 2x - 30 = 2(2x^2 - x - 15)$

Now try to factor $2x^2 - x - 15$.

$(2x - 3)(x + 5) = 2x^2 + 7x - 15$
   Incorrect

$(2x + 5)(x - 3) = 2x^2 - x - 15$
   Correct

The complete factored form is

$4x^2 - 2x - 30 = 2(2x + 5)(x - 3).$

(b) $18p^4 + 63p^3 + 27p^2$
$= 9p^2(2p^2 + 7p + 3).$

Now factor $2p^2 + 7p + 3$.

$(2p + 1)(p + 3) = 2p^2 + 7p + 3$
   Correct

The complete factored form is

$18p^4 + 63p^3 + 27p^2$
$= 9p^2(2p + 1)(p + 3).$

(c) $6a^2 + 3ab - 18b^2$

First factor out the greatest common factor, 3.

$6a^2 + 3ab - 18b^2$
$= 3(2a^2 + ab - 6b^2).$

Now try to factor $2a^2 + ab - 6b^2$.

$(2a + b)(a - 6b) = 2a^2 - 11ab - 6b^2$
   Incorrect

$(2a - 3b)(a + 2b) = 2a^2 + ab - 6b^2$
   Correct

The complete factored form is

$6a^2 + 3ab - 18b^2$
$= 3(2a - 3b)(a + 2b).$

6. 3 and 4 have a product of 12 and a sum of 7.

7. **(a)** $2m^2 + 7m + 3$

Find two integers whose product is $2(3) = 6$ and whose sum is 7. The integers are 1 and 6. Write the middle term, $7m$, as $1m + 6m$.

$$2m^2 + 7m + 3 = 2m^2 + 1m + 6m + 3$$
$$= m(2m + 1) + 3(2m + 1)$$
$$= (2m + 1)(m + 3)$$

(b) $5p^2 - 2p - 3$

Find two integers whose product is $5(-3) = -15$ and whose sum is -2. The integers are -5 and 3.

$$5p^2 - 2p - 3 = 5p^2 - 5p + 3p - 3$$
$$= 5p(p - 1) + 3(p - 1)$$
$$= (p - 1)(5p + 3)$$

(c) $15k^2 - k - 2$

Find two integers whose product is $15(-2) = -30$ and whose sum is -1. The integers are -6 and 5.

$$15k^2 - k - 2 = 15k^2 - 6k + 5k - 2$$
$$= 3k(5k - 2) + 1(5k - 2)$$
$$= (5k - 2)(3k + 1)$$

4.3 Section Exercises

3. $4y^2 + 17y - 15$

   (a) $(y + 5)(4y - 3)$
   $$= 4y^2 - 3y + 20y - 15$$
   $$= 4y^2 + 17y - 15$$

   (b) $(2y - 5)(2y + 3)$
   $$= 4y^2 + 6y - 10y - 15$$
   $$= 4y^2 - 4y - 15$$

Because
$$(y + 5)(4y - 3) = 4y^2 + 17y - 15,$$
the correct choice is (a).

7. $6a^2 + 7ab - 20b^2 = (3a - 4b)()$

The first term of the polynomial is $6a^2$ and the first term of the first factor is $3a$, so the first term of the second factor would be $6a^2/3a = 2a$. The last term of the polynomial is $-20b^2$ and the second term of the first factor is $-4b$, so the second term of the second factor would be $-20b^2/-4b = 5b$. Thus,

$$6a^2 + 7ab - 20b^2$$
$$= (3a - 4b)(2a + 5b).$$

11. $4z^3 - 10z^2 - 6z = 2z(2z^2 - 5z - 3)$
    $$= 2z(2z + 1)(z - 3)$$

The final factored form may also be written $2z(z - 3)(2z + 1)$.

Note: In Exercises 15–55, either the trial and error method or the grouping method can be used to factor each polynomial. In order to illustrate both methods, about half of the solutions use each method.

15. $2x^2 + 7x + 3$

Factor by trial and error. Possible factors of $2x^2$ are $2x$ and x.

Possible factors of 3 are 3 and 1 or 1 and 3.

$(2x + 3)(x + 1) = 2x^2 + 5x + 3$ *Incorrect*

$(2x + 1)(x + 3) = 2x^2 + 7x + 3$ *Correct*

19. $4r^2 + r - 3$

    Factor by trial and error. Possible factors of $4r^2$ are $4r$ and r, $2r$ and $2r$, or r and $4r$. Possible factors of -3 are -3 and 1 or 3 and -1.

    $(4r + 3)(r - 1) = 4r^2 - r - 3$ *Incorrect*

    The middle term differs only in sign from the correct product, so change the signs of the two factors.

    $(4r - 3)(r + 1) = 4r^2 + r - 3$ *Correct*

23. $8m^2 - 10m - 3$

    Factor by the grouping method. Look for two integers whose product is $8(-3)$ or -24 and whose sum is -10.

    The integers are -12 and 2.

    $8m^2 - 10m - 3$
    $= 8m^2 - 12m + 2m - 3$
    $= 4m(2m - 3) + 1(2m - 3)$ *Factor each group*
    $= (2m - 3)(4m + 1)$ *Factor out $2m - 3$*

27. $21m^2 + 13m + 2$

    Factor by grouping. Look for two integers whose product is $21 \cdot 2$ or 42 and whose sum is 13. The integers are 7 and 6.

    $21m^2 + 13m + 2$
    $= 21m^2 + 7m + 6m + 2$
    $= 7m(3m + 1) + 2(3m + 1)$ *Factor each group*
    $= (3m + 1)(7m + 2)$ *Factor out $3m + 1$*

31. $6b^2 + 7b + 2$

    Use the grouping method. Find two integers whose product is $(6)(2) = 12$ and whose sum is 7. The numbers are 3 and 4.

    $6b^2 + 7b + 2 = 6b^2 + 3b + 4b + 2$
    $= 3b(2b + 1) + 2(2b + 1)$
    $= (2b + 1)(3b + 2)$

35. $40m^2q + mq - 6q$

    First, factor out the greatest common factor, q.

    $40m^2q + mq - 6q = q(40m^2 + m - 6)$

    Now factor $40m^2 + m - 6$ by trial and error to obtain

    $40m^2 + m - 6 = (5m + 2)(8m - 3)$.

    The complete factoring is

    $40m^2q + mq - 6q = q(5m + 2)(8m - 3)$.

39. $15n^4 - 39n^3 + 18n^2$

Factor out the greatest common factor, $3n^2$.

$15n^4 - 39n^3 + 18n^2$
$= 3n^2(5n^2 - 13n + 6)$

Factor $5n^2 - 13n + 6$ by the trial and error method. Possible factors of $5n^2$ are $5n$ and n. Possible factors of 6 are -6 and -1, -3 and -2, -2 and -3, or -1 and -6.

$(5n - 6)(n - 1) = 5n^2 - 11n^2 + 6$
  Incorrect
$(5n - 3)(n - 2) = 5n^2 - 13n^2 + 6$
  Correct

Stop here. Thus,

$15n^4 - 39n^3 + 18n^2$
$= 3n^2(5n - 3)(n - 2)$.
  Include common factor

43. $15x^2y^2 - 7xy^2 - 4y^2$

Factor out the greatest common factor, y^2.

$15x^2y^2 - 7xy^2 - 4y^2$
$= y^2(15x^2 - 7x - 4)$

Factor $15x^2 - 7x - 4$ by the grouping method. Look for two integers whose product is $15(-4)$ or -60 and whose sum is -7. The integers are -12 and 5.

$15x^2y^2 - 7xy^2 - 4y^2$
$= y^2(15x^2 - 12x + 5x - 4)$
$= y^2[3x(5x - 4) + 1(5x - 4)]$
$= y^2(5x - 4)(3x + 1)$

47. $25a^2 + 25ab + 6b^2$

Use the grouping method. Find two integers whose product is $(25)(6) = 150$ and whose sum is 25. The numbers are 10 and 15.

$25a^2 + 25ab + 6b^2$
$= 25a^2 + 15ab + 10ab + 6b^2$
$= 5a(5a + 3b) + 2b(5a + 3b)$
$= (5a + 3b)(5a + 2b)$

51. $6m^6n + 7m^5n^2 + 2m^4n^3$

Factor out the greatest common factor, m^4n.

$6m^6n + 7m^5n^2 + 2m^4n^3$
$= m^4n(6m^2 + 7mn + 2n^2)$

Now factor $6m^2 + 7mn + 2n^2$ by trial and error.

$(3m + 2n)(2m + n) = 6m^2 + 7mn + 2n^2$
  Correct

Thus,

$6m^6n + 7m^5n^2 + 2m^4n^3$
$= m^4n(3m + 2n)(2m + n)$.

55. $16 + 16x + 3x^2$

Factor by the grouping method. Find two integers whose product is $(16)(3) = 48$ and whose sum is 16. The numbers are 4 and 12.

$16 + 16x + 3x^2$
$= 16 + 4x + 12x + 3x^2$
$= 4(4 + x) + 3x(4 + x)$
$= (4 + x)(4 + 3x)$

59. $-3x^2 - x + 4 = -1(3x^2 + x - 4)$
$= -1(3x + 4)(x - 1)$

63. $(x + 7)(3 - x) = 3x - x^2 + 21 - 7x$
    FOIL
$= -x^2 + 3x - 7x + 21$
    Commutative property
$= -x^2 - 4x + 21$
    Combine like terms

Thus, $(x + 7)(3 - x)$ is also a correct answer.

67. To multiply these two binomials, use the formula for the product of the sum and difference of two terms, $(a + b)(a - b) = a^2 - b^2$.

$\left(r^2 + \frac{1}{2}\right)\left(r^2 - \frac{1}{2}\right) = (r^2)^2 - \left(\frac{1}{2}\right)^2$

$= r^4 - \frac{1}{4}$

Section 4.4 Special Factorizations

4.4 Margin Exercises

1. **(a)** $p^2 - 100 = p^2 - 10^2$
$= (p + 10)(p - 10)$

(b) $9m^2 - 49 = (3m)^2 - 7^2$
$= (3m + 7)(3m - 7)$

(c) $64a^2 - 25 = (8a)^2 - 5^2$
$= (8a + 5)(8a - 5)$

(d) $x^2 + y^2$

This binomial is the sum of two squares, and the terms have no common factor. Unlike the difference of two squares, it cannot be factored. It is a prime polynomial.

2. **(a)** $50r^2 - 32 = 2(25r^2 - 16)$
    Factor out common factor, 2
$= 2[(5r)^2 - 4^2]$
$= 2(5r + 4)(5r - 4)$
    Difference of squares

(b) $27y^2 - 75 = 3(9y^2 - 25)$
    Factor out common factor, 3
$= 3[(3y)^2 - 5^2]$
$= 3(3y + 5)(3y - 5)$
    Difference of squares

(c) $k^4 - 49 = (k^2)^2 - 7^2$
$= (k^2 + 7)(k^2 - 7)$

(d) $81r^4 - 16$
$= (9r^2)^2 - 4^2$
$= (9r^2 + 4)(9r^2 - 4)$
$= (9r^2 + 4)[(3r)^2 - 2^2]$
$= (9r^2 + 4)(3r + 2)(3r - 2)$

3. **(a)** $p^2 + 14p + 49$
$= (p)^2 + 2 \cdot p \cdot 7 + (7)^2$
$= (p + 7)^2$

(b) $m^2 + 8m + 16$
$= (m)^2 + 2 \cdot m \cdot 4 + (4)^2$
$= (m + 4)^2$

(c) $x^2 + 2x + 1$
$= (x)^2 + 2 \cdot x \cdot 1 + (1)^2$
$= (x + 1)^2$

4. **(a)** $p^2 - 18p + 81$

$2 \cdot p \cdot 9 = 18p$, so this is a perfect trinomial, and

$$p^2 - 18p + 81 = (p - 9)^2.$$

(b) $16a^2 + 56a + 49$

$2 \cdot 4a \cdot 7 = 56a$, so this is a perfect square trinomial, and

$$16a^2 + 56a + 49 = (4a + 7)^2.$$

(c) $121p^2 + 110p + 100$

The middle term, $110p$, is not twice the product of the first and last terms of the binomial $(11p + 10)$ since $2 \cdot 11p \cdot 10 = 220p$. This is not a perfect square trinomial.

(d) $64x^2 - 48x + 9$

$2 \cdot 8x \cdot 3 = 48x$, so this is a perfect square trinomial, and

$$64x^2 - 48x + 9 = (8x - 3)^2.$$

4.4 Section Exercises

3. The following powers of x are all perfect squares: $x^2, x^4, x^6, x^8, x^{10}$. Based on this observation, we may make a conjecture (an educated guess) that if the power of a variable is divisible by 2 (with 0 remainder), then we have a perfect square.

7. $9r^2 - 4 = (3r)^2 - 2^2$
$ = (3r + 2)(3r - 2)$

11. $36x^2 - 16 = 4(9x^2 - 4)$
$ = 4[(3x)^2 - 2^2]$
$ = 4(3x + 2)(3x - 2)$

15. $16r^2 - 25a^2 = (4r)^2 - (5a)^2$
$ = (4r + 5a)(4r - 5a)$

19. $p^4 - 49 = (p^2)^2 - 7^2$
$ = (p^2 + 7)(p^2 - 7)$

23. $p^4 - 256 = (p^2)^2 - 16^2$
$ = (p^2 + 16)(p^2 - 16)$
$ = (p^2 + 16)(p^2 - 4^2)$
$ = (p^2 + 16)(p + 4)(p - 4)$

27. $w^2 + 2w + 1$

w^2 is a perfect square, and 1 is a perfect square since $1 \cdot 1 = 1$. The middle term is twice the product of the first and last terms:

$$2w = 2(w)(1).$$

Therefore,

$$w^2 + 2w + 1 = (w + 1)^2.$$

31. $t^2 + t + \dfrac{1}{4}$

t^2 is a perfect square, and 1/4 is a perfect square since $1/2 \cdot 1/2 = 1/4$. The middle term is twice the product of the first and last terms:

$$t = 2(t)\left(\dfrac{1}{2}\right)$$

Chapter 4 Factoring

Therefore,

$$t^2 + t + \frac{1}{4} = \left(t + \frac{1}{2}\right)^2.$$

35. $2x^2 + 24x + 72$

First, factor out the greatest common factor, 2.

$$2x^2 + 24x + 72 = 2(x^2 + 12x + 36)$$

Now factor $x^2 + 12x + 36$ as a perfect square trinomial. x^2 is a perfect square. 36 is a perfect square since $6 \cdot 6 = 36$. The middle term is $12x = 2(x)(6)$.
Therefore,

$$2x^2 + 24x + 72 = 2(x + 6)^2.$$

39. $49x^2 - 28xy + 4y^2$

$49x^2$ is a perfect square since $7x \cdot 7x = 49x^2$. $4y^2$ is a perfect square since $2y \cdot 2y = 4y^2$. The middle term is $-28xy = -2(7x)(2y)$.
Therefore,

$$49x^2 - 28xy + 4y^2 = (7x - 2y)^2.$$

43. $-50h^2 + 40hy - 8y^2$
$= -2(25h^2 - 20hy + 4y^2)$

$25h^2$ is a perfect square since $5h \cdot 5h = 25h^2$. $4y^2$ is a perfect square since $2y \cdot 2y = 4y^2$. The middle term is

$$-20hy = 2(5h)(2y).$$

Therefore,

$$-50h^2 + 40hy - 8y^2 = -2(5h - 2y)^2.$$

47. $ay^2 - 12y + 4$
$= (3y - 2)^2$
$= (3y)^2 + 2(3y)(-2) + (-2)^2$
$= 9y^2 - 12y + 4$

By equating the coefficients of the y^2 terms, we see that the only possible value of a is 9.

51. $4z - 9 = 0$
$4z = 9$
$z = \frac{9}{4}$

Summary Exercises on Factoring

3. $14k^3 + 7k^2 - 70k$
$= 7k(2k^2 + k - 10)$ *Greatest common factor*
$= 7k(2k + 5)(k - 2)$

7. $49z^2 - 16y^2$
$= (7z)^2 - (4y)^2$
$= (7z + 4y)(7z - 4y)$ *Difference of two squares*

11. $10y^2 - 7yz - 6z^2$
$= 10y^2 - 12yz + 5yz - 6z^2$
$= 2y(5y - 6z) + z(5y - 6z)$
 Factor by grouping
$= (5y - 6z)(2y + z)$

15. $32z^3 + 56z^2 - 16z$
$= 8z(4z^2 + 7z - 2)$
 Greatest common factor
$= 8z(4z^2 + 8z - z - 2)$
$= 8z[4z(z + 2) - 1(z + 2)]$
 Factor by grouping
$= 8z(z + 2)(4z - 1)$

19. $y^2 - 4yk - 12k^2 = (y + 2k)(y - 6k)$

23. $p^2 - 17p + 66 = (p - 6)(p - 11)$

27. $z^2 - 3za - 10a^2 = (z - 5a)(z + 2a)$

31. $16r^2 + 24rm + 9m^2$
$= (4r)^2 + 2 \cdot 4r \cdot 3m + (3m)^2$
$= (4r + 3m)^2$ *Perfect square trinomial*

35. $16k^2 - 48k + 36$
$= 4(4k^2 - 12k + 9)$
$= 4[(2k)^2 - 2 \cdot 2k \cdot 3 + 3^2]$
$= 4(2k - 3)^2$ *Perfect square trinomial*

39. $5z^3 - 45z^2 + 70z$
$= 5z(z^2 - 9z + 14)$ *Greatest common factor*
$= 5z(z - 7)(z - 2)$

43. $6a^2 + 10a - 4$
$= 2(3a^2 + 5a - 2)$ *Greatest common factor*
$= 2(3a - 1)(a + 2)$

47. $125m^4 - 400m^3n + 195m^2n^2$
$= 5m^2(25m^2 - 80mn + 39n^2)$ *Greatest common factor*
$= 5m^2(25m^2 - 15mn - 65mn + 39n^2)$
$= 5m^2[5m(5m - 3n) - 13n(5m - 3n)]$ *Grouping*
$= 5m^2(5m - 3n)(5m - 13n)$

51. $24k^4p + 60k^3p^2 + 150k^2p^3$
$= 6k^2p(4k^2 + 10kp + 25p^2)$ *Greatest common factor*

55. $64p^2 - 100m^2$
$= 4(16p^2 - 25m^2)$ *Greatest common factor*
$= 4(4p + 5m)(4p - 5m)$ *Difference of two squares*

59. $a^2 + 8a + 16 = (a + 4)^2$ *Perfect square trinomial*

Section 4.5 Solving Quadratic Equations by Factoring

4.5 Margin Exercises

1. **(a)** $(x - 5)(x + 2) = 0$

   By the zero-factor property, $x - 5 = 0$ gives $x = 5$; $x + 2 = 0$ gives $x = -2$. The solutions are 5 and -2. Check both solutions by substituting first 5 and then -2 for x in the original equation.

   (b) $(3x - 2)(x + 6) = 0$

   By the zero-factor property, either $3x - 2 = 0$ or $x + 6 = 0$.
   Solve each equation.

   $3x - 2 = 0$ or $x + 6 = 0$
   $3x = 2$
   $x = \frac{2}{3}$ or $x = -6$

The solutions are 2/3 and −6. Check by substituting each solution for x in the original equation.

(c) $x(5x + 3) = 0$

By the zero-factor property, either $x = 0$ or $5x + 3 = 0$. Since the first equation is solved, we only need to solve the second.

$$5x + 3 = 0$$
$$5x = -3$$
$$x = -\frac{3}{5}$$

The solutions are 0 and −3/5. Check by substituting each solution for x in the original equation.

2. (a) $m^2 - 3m - 10 = 0$

First factor the equation to get

$$(m - 5)(m + 2) = 0.$$

Then set each factor equal to zero and solve.

$$m - 5 = 0 \quad \text{or} \quad m + 2 = 0$$
$$m = 5 \quad \text{or} \quad m = -2$$

The solutions are 5 and −2.

(b) $x^2 - 7x = 0$

First factor the equation to get $x(x - 7) = 0$. Then set each factor equal to zero and solve.

$$x = 0 \quad \text{or} \quad x - 7 = 0$$
$$x = 0 \quad \text{or} \quad x = 7$$

The solutions are 0 and 7.

(c) $r^2 + 2r = 8$

Bring all nonzero terms to the same side of the equals sign by subtracting 8 from each side.

$$r^2 + 2r - 8 = 0$$
$$(r - 2)(r + 4) = 0 \quad \textit{Factor}$$
$$r - 2 = 0 \quad \text{or} \quad r + 4 = 0 \quad \textit{Solve}$$
$$r = 2 \quad \text{or} \quad r = -4$$

The solutions are 2 and −4.

3. (a) $10a^2 - 5a - 15 = 0$
$$5(2a^2 - a - 3) = 0$$
$$5(2a - 3)(a + 1) = 0$$
$$5 = 0 \quad \text{or} \quad 2a - 3 = 0 \quad \text{or} \quad a + 1 = 0$$

The equation $5 = 0$ has no solution.

$$2a - 3 = 0 \quad \text{or} \quad a + 1 = 0$$
$$2a = 3$$
$$a = \frac{3}{2} \quad \text{or} \quad a = -1$$

The solutions are 3/2 and −1. Check them by substituting in the original equation.

(b)
$$4x^2 - 2x = 42$$
$$4x^2 - 2x - 42 = 0$$
$$2(2x^2 - x - 21) = 0$$
$$2(2x - 7)(x + 3) = 0$$
$$2 = 0 \quad \text{or} \quad 2x - 7 = 0 \quad \text{or} \quad x + 3 = 0$$

The equation $2 = 0$ has no solution.

$$2x - 7 = 0 \quad \text{or} \quad x + 3 = 0$$
$$2x = 7$$
$$x = \frac{7}{2} \quad \text{or} \quad x = -3$$

The solutions are 7/2 and −3. Check them by substituting in the original equation.

4. (a)
$$49m^2 - 9 = 0$$
$$(7m + 3)(7m - 3) = 0$$
$$7m + 3 = 0 \quad \text{or} \quad 7m - 3 = 0$$
$$7m = -3 \qquad\qquad 7m = 3$$
$$m = -\frac{3}{7} \quad \text{or} \quad m = \frac{3}{7}$$

The solutions are $-3/7$ and $3/7$.

(b)
$$p(4p + 7) = 2$$

Get all terms on one side of the equals sign, with 0 on the other side.

$$4p^2 + 7p = 2$$
$$4p^2 + 7p - 2 = 0$$
$$(4p - 1)(p + 2) = 0$$
$$4p - 1 = 0 \quad \text{or} \quad p + 2 = 0$$
$$4p = 1$$
$$p = \frac{1}{4} \quad \text{or} \quad p = -2$$

The solutions are $1/4$ and -2.

(c) $m^2 = 3m$

Get all terms on one side of the equals sign, with 0 on the other side.

$$m^2 - 3m = 0$$

Factor the binomial.

$$m(m - 3) = 0$$

Set each factor equal to 0 and solve the resulting equations.

$$m = 0 \quad \text{or} \quad m - 3 = 0$$
$$m = 0 \quad \text{or} \quad m = 3$$

The solutions are 0 and 3.

5. (a)
$$r^3 - 16r = 0$$
$$r(r^2 - 16) = 0$$
$$r(r - 4)(r + 4) = 0$$
$$r = 0 \quad \text{or} \quad r - 4 = 0 \quad \text{or} \quad r + 4 = 0$$
$$r = 0 \quad \text{or} \quad r = 4 \quad \text{or} \quad r = -4$$

Check by substituting 0, 4, and -4 in the original equation.

(b)
$$x^3 - 3x^2 - 18x = 0$$
$$x(x^2 - 3x - 18) = 0$$
$$x(x - 6)(x + 3) = 0$$
$$x = 0 \quad \text{or} \quad x - 6 = 0 \quad \text{or} \quad x + 3 = 0$$
$$x = 0 \quad \text{or} \quad x = 6 \quad \text{or} \quad x = -3$$

Check by substituting 0, 6, and -3 in the original equation.

6. (a)
$$(m + 3)(m^2 - 11m + 10) = 0$$
$$(m + 3)(m - 10)(m - 1) = 0$$
$$m + 3 = 0 \quad \text{or} \quad m - 10 = 0 \quad \text{or} \quad m - 1 = 0$$
$$m = -3 \quad \text{or} \quad m = 10 \quad \text{or} \quad m = 1$$

Check by substituting -3, 10, and 1 in the original equation.

(b)
$$(2x + 5)(4x^2 - 9) = 0$$
$$(2x + 5)(2x + 3)(2x - 3) = 0$$
$$2x + 5 = 0 \quad \text{or} \quad 2x + 3 = 0 \quad \text{or} \quad 2x - 3 = 0$$
$$2x = -5 \qquad\qquad 2x = -3 \qquad\qquad 2x = 3$$
$$x = -\frac{5}{2} \quad \text{or} \quad x = -\frac{3}{2} \quad \text{or} \quad x = \frac{3}{2}$$

Check by substituting $-5/2$, $-3/2$, and $3/2$ in the original equation.

148   Chapter 4   Factoring

4.5   Section Exercises

For all equations in this section, answers should be checked by substituting into the original equation. These checks will be shown here for only a few of the exercises.

3. $(3k + 8)(k + 7) = 0$

By the zero-factor property, the only way that the product of these two factors can be zero is if at least one of the factors is zero.

$$3k + 8 = 0 \quad \text{or} \quad k + 7 = 0$$

Solve each of these linear equations.

$$3k = -8$$
$$k = -\frac{8}{3} \quad \text{or} \quad k = -7$$

The solutions are $-8/3$ are -7.

7. $2x(3x - 4) = 0$

Set each factor equal to zero and solve the resulting linear equations.

$$2x = 0 \quad \text{or} \quad 3x - 4 = 0$$
$$3x = 4$$
$$x = 0 \quad \text{or} \quad x = \frac{4}{3}$$

The solutions are 0 and 4/3.

11. 9 is called a *double solution* for $(x - 9)^2 = 0$ because it occurs twice when the equation is solved.

$$(x - 9)^2 = 0$$
$$(x - 9)(x - 9) = 0$$
$$x - 9 = 0 \quad \text{or} \quad x - 9 = 0$$
$$x = 9 \quad \text{or} \quad x = 9$$

15. $y^2 - 3y + 2 = 0$

Factor the polynomial $y^2 - 3y + 2 = 0$.

$$(y - 1)(y - 2) = 0$$

Set each factor equal to 0.

$$y - 1 = 0 \quad \text{or} \quad y - 2 = 0$$

Solve these linear equations.

$$y = 1 \quad \text{or} \quad y = 2$$

The solutions are 1 and 2.

19. $x^2 = 3 + 2x$

Write the equation in standard form.

$$x^2 - 2x - 3 = 0$$

Factor the polynomial.

$$(x - 3)(x + 1) = 0$$

Set each factor equal to 0.

$$x - 3 = 0 \quad \text{or} \quad x + 1 = 0$$

Solve the linear equations.

$$x = 3 \quad \text{or} \quad x = -1$$

Check

$$x^2 = 3 + 2x$$
$$3^2 = 3 + 2(3) \quad ? \quad \text{Let } x = 3$$
$$9 = 3 + 6 \quad ?$$
$$9 = 9 \qquad\qquad \text{True}$$

$$x^2 = 3 + 2x$$
$$(-1)^2 = 3 + 2(-1) \quad ? \quad \text{Let } x = -1$$
$$1 = 3 - 2 \quad ?$$
$$1 = 1 \qquad\qquad \text{True}$$

The solutions are -1 and 3.

23. $m^2 + 8m + 16 = 0$

Factor $m^2 + 8m + 16$ as a perfect square trinomial.

$$(m + 4)^2 = 0$$

Set the factor $m + 4$ equal to 0 and solve.

$$m + 4 = 0$$
$$m = -4$$

The only solution is -4.

27.
$$6p^2 = 4 - 5p$$
$$6p^2 + 5p - 4 = 0$$
$$(3p + 4)(2p - 1) = 0$$
$$3p + 4 = 0 \text{ or } 2p - 1 = 0$$
$$3p = -4 \quad\quad 2p = 1$$
$$p = -\tfrac{4}{3} \text{ or } \quad p = \tfrac{1}{2}$$

The solutions are $-4/3$ and $1/2$.

31. $y^2 - 9 = 0$

Factor the polynomial as the difference of two squares; then set each factor equal to 0 and solve.

$$(y + 3)(y - 3) = 0$$
$$y + 3 = 0 \text{ or } y - 3 = 0$$
$$y = -3 \text{ or } \quad y = 3$$

The solutions are -3 and 3.

35.
$$n^2 = 121$$
$$n^2 - 121 = 0$$
$$(n + 11)(n - 11) = 0$$
$$n + 11 = 0 \text{ or } n - 11 = 0$$
$$n = -11 \text{ or } \quad n = 11$$

The solutions are -11 and 11.

39.
$$x^2 = 7x$$
$$x^2 - 7x = 0$$
$$x(x - 7) = 0$$
$$x = 0 \text{ or } x - 7 = 0$$
$$x = 0 \text{ or } \quad x = 7$$

The solutions are 0 and 7.

43.
$$g(g - 7) = -10$$
$$g^2 - 7g = -10$$
$$g^2 - 7g + 10 = 0$$
$$(g - 2)(g - 5) = 0$$
$$g - 2 = 0 \text{ or } g - 5 = 0$$
$$g = 2 \text{ or } \quad g = 5$$

The solutions are 2 and 5.

47.
$$2(y^2 - 66) = -13y$$
$$2y^2 - 132 = -13y$$
$$2y^2 + 13y - 132 = 0$$
$$(2y - 11)(y + 12) = 0$$
$$2y - 11 = 0 \text{ or } y + 12 = 0$$
$$2y = 11$$
$$y = \tfrac{11}{2} \text{ or } \quad y = -12$$

Check

$$2(y^2 - 66) = -13y$$
$$2\left[\left(\tfrac{11}{2}\right)^2 - 66\right] = -13\left(\tfrac{11}{2}\right) \text{ ? } \quad \text{Let } y = 11/2$$
$$2\left(\tfrac{121}{4} - 66\right) = -\tfrac{143}{2} \text{ ?}$$
$$2\left(\tfrac{121}{4} - \tfrac{264}{4}\right) = -\tfrac{143}{2} \text{ ?}$$
$$2\left(-\tfrac{143}{4}\right) = -\tfrac{143}{2} \text{ ?}$$
$$-\tfrac{143}{2} = -\tfrac{143}{2} \quad\quad \text{True}$$

$2(y^2 - 66) = -13y$

$2[(-12)^2 - 66] = -13(-12)$?

Let $y = -12$

$2(144 - 66) = 156$?

$2(78) = 156$?

$156 = 156$ *True*

The solutions are -12 and $11/2$.

51. $(2r + 5)(3r^2 - 16r + 5) = 0$

Begin by factoring $3r^2 - 16r + 5$.

$(2r + 5)(3r - 1)(r - 5) = 0$

Now use the zero-factor property to set each of the three factors equal to 0; then solve the resulting equations.

$2r + 5 = 0$ or $3r - 1 = 0$ or $r - 5 = 0$

$2r = -5$ $3r = 1$

$r = -\frac{5}{2}$ or $r = \frac{1}{3}$ or $r = 5$

The solutions are $-5/2$, $1/3$, and 5.

55. $9y^3 - 49y = 0$

To factor the polynomial, begin by factoring out the greatest common factor.

$y(9y^2 - 49) = 0$

Now factor $9y^2 - 49$ as the difference of two squares.

$y(3y + 7)(3y - 7) = 0$

Set each of the three factors equal to 0 and solve.

$y = 0$ or $3y + 7 = 0$ or $3y - 7 = 0$

                $3y = -7$ $3y = 7$

$y = 0$ or $y = -\frac{7}{3}$ or $y = \frac{7}{3}$

The solutions are $-7/3$, 0, and $7/3$.

Section 4.6 Applications of Quadratic Equations

4.6 Margin Exercises

1. Let x = the width of the rectangle;
   $x + 2$ = the length.

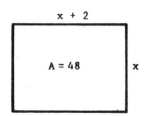

Use the formula for the area of a rectangle, $A = LW$.

Substitute 48 for A, $x + 2$ for L, and x for W.

$A = LW$

$48 = (x + 2)x$

$48 = x^2 + 2x$

$0 = x^2 + 2x - 48$ *Subtract 48*

$0 = (x - 6)(x + 8)$ *Factor*

$x - 6 = 0$ or $x + 8 = 0$

$x = 6$ or $x = -8$

Discard the solution -8 since the width cannot be negative. The width is 6 meters and the length is

$x + 2 = 6 + 2 = 8$ meters.

2. Let x = the length of the yard;
   $x - 4$ = the width of the yard.

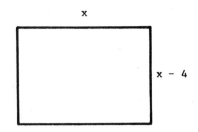

Use A = LW and P = 2L + 2W to write the equation. Substitute x for L and x − 4 for W.

The area	is	92	more than	the perimeter.
↓	↓	↓	↓	↓
x(x − 4)	=	92	+	2x + 2(x − 4)

Solve this equation.

$$x^2 - 4x = 92 + 2x + 2x - 8$$
$$x^2 - 4x = 84 + 4x$$
$$x^2 - 8x = 84 \quad \text{Subtract } 4x$$
$$x^2 - 8x - 84 = 0 \quad \text{Subtract } 84$$
$$(x - 14)(x + 6) = 0 \quad \text{Factor}$$
$$x - 14 = 0 \quad \text{or} \quad x + 6 = 0$$
$$x = 14 \quad \text{or} \quad x = -6$$

Discard −6. The length is 14 meters. The width is

$$x - 4 = 14 - 4 = 10 \text{ meters.}$$

3. Let x = the first integer;
   x + 2 = the next even integer.

The product	is	4	more than	two times	their sum.
↓	↓	↓	↓	↓	↓
x(x + 2)	=	4	+	2 ·	[x + (x + 2)]

Solve this equation.

$$x^2 + 2x = 4 + 2(2x + 2)$$
$$x^2 + 2x = 4 + 4x + 4$$
$$x^2 - 2x - 8 = 0$$
$$(x - 4)(x + 2) = 0$$

$$x - 4 = 0 \quad \text{or} \quad x + 2 = 0$$
$$x = 4 \quad \text{or} \quad x = -2$$

We need to find two consecutive even integers. If x = 4 is the first, then the second is

$$x + 2 = 4 + 2 = 6.$$

If x = −2 is the first, then the second is

$$x + 2 = -2 + 2 = 0.$$

The two integers are 4 and 6, or they are −2 and 0.

4. Let x = the length of the longer leg of the right triangle.
   x + 3 = the length of the hypotenuse;
   x − 3 = the length of the shorter leg.

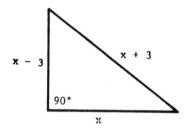

Use the Pythagorean formula, substituting x − 3 for a, x for b, and x + 3 for c.

$$a^2 + b^2 = c^2$$
$$(x - 3)^2 + x^2 = (x + 3)^2$$
$$x^2 - 6x + 9 + x^2 = x^2 + 6x + 9$$
$$2x^2 - 6x + 9 = x^2 + 6x + 9$$
$$x^2 - 12x = 0$$
$$x(x - 12) = 0$$
$$x = 0 \quad \text{or} \quad x - 12 = 0$$
$$x = 0 \quad \text{or} \quad x = 12$$

Discard 0 since the length of a side of a triangle cannot be zero. The length of the longer leg is 12 inches, the length of the hypotenuse is x + 3 = 12 + 3 = 15 inches, and the length of the shorter leg is x − 3 = 12 − 3 = 9 inches.

4.6 Section Exercises

3. A = bh; A = 45, b = 2x + 1, h = x + 1

   (a) A = bh

   45 = (2x + 1)(x + 1)

   (b) $45 = 2x^2 + 3x + 1$

   $0 = 2x^2 + 3x - 44$

   $0 = (2x + 11)(x - 4)$

   2x + 11 = 0   or   x − 4 = 0

   2x = −11

   $x = -\frac{11}{2}$ or x = 4

   (c) Substitute these solutions for x in the expressions 2x + 1 and x + 1 to find the values of b and h.

   $b = 2x + 1 = 2(-\frac{11}{2}) + 1$

   = −11 + 1 = −10

   or b = 2x + 1 = 2(4) + 1

   = 8 + 1 = 9

   We must discard the first solution because the base of a parallelogram cannot have a negative length.

Since b = −11/2 will not give a realistic answer for the base, we only need to substitute 4 for x to compute the height.

$$h = x + 1 = 4 + 1 = 5$$

The base is 9 units and the height is 5 units.

7. Let   x = the width of the shell.
   Then x + 3 = the length of the shell.

   Substitute 28 for the area, x for the width, and x + 3 for the length in the formula for the area for a rectangle.

   A = LW

   28 = (x + 3)x

   $28 = x^2 + 3x$

   $0 = x^2 + 3x - 28$

   0 = (x + 7)(x − 4)

   x + 7 = 0 or x − 4 = 0

   x = −7 or   x = 4

   The width of a rectangle cannot be negative, so we reject −7. The width of the shell is 4 inches, and the length is 4 + 3 = 7 inches.

11. Let   x = the length of the side of the square sign.
    Then x − 2 = the length of the side of the square poster.

    The formula for the area of a square is $A = s^2$, the area of the poster is x^2 and the area of the sign is $(x - 2)^2$.

$$x^2 - (x - 2)^2 = 32$$
$$x^2 - (x^2 - 4x + 4) = 32$$
$$x^2 - x^2 + 4x - 4 = 32$$
$$4x - 4 = 32$$
$$4x = 36$$
$$x = 9$$

The length of the side of the sign is 9 feet. The length of the side of the poster is $9 - 2 = 7$ feet.

15. Let x = the width of the aquarium.

    Then $x + 3$ = the height of the aquarium.

    Use the formula for the volume of a rectangular box.

    $$V = LWH$$
    $$2730 = 21x(x + 3)$$
    $$130 = x(x + 3)$$
    $$130 = x^2 + 3x$$
    $$0 = x^2 + 3x - 130$$
    $$0 = (x + 13)(x - 10)$$
    $$x + 13 = 0 \quad \text{or} \quad x - 10 = 0$$
    $$x = -13 \quad \text{or} \quad x = 10$$

    We discard -13 because the width cannot be negative. The width is 10 inches. The height is $10 + 3 = 13$ inches.

19. Let x = the smallest odd integer;

    $x + 2$ = the next smallest odd integer;

    $x + 4$ = the largest odd integer.

$$3[x + (x + 2) + (x + 4)] = x(x + 2) + 18$$
$$3(3x + 6) = x^2 + 2x + 18$$
$$9x + 18 = x^2 + 2x + 18$$
$$0 = x^2 - 7x$$
$$0 = x(x - 7)$$
$$x = 0 \quad \text{or} \quad x - 7 = 0$$
$$x = 0 \quad \text{or} \quad x = 7$$

We must discard 0 because it is even and the problem requires the integers to be odd. If $x = 7$, $x + 2 = 9$, and $x + 4 = 11$. The three integers are 7, 9, and 11.

23. Let x = Alan's distance from home;

    $x + 1$ = the distance between Wei-Jen and Alan.

    Refer to the diagram in the textbook. Use the Pythagorean formula.

    $$a^2 + b^2 = c^2$$
    $$x^2 + 5^2 = (x + 1)^2 \quad \text{Let } a = x, b = 5, c = x + 1$$
    $$x^2 + 25 = x^2 + 2x + 1$$
    $$x^2 - x^2 - 2x + 25 - 1 = 0$$
    $$-2x = -24$$
    $$x = 12$$

    Alan is 12 miles from home.

27. Let x = the numerical value of the area of the base;

    $x - 10$ = the height.

    Use the formula for the volume of a pyramid.

    $$V = \tfrac{1}{3}Bh$$

    $$32 = \tfrac{1}{3}x(x - 10) \quad \text{Let } V = 32, B = x, h = x - 10$$

154   Chapter 4   Factoring

$96 = x(x - 10)$ *Multiply by 3*
$96 = x^2 - 10x$ *Distributive property*
$0 = x^2 - 10x - 96$ *Standard form*
$0 = (x - 16)(x + 6)$ *Factor*
$x - 16 = 0$ or $x + 6 = 0$
$x = 16$ or $x = -6$

Since area cannot be negative, the area of the base is 16 square meters. If $x = 16$, $x - 10 = 6$, so the height of the pyramid is 6 meters.

31. There are two possible answers in Exercise 30(a) because the object will have a height of 48 feet twice, the first time as it is going up, and the second time when it is on its way down.

35. $\dfrac{26}{156} = \dfrac{1 \cdot 26}{6 \cdot 26} = \dfrac{1}{6}$

Chapter 4   Review Exercises

1. $7t + 14 = 7 \cdot t + 7 \cdot 2 = 7(t + 2)$

2. $60z^3 + 30z = 30z \cdot 2z^2 + 30z \cdot 1$
 $= 30z(2z^2 + 1)$

3. $35x^3 + 70x^2 = 35x^2 \cdot x + 35x^2 \cdot 2$
 $= 35x^2(x + 2)$

4. $2xy - 8y + 3x - 12$
 $= 2y(x - 4) + 3(x - 4)$
 $= (x - 4)(2y + 3)$

5. $100m^2n^3 - 50m^3n^4 + 150m^2n^2$
 $= 50m^2n^2(2n) + 50m^2n^2(-mn^2)$
 $\quad + 50m^2n^2(3)$
 $= 50m^2n^2(2n - mn^2 + 3)$

6. $6y^2 + 9y + 4y + 6$
 $= 3y(2y + 3) + 2(2y + 3)$
 $= (2y + 3)(3y + 2)$

7. $x^2 + 5x + 6$

 Find two integers whose product is 6 and whose sum is 5. The integers are 3 and 2. Thus,
 $x^2 + 5x + 6 = (x + 3)(x + 2)$.

8. $y^2 - 13y + 40$

 Find two integers whose product is 40 and whose sum if -13.

Factors of 40	Sum of factors
$-1, 40$	$(-1) + (-40) = -41$
$-2, -20$	$(-2) + (-20) = -22$
$-4, -10$	$(-4) + (-10) = -14$
$-5, -8$	$(-5) + (-8) = -13$

 The integers are -5 and -8, so
 $y^2 - 13y + 40 = (y - 5)(y - 8)$.

9. $q^2 + 6p - 27$

   Find two integers whose product is -27 and whose sum is 6. The integers are -3 and 9, so
   $$q^2 + 6q - 27 = (q - 3)(q + 9).$$

10. $r^2 - r - 56$

    Find two integers whose product is -56 and whose sum is -1. The integers are 7 and -8, so
    $$r^2 - r - 56 = (r + 7)(r - 8).$$

11. $r^2 - 4rs - 96s^2$

    Find two expressions whose product is $-96s^2$ and whose sum is $-4s$. The expressions are $8s$ and $-12s$, so
    $$r^2 - 4rs - 96s^2 = (r + 8s)(r - 12s).$$

12. $p^2 + 2pq - 120q^2$

    Find two expressions whose product is $-120q^2$ and whose sum is $2q$. The expressions are $12q$ and $-10q$, so
    $$p^2 + 2pq - 120q^2 = (p + 12q)(p - 10q).$$

13. $8p^3 - 24p^2 - 80p$

    First, factor out the greatest common factor, 8.
    $$8p^3 - 24p^2 - 80p = 8p(p^2 - 3p - 10)$$
    Now factor $p^2 - 3p - 10$.
    $$p^2 - 3p - 10 = (p + 2)(p - 5)$$

    The complete factored form is
    $$8p^3 - 24p^2 - 80p$$
    $$= 8p(p + 2)(p - 5).$$

14. $3x^4 + 30x^3 + 48x^2$
    $$= 3x^2(x^2 + 10x + 16)$$
    $$= 3x^2(x + 2)(x + 8)$$

15. $m^2 - 3mn - 18n^2$

    Find two expressions whose product is $-18n^2$ and whose sum is $-3n$. The expressions are $3n$ and $-6n$, so
    $$m^2 - 3mn - 18n^2$$
    $$= (m + 3n)(m - 6n).$$

16. $y^2 - 8yz + 15z^2$

    Find two expressions whose product is $15z^2$ and whose sum is $-8z$. The expressions are $-3z$ and $-5z$, so
    $$y^2 - 8yz + 15z^2 = (y - 3z)(y - 5z).$$

17. $p^7 - p^6q - 2p^5q^2 = p^5(p^2 - pq - 2q^2)$
    $$= p^5(p + q)(p - 2q)$$

18. $3r^5 - 6r^4s - 45r^3s^2$
    $$= 3r^3(r^2 - 2rs - 15s^2)$$
    $$= 3r^3(r + 3s)(r - 5s)$$

19. $x^2 + x + 1$

    There is no pair of integers whose product is 1 and whose sum is 1, so $x^2 + x + 1$ is prime.

20. $3x^2 + 6x + 6 = 3(x^2 + 2x + 2)$

    There is no pair of integers whose product is 2 and whose sum is 2, so $x^2 + 2x + 2$ is prime. Thus, the complete factored form is

    $3x^2 + 6x + 6 = 3(x^2 + 2x + 2)$.

21. To begin factoring $6r^2 - 5r - 6$, the possible first terms of the two binomial factors are r and 6r, or 2r and 3r.

22. When factoring $2z^3 + 9z^2 - 5z$, the first step is to factor out the common factor z.

In Exercises 23-30, either the trial and error method or the grouping method can be used to factor each polynomial.

23. $2k^2 - 5k + 2$

    Factor by trial and error.

    $2k^2 - 5k + 2 = (2k - 1)(k - 2)$

24. $3r^2 + 11r - 4$

    Factor by grouping. Look for two integers whose product is $3(-4) = -12$ and whose sum is 11. The integers are 12 and -1.

    $3r^2 + 11r - 4 = 3r^2 + 12r - r - 4$
    $ = 3r(r + 4) - 1(r + 4)$
    $ = (r + 4)(3r - 1)$

25. $6r^2 - 5r - 6$

    Factor by grouping. Find two integers whose product is -6 and whose sum is -5. The integers are -9 and 4.

    $6r^2 - 5r - 6 = 6r^2 - 9r + 4r - 6$
    $ = 3r(2r - 3) + 2(2r - 3)$
    $ = (2r - 3)(3r + 2)$

26. $10z^2 - 3z - 1$

    Factor by trial and error.

    $10z^2 - 3z - 1 = (5z + 1)(2z - 1)$

27. $8v^2 + 17v - 21$

    Factor by grouping. Look for two integers whose product is -21 and whose sum is 17. The integers are 24 and -7.

    $8v^2 + 17v - 21 = 8v^2 + 24v - 7v - 21$
    $ = 8v(v + 3) - 7(v + 3)$
    $ = (v + 3)(8v - 7)$

28. $24x^5 - 20x^4 + 4x^3$

    Factor out the greatest common factor, $4x^3$. Then complete the factoring by trial and error.

    $24x^5 - 20x^4 + 4x^3$
    $= 4x^3(6x^2 - 5x + 1)$
    $= 4x^3(3x - 1)(2x - 1)$

29. $-6x^2 + 3x + 30 = -3(2x^2 - x - 10)$
    $ = -3(2x - 5)(x + 2)$

30. $10r^3s + 17r^2s^2 + 6rs^3$
    $= rs(10r^2 + 17rs + 6s^2)$
    $= rs(5r + 6s)(2r + s)$

31. Only (b) $4x^2y^2 - 25z^2$ is the difference of two squares. In (a), 32 is not a perfect square. In (c), we have a sum, not a difference. In (d), y^3 is not a square. The correct choice is (b).

32. Only (d) $x^2 - 20x + 100$ is a perfect square trinomial because $x^2 = x \cdot x$, $100 = 10 \cdot 10$, and $-20x = -2(x)(10)$.

In Exercises 33–35, use the rule for factoring a difference of two squares.

33. $n^2 - 49 = n^2 - 7^2 = (n + 7)(n - 7)$

34. $25b^2 - 121 = (5b)^2 - 11^2$
    $= (5b + 11)(5b - 11)$

35. $49y^2 - 25w^2 = (7y)^2 - (5w)^2$
    $= (7y + 5w)(7y - 5w)$

36. $144p^2 - 36q^2 = 36(4p^2 - q^2)$
    $= 36[(2p)^2 - q^2]$
    $= 36(2p + q)(2p - q)$

37. $x^2 + 100$

    This polynomial is prime because it is the sum of two squares and the two terms have no common factor.

In Exercises 38–42, use the rules for factoring a perfect square trinomial.

38. $z^2 + 10z + 25 = z^2 + 2(5)(z) + 5^2$
    $= (z + 5)^2$

39. $r^2 - 12r + 36 = r^2 - 2(6)(r) + 6^2$
    $= (r - 6)^2$

40. $9t^2 - 42t + 49 = (3t)^2 - 2(3t)(7) + 7^2$
    $= (3t - 7)^2$

41. $16m^2 + 40mn + 25n^2$
    $= (4m)^2 + 2(4m)(5n) + (5n)^2$
    $= (4m + 5n)^2$

42. $54x^3 - 72x^2 + 24x$
    $= 6x(9x^2 - 12x + 4)$
    $= 6x[(3x)^2 - 2(3x)(2) + 2^2]$
    $= 6x(3x - 2)^2$

In Exercises 43–56, all solutions should be checked by substituting in the original equations. The checks will not be shown here.

43. $(4t + 3)(t - 1) = 0$
    $4t + 3 = 0$ or $t - 1 = 0$
    $4t = -3$
    $t = -\frac{3}{4}$ or $t = 1$

    The solutions are $-3/4$ and 1.

44. $(x + 7)(x - 4)(x + 3) = 0$
    $x + 7 = 0$ or $x - 4 = 0$ or $x + 3 = 0$
    $x = -7$ or $x = 4$ or $x = -3$

    The solutions are -7, 4, and -3.

45. $x(2x - 5) = 0$

$x = 0$ or $2x - 5 = 0$

$\phantom{x = 0 \text{ or }} 2x = 5$

$x = 0$ or $x = \dfrac{5}{2}$

The solutions are 0 and 5/2.

46. $z^2 + 4z + 3 = 0$

$(z + 3)(z + 1) = 0$

$z + 3 = 0$ or $z + 1 = 0$

$z = -3$ or $z = -1$

The solutions are -3 and -1.

47. $m^2 - 5m + 4 = 0$

$(m - 1)(m - 4) = 0$

$m - 1 = 0$ or $m - 4 = 0$

$m = 1$ or $m = 4$

The solutions are 1 and 4.

48. $x^2 = -15 + 8x$

$x^2 - 8x + 15 = 0$

$(x - 3)(x - 5) = 0$

$x - 3 = 0$ or $x - 5 = 0$

$x = 3$ or $x = 5$

The solutions are 3 and 5.

49. $3z^2 - 11z - 20 = 0$

$(3z + 4)(z - 5) = 0$

$3z + 4 = 0$ or $z - 5 = 0$

$3z = -4$

$z = -\dfrac{4}{3}$ or $z = 5$

The solutions are $-4/3$ and 5.

50. $81t^2 - 64 = 0$

$(9t + 8)(9t - 8) = 0$

$9t + 8 = 0$ or $9t - 8 = 0$

$9t = -8 9t = 8$

$t = -\dfrac{8}{9}$ or $t = \dfrac{8}{9}$

The solutions are $-8/9$ and $8/9$.

51. $y^2 = 8y$

$y^2 - 8y = 0$

$y(y - 8) = 0$

$y = 0$ or $y - 8 = 0$

$y = 0$ or $y = 8$

The solutions are 0 and 8.

52. $n(n - 5) = 6$

$n^2 - 5n = 6$

$n^2 - 5n - 6 = 0$

$(n + 1)(n - 6) = 0$

$n + 1 = 0$ or $n - 6 = 0$

$n = -1$ or $n = 6$

The solutions are -1 and 6.

53. $t^2 - 14t + 49 = 0$

$(t - 7)^2 = 0$

$t - 7 = 0$

$t = 7$

The only solution is 7.

54. $t^2 = 12(t - 3)$

$t^2 = 12t - 36$

$t^2 - 12t + 36 = 0$

$(t - 6)^2 = 0$

$t - 6 = 0$

$t = 6$

The only solution is 6.

55. $(5z + 2)(z^2 + 3z + 2) = 0$
$(5z + 2)(z + 2)(z + 1) = 0$
$5z + 2 = 0$ or $z + 2 = 0$ or $z + 1 = 0$
$5z = -2$
$z = -\frac{2}{5}$ or $z = -2$ or $x = -1$

The solutions are $-2/5$, -2, and -1.

56. $x^2 = 9$
$x^2 - 9 = 0$
$(x + 3)(x - 3) = 0$
$x + 3 = 0$ or $x - 3 = 0$
$x = -3$ or $x = 3$

The solutions are -3 and 3.

57. Let x = The width of the rectangle.
Then $x + 6$ = length.

$A = LW$
$40 = (x + 6)x$
$40 = x^2 + 6x$
$x^2 + 6x - 40 = 0$
$(x + 10)(x - 4) = 0$
$x + 10 = 0$ or $x - 4 = 0$
$x = -10$ or $x = 4$

Reject -10 since the width cannot be negative. The width of the rectangle is 4 meters and the length is $4 + 6$ or 10 meters.

58. Let x = the width of the rectangle.
Then $3x$ = the length.

Use $A = LW$.

The width increased by 3	times	the same length
↓	↓	↓
$(x + 3)$	·	$3x$

would be	an area of 30.
↓	↓
=	30

Solve the equation.

$(x + 3)(3x) = 30$
$3x^2 + 9x = 30$
$3x^2 + 9x - 30 = 0$
$3(x^2 + 3x - 10) = 0$
$x^2 + 3x - 10 = 0$
$(x + 5)(x - 2) = 0$
$x + 5 = 0$ or $x - 2 = 0$
$x = -5$ or $x = 2$

Reject -5. The width of the original rectangle is 2 meters and the length is $3(2)$ or 6 meters.

59. Let x = the width of the rectangle.
Then $x + 2$ = the length.

Use $A = LW$.
$A = (x + 2)x$
Use $P = 2L + 2W$.
$P = 2(x + 2) + 2x$

The area	is numerically	44 more than	the perimeter.
↓	↓	↓	↓
$(x + 2)x$	=	$44 + 2(x + 2) + 2x$	

Solve the equation.

$$(x + 2)x = 44 + 2(x + 2) + 2x$$
$$x^2 + 2x = 44 + 2x + 4 + 2x$$
$$x^2 - 2x - 48 = 0$$
$$(x - 8)(x + 6) = 0$$
$$x - 8 = 0 \text{ or } x + 6 = 0$$
$$x = 8 \text{ or } x = -6$$

Reject −6. The width is 8 centimeters and the length is 8 + 2 or 10 centimeters.

60. Let x = the length of the box.
Then $x - 1$ = the height.

$$V = LWH$$
$$120 = x(4)(x - 1)$$
$$120 = 4x^2 - 4x$$
$$4x^2 - 4x - 120 = 0$$
$$4(x^2 - x - 30) = 0$$
$$x^2 - x - 30 = 0$$
$$(x - 6)(x + 5) = 0$$
$$x - 6 = 0 \text{ or } x + 5 = 0$$
$$x = 6 \text{ or } x = -5$$

Reject −5. The length of the box is 6 meters and the height is 6 − 1 or 5 meters.

61. Let x = the width of the rectangle.
Then $x + 4$ = the length.

Use $A = LW$.
$$A = (x + 4)x$$
Use $P = 2L + 2W$.
$$P = 2(x + 4) + 2x$$

The area is 1 more than the perimeter.
↓ ↓ ↓ ↓ ↓
$(x + 4)x = 1 + 2(x + 4) + 2x$

Solve the equation.

$$(x + 4)x = 1 + 2(x + 4) + 2x$$
$$x^2 + 4x = 1 + 2x + 8 + 2x$$
$$x^2 + 4x = 9 + 4x$$
$$x^2 - 9 = 0$$
$$(x + 3)(x - 3) = 0$$
$$x + 3 = 0 \text{ or } x - 3 = 0$$
$$x = -3 \text{ or } x = 3$$

Reject −3. The width is 3 feet and the length is 3 + 4 or 7 feet.

62. Let x = the length of the rectangle.
Then $x - 5$ = the width.

Use $A = LW$.
$$A = x(x - 5)$$
Use $P = 2L + 2W$.
$$P = 2x + 2(x - 5)$$

The area is 10 more than the perimeter.
↓ ↓ ↓ ↓
$x(x - 5) = 10 + [2x + 2(x - 5)]$

Solve the equation.

$$x(x - 5) = 10 + [2x + 2(x - 5)]$$
$$x^2 - 5x = 10 + 2x + 2x - 10$$
$$x^2 - 5x = 4x$$
$$x^2 - 9x = 0$$
$$x(x - 9) = 0$$
$$x = 0 \text{ or } x - 9 = 0$$

Reject 0. The length is 9 inches and the width is 9 − 5 or 4 inches.

63. Let $\quad x =$ the first integer.
Then $x + 1 =$ the second integer.

$$x(x + 1) = x + (x + 1) + 29$$
$$x^2 + x = x + x + 1 + 29$$
$$x^2 + x = 2x + 30$$
$$x^2 - x - 30 = 0$$
$$(x + 5)(x - 6) = 0$$
$$x + 5 = 0 \quad \text{or} \quad x - 6 = 0$$
$$x = -5 \text{ or} \quad\quad x = 6$$

If $x = -5$, $x + 1 = -4$.
If $x = 6$, $x + 1 = 7$.
The two consecutive integers are -5 and -4, or 6 and 7.

64. Let $\quad x =$ the length of the shortest side.
Then $x + 2 =$ the length of the middle side
and $\quad x + 4 =$ the length of the hypotenuse.

Use the Pythagorean formula with $a = x$, $b = x + 2$, and $c = x + 4$.

$$x^2 + (x + 2)^2 = (x + 4)^2$$
$$x^2 + x^2 + 4x + 4 = x^2 + 8x + 16$$
$$2x^2 + 4x + 4 = x^2 + 8x + 16$$
$$x^2 - 4x - 12 = 0$$
$$(x - 6)(x + 2) = 0$$
$$x - 6 = 0 \quad \text{or} \quad x + 2 = 0$$
$$x = 6 \quad \text{or} \quad\quad x = -2$$

Reject -2 because the length of a side cannot be negative.
If $x = 6$, $x + 2 = 8$ and $x + 4 = 10$.
The length of the sides are 6 inches, 8 inches, and 10 inches.

65. $h = 128t - 16t^2$
$ h = 128(1) - 16(1)^2 \quad$ *Let $t = 1$*
$ = 128 - 16$
$ = 112$

After 1 second, the height is 112 feet.

66. $h = 128t - 16t^2$
$ h = 128(2) - 16(2)^2 \quad$ *Let $t = 2$*
$ = 256 - 16(4)$
$ = 256 - 64$
$ = 192$

After 2 seconds, the height is 192 feet.

67. $h = 128t - 16t^2$
$ = 128(4) - 16(4)^2 \quad$ *Let $t = 4$*
$ = 512 - 256$
$ = 256$

After 4 seconds, the height is 256 feet.

68. The object hits the ground when $h = 0$.

$$h = 128t - 16t^2$$
$$0 = 128t - 16t^2 \quad \textit{Let } h = 0$$
$$0 = 16t(8 - t)$$
$$16t = 0 \quad \text{or} \quad 8 - t = 0$$
$$t = 0 \quad \text{or} \quad\quad 8 = t$$

The solution $t = 0$ represents the time before the object is thrown. The object returns to the ground after 8 seconds.

69. Draw a sketch.

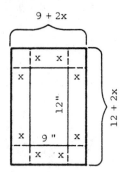

Let x represent the width of the border.
Then 9 + 2x represents the width of the finished mat and picture, and 12 + 2x represents the length.
Use A = LW.
Substitute A = 208, L = 12 + 2x, and W = 9 + 2x.

$$A = LW$$
$$208 = (12 + 2x)(9 + 2x)$$
$$208 = 108 + 42x + 4x^2$$
$$4x^2 + 42x + 108 - 208 = 0$$
$$4x^2 + 42x - 100 = 0$$
$$2(2x^2 + 21x - 50) = 0$$
$$2(2x + 25)(x - 2) = 0$$
$$2x + 25 = 0 \quad \text{or} \quad x - 2 = 0$$
$$2x = -25$$
$$x = -\frac{25}{2} \quad \text{or} \quad x = 2$$

Reject −25/2 since the width cannot be negative. The width of the border is 2 inches.

70. Draw a sketch.

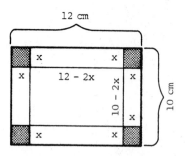

Let x represent the length of a side of each cutout square.
Then 12 − 2x represents the length of the bottom of the box and 10 − 2x represents the width.
Use A = LW.
Substitute A = 48, L = 12 − 2x, and W = 10 − 2x.

$$A = L \cdot W$$
$$48 = (12 - 2x)(10 - 2x)$$
$$48 = 120 - 44x + 4x^2$$
$$4x^2 - 44x + 120 - 48 = 0$$
$$4x^2 - 44x + 72 = 0$$
$$4(x^2 - 11x + 18) = 0$$
$$4(x - 9)(x - 2) = 0$$
$$x - 9 = 0 \quad \text{or} \quad x - 2 = 0$$
$$x = 9 \quad \text{or} \quad x = 2$$

Reject 9 because if x = 9, then 12 − 2x = −6, and a length cannot be negative. The length of a side of the cutout squares is 2 centimeters.

71. $z^2 - 11zx + 10x^2 = (z - x)(z - 10x)$

72. $3k^2 + 11k + 10$

Two integers with product $3(10) = 30$ and sum 11 are 5 and 6.

$= 3k^2 + 5k + 6k + 10$
$= k(3k + 5) + 2(3k + 5)$
$= (3k + 5)(k + 2)$

73. $15m^2 + 20mp - 12mp - 16p^2$
$= 5m(3m + 4p) - 4p(3m + 4p)$
   Factor by grouping
$= (3m + 4p)(5m - 4p)$

74. $y^4 - 625$
$= (y^2)^2 - 25^2$
$= (y^2 + 25)(y^2 - 25)$
   Difference of two squares
$= (y^2 + 25)(y + 5)(y - 5)$
   Difference of two squares

75. $6m^3 - 21m^2 - 45m$
$= 3m(2m^2 - 7m - 15)$
   Greatest common factor
$= 3m(2m^2 - 10m + 3m - 15)$
$= 3m[2m(m - 5) + 3(m - 5)]$
   Factor by grouping
$= 3m(m - 5)(2m + 3)$

76. $24ab^3c^2 - 56a^2bc^3 + 72a^2b^2c$
   Greatest common factor
$= 8abc(3b^2c - 7ac^2 + 9ab)$

77. $25a^2 + 15ab + 9b^2$ is a prime polynomial.

78. $12x^2yz^3 + 12xy^2z - 30x^3y^2z^4$
$= 6xyz(2xz^2 + 2y - 5x^2yz^3)$
   Greatest common factor

79. $2a^5 - 8a^4 - 24a^3$
$= 2a^3(a^2 - 4a - 12)$
   Greatest common factor
$= 2a^3(a - 6)(a + 2)$

80. $12r^2 + 18rq - 10rq - 15q^2$
$= 6r(2r + 3q) - 5q(2r + 3q)$
   Factor by grouping
$= (2r + 3q)(6r - 5q)$

81. $100a^2 - 9 = (10a)^2 - 3^2$
$= (10a + 3)(10a - 3)$
   Difference of two squares

82. $49t^2 + 56t + 16$
$= (7t)^2 + 2(7t)(4) + 4^2$
$= (7t + 4)^2$

83. $t(t - 7) = 0$
$t = 0$ or $t - 7 = 0$
$t = 0$ or $t = 7$

The solutions are 0 and 7.

84. $x(x + 3) = 0$
$x^2 + 3x = 10$
$x^2 + 3x - 10 = 0$
$(x + 5)(x - 2) = 0$
$x + 5 = 0$ or $x - 2 = 0$
$x = -5$ or $x = 2$

The solutions are -5 and 2.

85. $25x^2 + 20x + 4 = 0$
    $(5x + 2)^2 = 0$
    $5x + 2 = 0$
    $5x = -2$
    $x = -\dfrac{2}{5}$

   The only solution is -2/5.

86. Let x = the length of the shorter leg;
    $2x + 6$ = the length of the longer leg;
    $(2x + 6) + 3$ = the length of the hypotenuse.

   Use the Pythagorean formula
   $a^2 + b^2 = c^2$.

   Substitute $a = x$, $b = 2x + 6$, and $c = (2x + 6) + 3$.

   $x^2 + (2x + 6)^2 = [(2x + 6) + 3]^2$
   $x^2 + (2x + 6)^2 = (2x + 9)^2$
   $x^2 + 4x^2 + 24x + 36 = 4x^2 + 36x + 81$
   $5x^2 + 24x + 36 = 4x^2 + 36x + 81$
   $x^2 - 12x - 45 = 0$
   $(x - 15)(x + 3) = 0$
   $x - 15 = 0$ or $x + 3 = 0$
   $x = 15$ or $x = -3$

   Reject -3 because a length cannot be negative. The sides of the lot are 15 meters, 2(15) + 6 = 36 meters, and 36 + 3 = 39 meters.

87. Let x = the width of the base.
    Then $x + 2$ = the length.

   The area of the base, B, is given by
   $A = LW$
   $B = x(x + 2)$.

   Use the formula for the volume of a pyramid, $V = \dfrac{1}{3}Bh$.

   Substitute $V = 48$, $B = x(x + 2)$, and $h = 6$.

   $V = \dfrac{1}{3} \cdot B \cdot h$
   $48 = \dfrac{1}{3}x(x + 2)(6)$
   $48 = 2x(x + 2)$
   $48 = 2x^2 + 4x$
   $2x^2 + 4x - 48 = 0$
   $2(x^2 + 2x - 24)$
   $(x + 6)(x - 4) = 0$
   $x + 6 = 0$ or $x - 4 = 0$
   $x = -6$ or $x = 4$

   Reject -6. The width of the base is 4 meters and the length is 4 + 2 or 6 meters.

88. Let x = the smallest integer;
    $x + 1$ = the next smallest integer;
    $x + 2$ = the largest integer.

   $x(x + 1) = 23 + (x + 2)$
   $x^2 + x = 23 + x + 2$
   $x^2 + x = 25 + x$
   $x^2 - 25 = 0$
   $(x + 5)(x - 5) = 0$
   $x + 5 = 0$ or $x - 5 = 0$
   $x = -5$ or $x = 5$

   If $x = -5$, then $x + 1 = -4$ and $x + 2 = -3$.
   If $x = 5$, then $x + 1 = 6$ and $x + 2 = 7$.
   The integers are -5, -4, and -3, or 5, 6, and 7.

89. Let x = the first even integer;
x + 2 = the next even integer.

$$x + (x + 2) = x(x + 2) - 34$$
$$x + x + 2 = x^2 + 2x - 34$$
$$2x + 2 = x^2 + 2x - 34$$
$$0 = x^2 - 36$$
$$0 = (x + 6)(x - 6)$$
$$x + 6 = 0 \text{ or } x - 6 = 0$$
$$x = -6 \text{ or } x = 6$$

If x = -6, then x + 2 = -4.
If x = 6, then x + 2 = 8.
The integers are -6 and -4, or 6 and 8.

90. Let x = the width of the house;
x + 7 = the length of the house.

Use A = LW. Substitute 170 for A, x + 7 for L, and x for W.

$$170 = (x + 7)(x)$$
$$170 = x^2 + 7x$$
$$0 = x^2 + 7x - 170$$
$$0 = (x + 17)(x - 10)$$
$$x + 17 = 0 \text{ or } x - 10 = 0$$
$$x = -17 \text{ or } x = 10$$

Discard -17 because the width cannot be negative.

W = x = 10
L = x + 7 = 10 + 7 = 17

The width is 10 meters and the length is 17 meters.

91. Let b = the base of the sail;
b + 4 = the height of the sail.

Use the formula for the area of a triangle.

$$A = \tfrac{1}{2}bh$$
$$30 = \tfrac{1}{2}(b)(b + 4) \quad \text{Let } A = 30,\ h = b + 4$$
$$60 = b(b + 4) \quad \text{Multiply by 2}$$
$$60 = b^2 + 4b \quad \text{Distributive property}$$
$$0 = b^2 + 4b - 60 \quad \text{Standard form}$$
$$0 = (b + 10)(b - 6) \quad \text{Factor}$$

Discard -10 since the base of a triangle cannot be negative. The base of the triangular sail is 6 meters.

Chapter 4 Test

1. $2x^2 - 2x - 24 = 2(x^2 - x - 12)$
$= 2(x + 3)(x - 4)$

The correct completely factored form is choice (d). Note that the factored forms (a) $(2x + 6)(x - 4)$ and (b) $(x + 3)(2x - 8)$ also can be multiplied to give a product of $2x^2 - 2x - 24$, but neither of these is completely factored because $2x + 6$ and $2x - 8$ both contain a common factor of 2.

2. $12x^2 - 30x = 6x(2x - 5)$

3. $2m^3n^2 + 3m^3n - 5m^2n^2$
$= m^2n(2mn + 3m - 5n)$

4. $x^2 - 5x - 24$

Find two integers whose product is -24 and whose sum is -5. The integers are 3 and -8.

$x^2 - 5x - 24 = (x + 3)(x - 8)$

5. $x^2 - 9x + 14$

   Find two integers whose product is 14 and whose sum is -9. The integers are -7 and -2.

   $x^2 - 9x + 14 = (x - 7)(x - 2)$

6. $2x^2 + x - 3$

   Factor by trial and error.

   $2x^2 + x - 3 = (2x + 3)(x - 1)$

7. $6x^2 - 19x - 7$

   To factor by grouping, find two integers whose product is $6(-7) = -42$ and whose sum is -19. The integers are -21 and 2.

   $6x^2 - 19x - 7 = 6x^2 - 21x + 2x - 7$
   $= 3x(2x - 7) + 1(2x - 7)$
   $= (2x - 7)(3x + 1)$

8. $3x^2 - 12x - 15 = 3(x^2 - 4x - 5)$
   $= 3(x + 1)(x - 5)$

9. $10z^2 - 17z + 3$
   $= 10z^2 - 15z - 2z + 3$
   $= 5z(2z - 3) - 1(2z - 3)$
   $= (2z - 3)(5z - 1)$

10. $t^2 + 2t + 3$

    We cannot find two integers whose product is 3 and whose sum is 2. This polynomial is prime.

11. $x^2 + 36$

    This polynomial is prime because the sum of two squares cannot be factored and the two terms have no common factor.

12. $y^2 - 49 = y^2 - 7^2 = (y + 7)(y - 7)$

13. $9y^2 - 64 = (3y)^2 - 8^2$
    $= (3y + 8)(3y - 8)$

14. $x^2 + 16x + 64 = x^2 + 2(8)(x) + 8^2$
    $= (x + 8)^2$

15. $4x^2 - 28xy + 49y^2$
    $= (2x)^2 - 2(2x)(7y) + (7y)^2$
    $= (2x - 7y)^2$

16. $-2x^2 - 4x - 2$
    $= -2(x^2 + 2x + 1)$
    $= -2(x^2 + 2 \cdot x \cdot 1 + 1^2)$
    $= -2(x + 1)^2$

17. $6t^4 + 3t^3 - 108t^2$
    $= 3t^2(2t^2 + t - 36)$
    $= 3t^2(2t + 9)(t - 4)$

18. $4r^2 + 10rt + 25t^2$

    Note that $(2r + 5t)^2 = 4r^2 + 20rt + 25t^2$, not $4r^2 + 10rt + 25t^2$, so the given polynomial is not a perfect square trinomial. This polynomial is prime because there are no two integers whose product is $(4)(25) = 100$ and whose sum is 10.

19. $4t^3 + 32t^2 + 64t = 4t(t^2 + 8t + 16)$
$ = 4t(t + 4)^2$

20. $x^4 - 81 = (x^2)^2 - 9^2$
$ = (x^2 + 9)(x^2 - 9)$
$ = (x^2 + 9)(x^2 - 3^2)$
$ = (x^2 + 9)(x + 3)(x - 3)$

21. $(p + 3)(p + 3)$ is not the correct factorization of $p^2 + 9$ because

$(p + 3)(p + 3) = p^2 + 6p + 9$
$ \neq p^2 + 9.$

22. $(x + 3)(x - 9) = 0$
$x + 3 = 0$ or $x - 9 = 0$
$x = -3$ or $x = 9$

The solutions are -3 and 9.

23. $2r^2 - 13r + 6 = 0$
$(2r - 1)(r - 6) = 0$
$2r - 1 = 0$ or $r - 6 = 0$
$2r = 1$
$r = \frac{1}{2}$ or $r = 6$

The solutions are $1/2$ and 6.

24. $25x^2 - 4 = 0$
$(5x + 2)(5x - 2) = 0$
$5x + 2 = 0$ or $5x - 2 = 0$
$5x = -2$ or $5x = 2$
$x = -\frac{2}{5}$ or $x = \frac{2}{5}$

The solutions are $-2/5$ and $2/5$.

25. $x(x - 20) = -100$
$x^2 - 20x = -100$
$x^2 - 20x + 100 = 0$
$(x - 10)^2 = 0$
$x - 10 = 0$
$x = 10$

The only solution is 10.

26. $t^2 = 3t$
$t^2 - 3t = 0$
$t(t - 3) = 0$
$t = 0$ or $t - 3 = 0$
$t = 0$ or $t = 3$

The solutions are 0 and 3.

27. $x^2 = \frac{4}{9}$

$x^2 - \frac{4}{9} = 0$

$\left(x + \frac{2}{3}\right)\left(x - \frac{2}{3}\right) = 0$

$x + \frac{2}{3} = 0$ or $x - \frac{2}{3} = 0$

$x = -\frac{2}{3}$ or $x = \frac{2}{3}$

$x = 2/3$ is not the correct response because $-2/3$ is also a solution of the given equation.

28. Let x = the width of the flower bed;

$2x - 3$ = the length of the flower bed.

Use the formula $A = LW$.

$x(2x - 3) = 54$
$2x^2 - 3x = 54$
$2x^2 - 3x - 54 = 0$
$(2x + 9)(x - 6) = 0$

2x + 9 = 0   or   x − 6 = 0
    2x = −9
    $x = -\frac{9}{2}$ or   x = 6

Reject −9/2.
If x = 6, 2x − 3 = 2(6) − 3 = 9.
The dimensions of the flower bed are 6 feet by 9 feet.

29. Let x = the length of the stud;
    3x − 7 = the length of the brace.

    The figure shows that a right triangle is formed with the brace as the hypotenuse. Use the Pythagorean formula.

    $x^2 + 15^2 = (3x - 7)^2$
    $x^2 + 225 = 9x^2 - 42x + 49$
    $0 = 8x^2 - 42x - 176$
    $0 = 2(4x^2 - 21x - 88)$
    $0 = 2(4x + 11)(x - 8)$
    4x + 11 = 0   or   x − 8 = 0
        4x = −11
        $x = -\frac{11}{4}$ or   x = 8

Reject −11/4.
If x = 8, 3x − 7 = 17, so the brace should be 17 feet long.

30. Let x = the first integer;
    x + 1 = the second integer.

    $[x + (x + 1)]^2 = x + 11$
    $(2x + 1)^2 = x + 11$
    $4x^2 + 4x + 1 = x + 11$
    $4x^2 + 3x - 10 = 0$
    $(4x - 5)(x + 2) = 0$

4x − 5 = 0   or   x + 2 = 0
    4x = 5
    $x = \frac{5}{4}$ or   x = −2

Reject 5/4 because it is not an integer.
If x = −2, x + 1 = −1. The integers are −2 and −1.

Cumulative Review: Chapters R–4

1. 3x + 2(x − 4) = 4(x − 2)
   3x + 2x − 8 = 4x − 8
   5x − 8 = 4x − 8
   x − 8 = −8
   x = 0

2. .3x + .9x = .06

   Multiply both sides by 100 to clear decimals.

   100(.3x + .9x) = 100(.06)
   30x + 90x = 6
   120x = 6
   $x = \frac{6}{120} = \frac{1}{20} = .05$

3. $\frac{2}{3}y - \frac{1}{2}(y - 4) = 3$

   To clear fractions, multiply both sides by the least common denominator, which is 6.

$$6\left[\frac{2}{3}y - \frac{1}{2}(y - 4)\right] = 6(3)$$
$$4y - 3(y - 4) = 18$$
$$4y - 3y + 12 = 18$$
$$y + 12 = 18$$
$$y = 6$$

4. $$A = P + Prt$$
$$A = P(1 + rt)$$
$$\frac{P}{1 + rt} = \frac{P(1 + rt)}{1 + rt}$$
$$\frac{A}{1 + rt} = P$$

5. Let x = the unknown number.
$$4x + (x + 12) = 3$$
$$4x + x + 12 = 3$$
$$5x + 12 = 3$$
$$5x = -9$$
$$x = -\frac{9}{5}$$

The number is $-9/5$.

6. Let x = the number of tranquilizer prescriptions;
$\frac{2}{3}x$ = the number of antibiotic prescriptions.
$$x + \frac{2}{3}x = 90$$
$$3\left(x + \frac{2}{3}x\right) = 90 \cdot 3$$
$$3x + 2x = 270$$
$$5x = 270$$
$$x = 54$$

There were 54 tranquilizer prescriptions and 36 antibiotics prescriptions.

7. Let x = Louise's salary before the deduction.
$$x - .06x = 423$$
$$.94x = 423$$
$$x = \frac{423}{.94} = 450$$

Louise earns $450 per week.

8. Let x = number of pounds of gravel;
$3x$ = number of pounds of cement.
$$x + 3x = 140$$
$$4x = 140$$
$$x = 35$$

There are 35 pounds of gravel.

9. $2^{-3} \cdot 2^5 = 2^{-3+5} = 2^2 = 4$

10. $\left(\frac{3}{4}\right)^{-2} = \left(\frac{4}{3}\right)^2 = \frac{16}{9}$

11. $$\frac{6^5 \cdot 6^{-2}}{6^3} = \frac{6^{5+(-2)}}{6^3}$$
$$= \frac{6^3}{6^3} = 1$$

12. $$\left(\frac{4^{-3} \cdot 4^4}{4^5}\right)^{-1} = \frac{4^5}{4^{-3} \cdot 4^4}$$
$$= \frac{4^5}{4^1} = 4^4 = 256$$

13. $$\frac{(p^2)^3 p^{-4}}{(p^{-3})^{-1}p} = \frac{p^{2 \cdot 3} p^{-4}}{p^{(-3)(-1)}p}$$
$$= \frac{p^6 p^{-4}}{p^3 p^1}$$
$$= \frac{p^{6-4}}{p^{3+1}} = \frac{p^2}{p^4}$$
$$= \frac{1}{p^2}$$

14. $\dfrac{(m^{-2})^3 m}{m^5 m^{-4}} = \dfrac{m^{-2 \cdot 3} m}{m^5 m^{-4}}$

$= \dfrac{m^{-6} m^1}{m^5 m^{-4}} = \dfrac{m^{-6+1}}{m^{5-4}}$

$= \dfrac{m^{-5}}{m^1} = m^{-5-1}$

$= \dfrac{1}{m^6}$

15. $(2k^2 + 4k) - (5k^2 - 2) - (k^2 + 8k - 6)$
$= (2k^2 + 4k) + (-5k^2 - 2)$
$\quad + (-k^2 - 8k + 6)$
$= 2k^2 + 4k - 5k^2 + 2 - k^2 - 8k + 6$
$= -4k^2 - 4k + 8$

16. $3m^3(2m^5 - 5m^3 + m)$
$= 3m^3 \cdot 2m^5 - 3m^3 \cdot 5m^3 + 3m^3 \cdot m$
$= 6m^8 - 15m^6 + 3m^4$

17. $(y^2 + 3y + 5)(3y - 1)$

Multiply vertically.

$$\begin{array}{r} y^2 + 3y + 5 \\ 3y - 1 \\ \hline -y^2 - 3y - 5 \\ 3y^3 + 9y^2 + 15y \\ \hline 3y^3 + 8y^2 + 12y - 5 \end{array}$$

18. $(3p + 2)^2 = (3p)^2 + 2(3p)(2) + (2)^2$
$ = 9p^2 + 12p + 4$

19. $(2p + 3q)(2p - 3q) = (2p)^2 - (3q)^2$
$ = 4p^2 - 9q^2$

20. $(9x + 6)(5x - 3)$
$= 45x^2 - 27x + 30x - 18$
$= 45x^2 + 3x - 18$

21. $\dfrac{8x^4 + 12x^3 - 6x^2 + 20x}{2x}$

$= \dfrac{8x^4}{2x} + \dfrac{12x^3}{2x} - \dfrac{6x^2}{2x} + \dfrac{20x}{2x}$

$= 4x^3 + 6x^2 - 3x + 10$

22. $(12p^3 + 2p^2 - 12p + 4) \div (2p - 2)$

$$\begin{array}{r} 6p^2 + 7p + 1 \\ 2p - 2 \overline{\smash{\big)}\, 12p^3 + 2p^2 - 12p + 4} \\ \underline{12p^3 - 12p^2} \\ 14p^2 - 12p \\ \underline{14p^2 - 14p} \\ 2p + 4 \\ \underline{2p - 2} \\ 6 \end{array}$$

The remainder is 6. The answer is

$6p^2 + 7p + 1 + \dfrac{6}{2p - 2}.$

23. $2a^2 + 7a - 4$

Factor by trial and error.

$2a^2 + 7a - 4 = (a + 4)(2a - 1)$

24. $10m^2 + 19m + 6$

To factor by grouping, find two integers whose product is $10(6) = 60$ and whose sum is 19. The integers are 15 and 4.

$10m^2 + 19m + 6$
$= 10m^2 + 15m + 4m + 6$
$= 5m(2m + 3) + 2(2m + 3)$
$= (2m + 3)(5m + 2)$

25. $15x^2 - xy - 6y^2$
$= 15x^2 - 10xy + 9xy - 6y^2$
$= 5x(3x - 2y) + 3y(3x - 2y)$
$= (3x - 2y)(5x + 3y).$

26. $8t^2 + 10tv + 3v^2$
    $= 8t^2 + 6tv + 4tv + 3v^2$
    $= 2t(4t + 3v) + v(4t + 3)$
    $= (4t + 3v)(2t + v)$

27. $9x^2 + 6x + 1 = (3x)^2 + 2(3x)(1) + 1^2$
    $= (3x + 1)^2$

28. $4p^2 - 12p + 9$
    $= (2p)^2 - 2(2p)(3) + 3^2$
    $= (2p - 3)^2$

29. $-32t^2 - 112tz - 98z^2$
    $= -2(16t^2 + 56tz + 49z^2)$
    $= -2[(4t)^2 + 2(4t)(7z) + (7z)^2]$
    $= -2(4t + 7z)^2$

30. $25r^2 - 81t^2 = (5r)^2 - (9t)^2$
    $= (5r + 9t)(5r - 9t)$

31. $100x^2 + 25 = 25(4x^2 + 1)$

32. $6a^2m + am - 2m = m(6a^2 + a - 2)$
    $= m(2a - 1)(3a + 2)$

33. $2pq + 6p^3q + 8p^2q$
    $= 2pq(1 + 3p^2 + 4p)$
    $= 2pq(3p^2 + 4p + 1)$
    $= 2pq(3p + 1)(p + 1)$

34. $2ax - 2bx + ay - by$
    $= (2ax - 2bx) + (ay - by)$
        Group like terms
    $= 2x(a - b) + y(a - b)$
        Factor each group
    $= (a - b)(2x + y)$ *Factor out $a - b$*

35. $(2p - 3)(p + 2)(p - 6) = 0$

    Set each factor equal to 0 and solve the resulting linear equations.

    $2p - 3 = 0$ or $p + 2 = 0$ or $p - 6 = 0$
    $2p = 3$
    $p = \dfrac{3}{2}$ or $\quad p = -2$ or $\quad p = 6$

    The solutions are 3/2, -2, and 6.

36. $\quad 6m^2 + m - 2 = 0$
    $(3m + 2)(2m - 1) = 0$
    $3m + 2 = 0$ or $2m - 1 = 0$
    $\quad 3m = -2 \quad\quad\quad 2m = 1$
    $\quad m = -\dfrac{2}{3}$ or $\quad m = \dfrac{1}{2}$

    The solutions are -2/3 and 1/2.

37. $\quad 8x^2 = 64x$
    $8x^2 - 64x = 0$
    $8x(x - 8) = 0$
    $x = 0$ or $x - 8 = 0$
    $x = 0$ or $\quad x = 8$

    The solutions are 0 and 8.

38. Let x = the width of the rectangle.
    $2x - 1$ = the length of the rectangle.
    $x(2x - 1) = 15$
    $2x^2 - x = 15$
    $2x^2 - x - 15 = 0$
    $(2x + 5)(x - 3) = 0$
    $2x + 5 = 0$ or $x - 3 = 0$
    $2x = -5$
    $x = -\dfrac{5}{3}$ or $\quad x = 3$

Reject −5/2 because the width cannot be negative. The width is 3 centimeters.

39. Let x = the first even integer;
x + 2 = the next consecutive even integer.

$$x^2 + (x + 2)^2 = 5(x + 2)$$
$$x^2 + x^2 + 4x + 4 = 5x + 10$$
$$2x^2 + 4x + 4 = 5x + 10$$
$$2x^2 - x - 6 = 0$$
$$(2x + 3)(x - 2) = 0$$
$$2x + 3 = 0 \quad \text{or} \quad x - 2 = 0$$
$$2x = -3$$
$$x = -\frac{3}{2} \quad \text{or} \quad x = 2$$

Reject −3/2 because it is not an integer. If x = 2, x + 2 = 4. The integers are 2 and 4.

40. Let x = the length of the shorter leg;
x + 7 = the length of the longer leg;
2x + 3 = the length of the hypotenuse.

$$x^2 + (x + 7)^2 = (2x + 3)^2$$
$$x^2 + x^2 + 14x + 49 = 4x^2 + 12x + 9$$
$$2x^2 + 14x + 49 = 4x^2 + 12x + 9$$
$$0 = 2x^2 - 2x - 40$$
$$0 = 2(x^2 - x - 20)$$
$$0 = (x - 5)(x + 4)$$
$$x - 5 = 0 \quad \text{or} \quad x + 4 = 0$$
$$x = 5 \quad \text{or} \quad x = -4$$

Reject −4 because the length of a leg cannot be negative.

If x = 5, x + 7 = 12, and 2x + 3 = 2(5) + 3 = 15.

The length of the sides are 5 meters, 12 meters, and 13 meters.

CHAPTER 5 RATIONAL EXPRESSIONS

Section 5.1 The Fundamental Property of Rational Expressions

5.1 Margin Exercises

1. (a) $\dfrac{x + 2}{x - 5}$

   Solve $x - 5 = 0$.
   Since $x = 5$ will make the denominator zero, the expression is undefined for 5.

   (b) $\dfrac{3r}{r^2 + 6r + 8}$

   Solve $r^2 + 6r + 8 = 0$.
   $(r + 4)(r + 2) = 0$, so the denominator is zero when either $r + 4 = 0$ or $r + 2 = 0$, that is, when $r = -4$ or $r = -2$. The expression is undefined for -4 and -2.

   (c) $\dfrac{-5m}{m^2 + 4}$

   Since $m^2 + 4$ cannot equal 0, there are no values of the variable that make the expression undefined.

2. (a) $\dfrac{x}{2x + 1} = \dfrac{3}{2(3) + 1}$ Let $x = 3$

   $= \dfrac{3}{6 + 1}$

   $= \dfrac{3}{7}$

   (b) $\dfrac{2x + 6}{x - 3} = \dfrac{2(3) + 6}{3 - 3}$ Let $x = 3$

   $= \dfrac{12}{0}$

   If we substitute 3 for x, the denominator is zero, so the quotient is undefined when $x = 3$.

3. (a) $\dfrac{5x^4}{15x^2} = \dfrac{(5x^2)(x^2)}{(5x^2)(3)}$ Factor

   $= \dfrac{x^2}{3}$ Fundamental property

   (b) $\dfrac{6p^3}{2p^2} = \dfrac{(2p^2)(3p)}{(2p^2)(1)}$ Factor

   $= \dfrac{3p}{1}$ Fundamental property

   $= 3p$

4. (a) $\dfrac{4y + 2}{6y + 3} = \dfrac{2(2y + 1)}{3(2y + 1)}$ Factor

   $= \dfrac{2}{3}$ Fundamental property

   (b) $\dfrac{8p + 8q}{5p + 5q} = \dfrac{8(p + q)}{5(p + q)}$ Factor

   $= \dfrac{8}{5}$ Fundamental property

5. (a) $\dfrac{x^2 + 4x + 4}{4x + 8}$

   $= \dfrac{(x + 2)(x + 2)}{4(x + 2)}$ Factor

   $= \dfrac{x + 2}{4}$ Fundamental property

   (b) $\dfrac{a^2 - b^2}{a^2 + 2ab + b^2}$

   $= \dfrac{(a - b)(a + b)}{(a + b)(a + b)}$ Factor

   $= \dfrac{a - b}{a + b}$ Fundamental property

6. (a) $\dfrac{5 - y}{y - 5}$

   Since $y - 5 = -1(-y + 5)$
   $= -1(5 - y)$,

   $5 - y$ and $y - 5$ are negatives (or opposites) of each other. Therefore,

   $\dfrac{5 - y}{y - 5} = -1.$

Chapter 5 Rational Expressions

(b) $\dfrac{m-n}{n-m}$

$$n - m = -1(-n + m)$$
$$= -1(m - n)$$

$m - n$ and $n - m$ are negatives of each other. Therefore,

$$\dfrac{m-n}{n-m} = -1.$$

(c) $\dfrac{9-k}{9+k}$

$$9 - k = -1(-9 + k) \ne -1(9 + k)$$

The expressions $9 - k$ and $9 + k$ are not negatives of each other. They do not have any common factors (other than 1), so the rational expression cannot be written in simpler form.

5.1 Section Exercises

3. $\dfrac{4x^2}{3x-5}$

To find the values for which this expression is undefined, set the denominator equal to zero and solve for x.

$$3x - 5 = 0$$
$$3x = 5$$
$$x = \dfrac{5}{3}$$

Because $x = 5/3$ will make the denominator zero, the given expression is undefined by $5/3$.

7. $\dfrac{3x}{x^2+2}$

This denominator cannot equal zero for any value of x because x^2 is always greater than or equal to zero, and adding 2 makes the sum greater than zero. Thus, the given rational expression is never undefined.

11. (a) $\dfrac{2x^2 - 4x}{3x} = \dfrac{2(2)^2 - 4(2)}{3(2)}$

$$\text{Let } x = 2$$

$$= \dfrac{2(4) - 4(2)}{3(2)}$$

$$= \dfrac{8 - 8}{6}$$

$$= \dfrac{0}{6} = 0$$

(b) $\dfrac{2x^2 - 4x}{3x} = \dfrac{2(-3)^2 - 4(-3)}{3(-3)}$

$$\text{Let } x = -3$$

$$= \dfrac{2(9) - (-12)}{-9}$$

$$= \dfrac{18 - (-12)}{-9}$$

$$= \dfrac{30}{-9} = -\dfrac{10}{3}$$

15. (a) $\dfrac{5x+2}{2x^2 + 11x + 12}$

$$= \dfrac{5(2) + 2}{2(2)^2 + 11(2) + 12}$$

$$\text{Let } x = 2$$

$$= \dfrac{10 + 2}{2(4) + 11(2) + 12}$$

$$= \dfrac{12}{8 + 22 + 12}$$

$$= \dfrac{12}{42} = \dfrac{2}{7}$$

(b) $\dfrac{5x + 2}{2x^2 + 11x + 12}$

$= \dfrac{5(-3) + 2}{2(-3)^2 + 11(-3) + 12}$ Let $x = -3$

$= \dfrac{-15 + 2}{2(9) + 11(-3) + 12}$

$= \dfrac{-13}{18 - 33 + 12}$

$= \dfrac{-13}{-3} = \dfrac{13}{3}$

19. $\dfrac{18r^3}{6r} = \dfrac{3r^2(6r)}{1(6r)}$ Factor

$= 3r^2$ Fundamental property

23. $\dfrac{(x + 1)(x - 1)}{(x + 1)^2} = \dfrac{(x + 1)(x - 1)}{(x + 1)(x + 1)}$

$= \dfrac{x - 1}{x + 1}$

27. $\dfrac{m^2 - n^2}{m + n} = \dfrac{(m + n)(m - n)}{m + n}$

$= m - n$

31. $\dfrac{3m^2 - 3m}{5m - 5} = \dfrac{3m(m - 1)}{5(m - 1)}$

$= \dfrac{3m}{5}$

35. $\dfrac{zw + 4z - 3w - 12}{zw + 4z + 5w + 20}$

$= \dfrac{z(w + 4) - 3(w + 4)}{z(w + 4) + 5(w + 4)}$

$= \dfrac{(w + 4)(z - 3)}{(w + 4)(z + 5)}$

$= \dfrac{z - 3}{z + 5}$

39. $\dfrac{6 - t}{t - 6} = \dfrac{-1(t - 6)}{1(t - 6)}$

$= \dfrac{-1}{1} = -1$

Note that $6 - t$ and $t - 6$ are negatives (or opposites), so we know that their quotient will be -1.

43. $\dfrac{q^2 - 4q}{4q - q^2} = \dfrac{q(q - 4)}{q(4 - q)}$

$= \dfrac{q - 4}{4 - q} = -1$

47. $\dfrac{2}{3} \cdot \dfrac{5}{6} = \dfrac{2 \cdot 5}{3 \cdot 6}$

$= \dfrac{10}{18} = \dfrac{5 \cdot 2}{9 \cdot 2}$

$= \dfrac{5}{9}$

51. $\dfrac{10}{3} \div \dfrac{5}{6} = \dfrac{10}{3} \cdot \dfrac{6}{5}$

$= \dfrac{10 \cdot 6}{3 \cdot 5}$

$= \dfrac{60}{15}$

$= \dfrac{4 \cdot 15}{1 \cdot 15} = 4$

Section 5.2 Multiplication and Division of Rational Expressions

5.2 Margin Exercises

1. (a) $\dfrac{3m^2}{2} \cdot \dfrac{10}{m} = \dfrac{3m^2 \cdot 10}{2 \cdot m}$

$= \dfrac{3 \cdot m \cdot m \cdot 2 \cdot 5}{m \cdot 2}$

$= 15m$

(b) $\dfrac{8p^2q}{3} \cdot \dfrac{9}{q^2p} = \dfrac{2 \cdot 4 \cdot p \cdot p \cdot q \cdot 3 \cdot 3}{3 \cdot q \cdot q \cdot p}$

$= \dfrac{24p}{q}$

2. (a) $\dfrac{a + b}{5} \cdot \dfrac{30}{2(a + b)} = \dfrac{30(a + b)}{10(a + b)} = 3$

(b) $\dfrac{3(p - q)}{p} \cdot \dfrac{q}{2(p - q)} = \dfrac{3q(p - q)}{2p(p - q)}$

$= \dfrac{3q}{2p}$

3. (a) $\dfrac{x^2 + 7x + 10}{3x + 6} \cdot \dfrac{6x - 6}{x^2 + 2x - 15}$

$= \dfrac{(x + 2)(x + 5)}{3(x + 2)} \cdot \dfrac{6(x - 1)}{(x + 5)(x - 3)}$ Factor

$= \dfrac{6(x + 2)(x + 5)(x - 1)}{3(x + 2)(x + 5)(x - 3)}$ Multiply

$= \dfrac{2(x - 1)}{x - 3}$ Lowest terms

(b) $\dfrac{m^2 + 4m - 5}{m + 5} \cdot \dfrac{m^2 + 8m + 15}{m - 1}$

$= \dfrac{(m - 1)(m + 5)}{m + 5} \cdot \dfrac{(m + 3)(m + 5)}{m - 1}$ Factor

$= \dfrac{(m - 1)(m + 5)(m + 3)(m + 5)}{(m + 5)(m - 1)}$ Multiply

$= (m + 3)(m + 5)$ Lowest terms

4. (a) The reciprocal of a rational expression is found by inverting the fraction, so the reciprocal of

$\dfrac{6b^5}{3r^2b}$

is

$\dfrac{3r^2b}{6b^5}$.

(b) The reciprocal of

$\dfrac{t^2 - 4t}{t^2 + 2t - 3}$

is

$\dfrac{t^2 + 2t - 3}{t^2 - 4t}$.

5. (a) $\dfrac{r}{r - 1} \div \dfrac{3r}{r + 4}$

$= \dfrac{r}{r - 1} \cdot \dfrac{r + 4}{3r}$ Multiply first expression by reciprocal of second

$= \dfrac{r(r + 4)}{3r(r - 1)}$ Multiply

$= \dfrac{r + 4}{3(r - 1)}$ Lowest terms

(b) $\dfrac{6x - 4}{3} \div \dfrac{15x - 10}{9}$

$= \dfrac{6x - 4}{3} \cdot \dfrac{9}{15x - 10}$ Multiply by reciprocal

$= \dfrac{2(3x - 2)}{3} \cdot \dfrac{9}{5(3x - 2)}$ Factor

$= \dfrac{18(3x - 2)}{15(3x - 2)}$ Multiply

$= \dfrac{6}{5}$ Lowest terms

6. (a) $\dfrac{5a^2b}{2} \div \dfrac{10ab^2}{8} = \dfrac{5a^2b}{2} \cdot \dfrac{8}{10ab^2}$ Multiply by reciprocal

$= \dfrac{40a^2b}{20ab^2}$ Multiply

$= \dfrac{2a}{b}$ Lowest terms

(b) $\dfrac{(3t)^2}{w} \div \dfrac{3t^2}{5w^4} = \dfrac{9t^2}{w} \cdot \dfrac{5w^4}{3t^2}$ Multiply by reciprocal

$= \dfrac{45t^2w^4}{3t^2w}$ Multiply

$= 15w^3$ Lowest terms

7. (a) $\dfrac{y^2 + 4y + 3}{y + 3} \div \dfrac{y^2 - 4y - 5}{y - 3}$

$= \dfrac{y^2 + 4y + 3}{y + 3} \cdot \dfrac{y - 3}{y^2 - 4y - 5}$ Multiply by reciprocal

$= \dfrac{(y + 3)(y + 1)}{y + 3} \cdot \dfrac{y - 3}{(y - 5)(y + 1)}$ Factor

$= \dfrac{(y + 3)(y + 1)(y - 3)}{(y + 3)(y - 5)(y + 1)}$ Multiply

$= \dfrac{y - 3}{y - 5}$ Lowest terms

(b) $\dfrac{4x(x+3)}{2x+1} \div \dfrac{-x^2(x+3)}{4x^2-1}$

$= \dfrac{4x(x+3)}{2x+1} \cdot \dfrac{4x^2-1}{-x^2(x+3)}$
 Multiply by reciprocal

$= \dfrac{4x(x+3)}{2x+1} \cdot \dfrac{(2x+1)(2x-1)}{-x^2(x+3)}$
 Factor

$= \dfrac{4x(x+3)(2x+1)(2x-1)}{-x^2(2x+1)(x+3)}$
 Multiply

$= -\dfrac{4(2x-1)}{x}$
 Lowest terms

8. (a) $\dfrac{ab-a^2}{a^2-1} \div \dfrac{b-a}{a^2+2a+1}$

$= \dfrac{ab-a^2}{a^2-1} \cdot \dfrac{a^2+2a+1}{b-a}$
 Multiply by reciprocal

$= \dfrac{a(b-a)}{(a+1)(a-1)} \cdot \dfrac{(a+1)(a+1)}{b-a}$
 Factor

$= \dfrac{a(b-a)(a+1)(a+1)}{(a+1)(a-1)(b-a)}$
 Multiply

$= \dfrac{a(a+1)}{a-1}$
 Lowest terms

(b) $\dfrac{x^2-y^2}{x^2-1} \div \dfrac{x^2+2xy+y^2}{x^2+x}$

$= \dfrac{x^2-y^2}{x^2-1} \cdot \dfrac{x^2+x}{x^2+2xy+y^2}$
 Multiply by reciprocal

$= \dfrac{(x-y)(x+y)}{(x-1)(x+1)} \cdot \dfrac{x(x+1)}{(x+y)(x+y)}$
 Factor

$= \dfrac{x(x-y)(x+y)(x+1)}{(x-1)(x+1)(x+y)(x+y)}$
 Multiply

$= \dfrac{x(x-y)}{(x-1)(x+y)}$
 Lowest terms

5.2 Section Exercises

3. $\dfrac{12x^4}{18x^3} \cdot \dfrac{-8x^5}{4x^2} = \dfrac{-96x^9}{72x^5}$
 Multiply numerators; multiply denominators

$= \dfrac{-4x^4(24x^5)}{3(24x^5)}$
 Group common factors

$= -\dfrac{4x^4}{3}$
 Lowest terms

7. The reciprocal of

$\dfrac{3p^3}{16q}$

is

$\dfrac{16q}{3p^3}.$

11. The reciprocal of

$\dfrac{z^2+7z+12}{z^2-9}$

is

$\dfrac{z^2-9}{z^2+7z+12}.$

15. $\dfrac{4t^4}{2t^5} \div \dfrac{(2t)^3}{-6} = \dfrac{4t^4}{2t^5} \cdot \dfrac{-6}{(2t)^3}$

$= \dfrac{4t^4}{2t^5} \cdot \dfrac{-6}{8t^3}$

$= \dfrac{-24t^4}{16t^8}$

$= \dfrac{-3(8t^4)}{2t^4(8t^4)}$

$= \dfrac{-3}{2t^4} = -\dfrac{3}{2t^4}$

178   Chapter 5   Rational Expressions

23. $\dfrac{2-t}{8} \div \dfrac{t-2}{6} = \dfrac{2-t}{8} \cdot \dfrac{6}{t-2}$

   Multiply by reciprocal

   $= \dfrac{6(2-t)}{8(t-2)}$

   Multiply numerators; multiply denominators

   $= \dfrac{6(-1)}{8}$

   $2 - t = -1(t - 2)$

   $= -\dfrac{3}{4}$ *Lowest terms*

27. $\dfrac{6(m-2)^2}{5(m+4)^2} \cdot \dfrac{15(m+4)}{2(2-m)}$

   $= \dfrac{2 \cdot 3(m-2)(m-2)}{5(m+4)(m+4)} \cdot \dfrac{3 \cdot 5(m+4)}{2(-1)(m-2)}$

   $= \dfrac{3(m-2)(3)}{(m+4)(-1)}$

   $= \dfrac{9(m-2)}{-1(m+4)}$

   $= \dfrac{-9(m-2)}{m+4}$

31. $\dfrac{2k^2 - k - 1}{2k^2 + 5k + 3} \div \dfrac{4k^2 - 1}{2k^2 + k - 3}$

   $= \dfrac{2k^2 - k - 1}{2k^2 + 5k + 3} \cdot \dfrac{2k^2 + k - 3}{4k^2 - 1}$

   $= \dfrac{(2k+1)(k-1)(2k+3)(k-1)}{(2k+3)(k+1)(2k+1)(2k-1)}$

   $= \dfrac{(k-1)(k-1)}{(k+1)(2k-1)}$

   $= \dfrac{(k-1)^2}{(k+1)(2k-1)}$

35. $\dfrac{m^2 + 2mp - 3p^2}{m^2 - 3mp + 2p^2} \div \dfrac{m^2 + 4mp + 3p^2}{m^2 + 2mp - 8p^2}$

   $= \dfrac{m^2 + 2mp - 3p^2}{m^2 - 3mp + 2p^2} \cdot \dfrac{m^2 + 2mp - 8p^2}{m^2 + 4mp + 3p^2}$

   $= \dfrac{(m+3p)(m-p)(m+4p)(m-2p)}{(m-2p)(m-p)(m+3p)(m+p)}$

   $= \dfrac{m+4p}{m+p}$

39. $18 = 2 \cdot 9$

   $= 2 \cdot 3 \cdot 3$

   $= 2 \cdot 3^2$

43. $24m, 18m^2, 6$

   Write each term in prime factored form.

   $24m = 2 \cdot 2 \cdot 2 \cdot 3 \cdot m = 2^3 \cdot 3 \cdot m$

   $18m^2 = 2 \cdot 3 \cdot 3 \cdot m^2 = 2 \cdot 3^2 \cdot m^2$

   $6 = 2 \cdot 3$

   The greatest common factor is $2 \cdot 3 = 6$.

Section 5.3   Least Common Denominators

5.3 Margin Exercises

1. **(a)** $\dfrac{7}{20p}, \dfrac{11}{30p}$

   Factor each denominator.

   $20p = 2 \cdot 2 \cdot 5 \cdot p = 2^2 \cdot 5 \cdot p$

   $30p = 2 \cdot 3 \cdot 5 \cdot p$

   Take each factor the greatest number of times it appears in any denominator; then multiply.

   $\text{LCD} = 2^2 \cdot 3 \cdot 5 \cdot p$

   $= 60p$

   (b) $\dfrac{9}{8m^4}, \dfrac{11}{12m^6}$

   Factor each denominator.

   $8m^4 = 2 \cdot 2 \cdot 2 \cdot m^4 = 2^3 \cdot m^4$

   $12m^6 = 2 \cdot 2 \cdot 3 \cdot m^6 = 2^2 \cdot 3 \cdot m^6$

   Take each factor the greatest number of times it appears in any denominator; then multiply.

$$\text{LCD} = 2^3 \cdot 3 \cdot m^6$$
$$= 24m^6$$

2. (a) $\dfrac{4}{16m^3 n}, \dfrac{5}{9m^5}$

   Factor each denominator.
   $$16m^3 n = 2 \cdot 2 \cdot 2 \cdot 2 \cdot m^3 \cdot n$$
   $$= 2^4 \cdot m^3 \cdot n$$
   $$9m^5 = 3 \cdot 3 \cdot m^5 = 3^2 \cdot m^5$$

   Take each factor the greatest number of times it appears in any denominator; then multiply.
   $$\text{LCD} = 2^4 \cdot 3^2 \cdot m^5 \cdot n$$
   $$= 144 m^5 n$$

   (b) $\dfrac{3}{25a^2}, \dfrac{2}{10a^3 b}$

   Factor each denominator.
   $$25a^2 = 5 \cdot 5 \cdot a^2 = 5^2 \cdot a^2$$
   $$10a^3 b = 2 \cdot 5 \cdot a^3 \cdot b$$

   Take each factor the greatest number of times it appears in any denominator; then multiply.
   $$\text{LCD} = 2 \cdot 5^2 \cdot a^3 \cdot b$$
   $$= 50 a^3 b$$

3. (a) $\dfrac{7}{3a}, \dfrac{5}{3a - 10}$

   Notice that each denominator is prime. Take each factor the greatest number of times it appears in any denominator; then multiply.
   $$\text{LCD} = 3 \cdot a(3a - 10)$$
   $$= 3a(3a - 10)$$

(b) $\dfrac{1}{12a}, \dfrac{5}{a^2 - 4a}$

Factor each denominator.
$$12a = 2 \cdot 2 \cdot 3 \cdot a$$
$$= 2^2 \cdot 3 \cdot a$$
$$a^2 - 4a = a(a - 4)$$
$$\text{LCD} = 2^2 \cdot 3 \cdot a(a - 4)$$
$$= 12a(a - 4)$$

(c) $\dfrac{2m}{m^2 - 3m + 2}, \dfrac{5m - 3}{m^2 + 3m - 10}$

Factor each denominator.
$$m^2 - 3m + 2 = (m - 2)(m - 1)$$
$$m^2 + 3m - 10 = (m + 5)(m - 2)$$
$$\text{LCD} = (m - 2)(m - 1)(m + 5)$$

(d) $\dfrac{6}{x - 4}, \dfrac{3x - 1}{4 - x}$

The expressions $x - 4$ and $4 - x$ are negatives of each other because
$$-(x - 4) = -x + 4 = 4 - x.$$

Therefore either $x - 4$ or $4 - x$ can be used as the LCD.

4. (a) $\dfrac{7k}{5} = \dfrac{}{30p}$

   Factor the denominator on the right; then compare it to the denominator on the left.
   $$\dfrac{7k}{5} = \dfrac{}{5 \cdot 6 \cdot p}$$

   The factors 6 and p are missing on the left, so multiply by 6p/6p.
   $$\dfrac{7k}{5} \cdot \dfrac{6p}{6p} = \dfrac{42kp}{30p}$$

(b) $\dfrac{9}{2a+5} = \dfrac{}{6a+15}$

Factor the denominator on the right.

$$\dfrac{9}{2a+5} = \dfrac{}{3(2a+5)}$$

The factor 3 is missing on the left, so multiply by 3/3.

$$\dfrac{9}{2a+5} \cdot \dfrac{3}{3} = \dfrac{27}{6a+15}$$

$$\dfrac{9}{2a+5} = \dfrac{9}{2a+5} \cdot \dfrac{3}{3}$$

$$= \dfrac{27}{3(2a+5)}$$

$$= \dfrac{27}{6a+15}$$

(c) $\dfrac{5k+1}{k^2+2k} = \dfrac{}{k(k+2)(k-1)}$

Factor the denominator on the left and compare denominators.

$$\dfrac{5k+1}{k(k+2)} = \dfrac{}{k(k+2)(k-1)}$$

$$\dfrac{5k+1}{k^2+2k} = \dfrac{5k+1}{k(k+2)} \cdot \dfrac{k-1}{k-1}$$

$$= \dfrac{(5k+1)(k-1)}{k(k+2)(k-1)}$$

5.3 Section Exercises

3. $\dfrac{2}{15}, \dfrac{3}{10}, \dfrac{7}{30}$

Write each denominator in prime factored form.

$$15 = 3 \cdot 5$$
$$10 = 2 \cdot 5$$
$$30 = 2 \cdot 3 \cdot 5$$

To find the least common denominator, take each different factor the *greatest* number of times it appears in *any* denominator.

$$\text{LCD} = 2 \cdot 3 \cdot 5 = 30$$

7. $\dfrac{5}{36q}, \dfrac{17}{24q}$

Factor each denominator.

$$36q = 2 \cdot 2 \cdot 3 \cdot 3 \cdot q$$
$$= 2^2 \cdot 3^2 \cdot q$$
$$24q = 2 \cdot 2 \cdot 2 \cdot 3 \cdot q$$
$$= 2^3 \cdot 3 \cdot q$$

Take each factor the greatest number of times it appears; then multiply.

$$\text{LCD} = 2^3 \cdot 3^2 \cdot q = 72q$$

11. If two denominators have a greatest common factor equal to 1, their least common denominator is the product of the two denominators.

15. $\dfrac{7}{5b-10}, \dfrac{11}{6b-12}$

Factor the denominators.

$$5b - 10 = 5(b-2)$$
$$6b - 12 = 6(b-2)$$
$$= 2 \cdot 3(b-2)$$
$$\text{LCD} = 2 \cdot 3 \cdot 5(b-2)$$
$$= 30(b-2)$$

19. $\dfrac{3}{k^2 + 5k}$, $\dfrac{2}{k^2 + 3k - 10}$

Factor each denominator.

$$k^2 + 5k = k(k + 5)$$
$$k^2 + 3k - 10 = (k + 5)(k - 2)$$
$$\text{LCD} = k(k + 5)(k - 2)$$

23. $\dfrac{4}{11} = \dfrac{}{55}$

$$\dfrac{4}{11} = \dfrac{4}{11} \cdot \dfrac{5}{5} = \dfrac{20}{55}$$

27. $\dfrac{13}{40y} = \dfrac{}{80y^3}$

$$\dfrac{13}{40y} = \dfrac{13}{40y} \cdot \dfrac{2y^2}{2y^2} = \dfrac{26y^2}{80y^3}$$

31. $\dfrac{5}{2(m + 3)} = \dfrac{}{8(m + 3)}$

$$\dfrac{5}{2(m + 3)} = \dfrac{5}{2(m + 3)} \cdot \dfrac{4}{4}$$
$$= \dfrac{20}{8(m + 3)}$$

35. $\dfrac{14}{z^2 - 3z} = \dfrac{}{z(z - 3)(z - 2)}$

$$\dfrac{14}{z^2 - 3z} = \dfrac{14}{z(z - 3)}$$
$$= \dfrac{14(z - 2)}{z(z - 3)(z - 2)}$$

39. $\dfrac{3}{4} + \dfrac{7}{4} = \dfrac{10}{4} = \dfrac{5}{2}$

43. $\dfrac{7}{5} - \dfrac{3}{4}$

$$\text{LCD} = 5 \cdot 4 = 20$$

$$\dfrac{7}{5} - \dfrac{3}{4} = \dfrac{7 \cdot 4}{5 \cdot 4} - \dfrac{3 \cdot 5}{4 \cdot 5}$$
$$= \dfrac{28}{20} - \dfrac{15}{20} = \dfrac{13}{20}$$

Section 5.4 Addition and Subtraction of Rational Expressions

5.4 Margin Exercises

1. (a) $\dfrac{3}{y + 4} + \dfrac{2}{y + 4} = \dfrac{3 + 2}{y + 4} = \dfrac{5}{y + 4}$

   (b) $\dfrac{x}{x + y} + \dfrac{1}{x + y} = \dfrac{x + 1}{x + y}$

   (c) $\dfrac{a}{a + b} + \dfrac{b}{a + b} = \dfrac{a + b}{a + b} = 1$

2. (a) $\dfrac{6}{5x} + \dfrac{9}{2x}$

   Step 1 $\text{LCD} = 5 \cdot 2 \cdot x = 10x$

   Step 2 $\dfrac{6}{5x} = \dfrac{6 \cdot 2}{5x \cdot 2} = \dfrac{12}{10x}$

   $\dfrac{9}{2x} = \dfrac{9 \cdot 5}{2x \cdot 5} = \dfrac{45}{10x}$

   Step 3 $\dfrac{6}{5x} + \dfrac{9}{2x} = \dfrac{12}{10x} + \dfrac{45}{10x}$

   $= \dfrac{12 + 45}{10x} = \dfrac{57}{10x}$

   (b) $\dfrac{m}{3n} + \dfrac{2}{7n}$

   Step 1 $\text{LCD} = 3 \cdot 7 \cdot n = 21n$

   Step 2 $\dfrac{m}{3n} = \dfrac{m \cdot 7}{3n \cdot 7} = \dfrac{7m}{21n}$

   $\dfrac{2}{7n} = \dfrac{2 \cdot 3}{7n \cdot 3} = \dfrac{6}{21n}$

   Step 3 $\dfrac{m}{3n} + \dfrac{2}{7n} = \dfrac{7m}{21n} + \dfrac{6}{21n}$

   $= \dfrac{7m + 6}{21n}$

182 Chapter 5 Rational Expressions

3. (a) $\dfrac{2p}{3p + 3} + \dfrac{5p}{2p + 2}$

Since $3p + 3 = 3(p + 1)$ and $2p + 2 = 2(p + 1)$, the LCD is $2 \cdot 3(p + 1) = 6(p + 1)$.

$\dfrac{2p}{3p + 3} + \dfrac{5p}{2p + 2} = \dfrac{2 \cdot 2p}{2(3p + 3)} + \dfrac{3 \cdot 5p}{3(2p + 2)}$

$= \dfrac{4p}{6p + 6} + \dfrac{15p}{6p + 6}$

$= \dfrac{4p + 15p}{6p + 6}$

$= \dfrac{19p}{6p + 6} = \dfrac{19p}{6(p + 1)}$

(b) $\dfrac{4}{y^2 - 1} + \dfrac{6}{y + 1}$

$= \dfrac{4}{(y + 1)(y - 1)} + \dfrac{6}{y + 1}$
Factor denominator

$= \dfrac{4}{(y + 1)(y - 1)} + \dfrac{6(y - 1)}{(y + 1)(y - 1)}$
LCD = $(y + 1)(y - 1)$

$= \dfrac{4 + 6(y - 1)}{(y + 1)(y - 1)}$
Add numerators

$= \dfrac{4 + 6y - 6}{(y + 1)(y - 1)}$
Distributive property

$= \dfrac{6y - 2}{(y + 1)(y - 1)}$
Combine terms

$= \dfrac{2(3y - 1)}{(y + 1)(y - 1)}$
Factor numerator

(c) $\dfrac{-2}{p + 1} + \dfrac{4p}{p^2 - 1}$

$= \dfrac{-2}{p + 1} + \dfrac{4p}{(p + 1)(p - 1)}$
Factor denominator

$= \dfrac{-2(p - 1)}{(p + 1)(p - 1)} + \dfrac{4p}{(p + 1)(p - 1)}$
LCD = $(p + 1)(p - 1)$

$= \dfrac{-2(p - 1) + 4p}{(p + 1)(p - 1)}$
Add numerators

$= \dfrac{-2p + 2 + 4p}{(p + 1)(p - 1)}$
Distributive property

$= \dfrac{2p + 2}{(p + 1)(p - 1)}$
Combine terms

$= \dfrac{2(p + 1)}{(p + 1)(p - 1)}$
Factor numerator

$= \dfrac{2}{p - 1}$
Lowest terms

4. (a) $\dfrac{2k}{k^2 - 5k + 4} + \dfrac{3}{k^2 - 1}$

$= \dfrac{2k}{(k - 4)(k - 1)} + \dfrac{3}{(k + 1)(k - 1)}$
Factor

$= \dfrac{2k(k + 1)}{(k - 4)(k - 1)(k + 1)}$

$+ \dfrac{3(k - 4)}{(k + 1)(k - 1)(k + 4)}$
LCD = $(k - 4)(k - 1)(k + 1)$

$= \dfrac{2k(k + 1) + 3(k - 4)}{(k - 4)(k - 1)(k + 1)}$
Add numerators

$= \dfrac{2k^2 + 2k + 3k - 12}{(k - 4)(k - 1)(k + 1)}$
Distributive property

$= \dfrac{2k^2 + 5k - 12}{(k - 4)(k - 1)(k + 1)}$
Combine terms

$= \dfrac{(2k - 3)(k + 4)}{(k - 4)(k - 1)(k + 1)}$
Factor numerator

(b) $\dfrac{4m}{m^2 + 3m + 2} + \dfrac{2m - 1}{m^2 + 6m + 5}$

$= \dfrac{4m}{(m + 2)(m + 1)} + \dfrac{2m - 1}{(m + 5)(m + 1)}$
Factor

$= \dfrac{4m(m + 5)}{(m + 2)(m + 1)(m + 5)}$

$+ \dfrac{(2m - 1)(m + 2)}{(m + 5)(m + 1)(m + 2)}$
LCD = $(m + 2)(m + 1)(m + 5)$

$$= \frac{4m^2 + 20m + 2m^2 + 3m - 2}{(m + 2)(m + 1)(m + 5)}$$

$$= \frac{6m^2 + 23m - 2}{(m + 2)(m + 1)(m + 5)}$$

5. (a) $\dfrac{3}{m^2} - \dfrac{2}{m^2} = \dfrac{3 - 2}{m^2} = \dfrac{1}{m^2}$

(b) $\dfrac{x}{2x + 3} - \dfrac{3x + 4}{2x + 3} = \dfrac{x - (3x + 4)}{2x + 3}$

$$= \frac{x - 3x - 4}{2x + 3}$$

$$= \frac{-2x - 4}{2x + 3}$$

$$= \frac{-2(x + 2)}{2x + 3}$$

6. (a) $\dfrac{1}{k + 4} - \dfrac{2}{k} = \dfrac{k \cdot 1}{k(k + 4)} - \dfrac{2(k + 4)}{k(k + 4)}$
LCD $= k(k + 4)$

$$= \frac{k - 2(k + 4)}{k(k + 4)}$$

$$= \frac{k - 2k - 8}{k(k + 4)}$$

$$= \frac{-k - 8}{k(k + 4)}$$

(b) $\dfrac{6}{a + 2} - \dfrac{1}{a - 3}$

$$= \frac{6(a - 3)}{(a + 2)(a - 3)} - \frac{1(a + 2)}{(a - 3)(a + 2)}$$
LCD $= (a + 2)(a - 3)$

$$= \frac{6(a - 3) - 1(a + 2)}{(a + 2)(a - 3)}$$

$$= \frac{6a - 18 - a - 2}{(a + 2)(a - 3)}$$

$$= \frac{5a - 20}{(a + 2)(a - 3)}$$

$$= \frac{5(a - 4)}{(a + 2)(a - 3)}$$

7. (a) $\dfrac{5}{x - 1} - \dfrac{3x}{1 - x}$

The denominators are negatives of each other, so either may be used as the common denominator. Let us choose $x - 1$.

$$\frac{5}{x - 1} - \frac{3x}{1 - x} = \frac{5}{x - 1} - \frac{-3x}{x - 1}$$

$$= \frac{5 - (-3x)}{x - 1}$$

$$= \frac{5 + 3x}{x - 1}$$

(b) $\dfrac{2y}{y - 2} - \dfrac{1 + y}{2 - y}$

The denominators are negatives of each other, so either may be used as the common denominator. Let us choose $y - 2$.

$$\frac{2y}{y - 2} - \frac{1 + y}{2 - y} = \frac{2y}{y - 2} - \frac{-1 - y}{y - 2}$$

$$= \frac{2y - (-1 - y)}{y - 2}$$

$$= \frac{2y + 1 + y}{y - 2}$$

$$= \frac{3y + 1}{y - 2}$$

8. (a) $\dfrac{4y}{y^2 - 1} - \dfrac{5}{y^2 + 2y + 1}$

$$= \frac{4y}{(y - 1)(y + 1)} - \frac{5}{(y + 1)(y + 1)}$$
Factor

$$= \frac{4y(y + 1)}{(y - 1)(y + 1)(y + 1)}$$

$$- \frac{5(y - 1)}{(y + 1)(y + 1)(y - 1)}$$
LCD $= (y - 1)(y + 1)(y + 1)$

$$= \frac{4y(y + 1) - 5(y - 1)}{(y - 1)(y + 1)(y + 1)}$$

$$= \frac{4y^2 + 4y - 5y + 5}{(y - 1)(y + 1)(y + 1)}$$

184 Chapter 5 Rational Expressions

$$= \frac{4y^2 - y + 5}{(y - 1)(y + 1)(y + 1)}$$

$$= \frac{4y^2 - y + 5}{(y - 1)(y + 1)^2}$$

(b) $\dfrac{3r}{r^2 - 5r} - \dfrac{4}{r^2 - 10r + 25}$

$$= \frac{3r}{r(r - 5)} - \frac{4}{(r - 5)(r - 5)}$$
Factor

$$= \frac{3}{r - 5} - \frac{4}{(r - 5)(r - 5)}$$

$$= \frac{3(r - 5)}{(r - 5)(r - 5)} - \frac{4}{(r - 5)(r - 5)}$$
LCD = $(r - 5)(r - 5)$

$$= \frac{3(r - 5) - 4}{(r - 5)(r - 5)}$$

$$= \frac{3r - 15 - 4}{(r - 5)(r - 5)}$$

$$= \frac{3r - 19}{(r - 5)^2}$$

9. (a) $\dfrac{2}{p^2 - 5p + 4} - \dfrac{3}{p^2 - 1}$

$$= \frac{2}{(p - 4)(p - 1)} - \frac{3}{(p + 1)(p - 1)}$$

$$= \frac{2(p + 1)}{(p - 4)(p - 1)(p + 1)}$$

$$- \frac{3(p - 4)}{(p + 1)(p - 1)(p - 4)}$$
LCD = $(p - 4)(p - 1)(p + 1)$

$$= \frac{(2p + 2) - (3p - 12)}{(p + 1)(p - 1)(p + 4)}$$

$$= \frac{2p + 2 - 3p + 12}{(p - 4)(p - 1)(p + 1)}$$

$$= \frac{14 - p}{(p - 4)(p - 1)(p + 1)}$$

(b) $\dfrac{q}{2q^2 + 5q - 3} - \dfrac{3q + 4}{3q^2 + 10q + 3}$

$$= \frac{q}{(2q - 1)(q + 3)} - \frac{3q + 4}{(q + 3)(3q + 1)}$$

$$= \frac{q(3q + 1)}{(2q - 1)(q + 3)(3q + 1)}$$

$$- \frac{(2q - 1)(3q + 4)}{(2q - 1)(q + 3)(3q + 1)}$$
LCD = $(2q - 1)(q + 3)(3q + 1)$

$$= \frac{(3q^2 + q) - (6q^2 + 5q - 4)}{(2q - 1)(q + 3)(3q + 1)}$$

$$= \frac{3q^2 + q - 6q^2 - 5q + 4}{(2q - 1)(q + 3)(3q + 1)}$$

$$= \frac{-3q^2 - 4q + 4}{(2q - 1)(q + 3)(3q + 1)}$$

$$= \frac{(-3q + 2)(q + 2)}{(2q - 1)(q + 3)(3q + 1)}$$

5.4 Section Exercises

3. $\dfrac{4}{m} + \dfrac{7}{m} = \dfrac{4 + 7}{m} = \dfrac{11}{m}$

7. $\dfrac{x^2}{x + 5} + \dfrac{5x}{x + 5} = \dfrac{x^2 + 5x}{x + 5}$
   Add numerators

$$= \frac{x(x + 5)}{x + 5}$$
Factor numerator

$$= x \quad \text{Lowest terms}$$

15. $\dfrac{5}{7} - \dfrac{r}{2} = \dfrac{5}{7} \cdot \dfrac{2}{2} - \dfrac{r}{2} \cdot \dfrac{7}{7}$ LCD = 14

$$= \frac{10}{14} - \frac{7r}{14}$$

$$= \frac{10 - 7r}{14}$$

19. $\dfrac{5 + 5k}{4} + \dfrac{1 + k}{8} = \dfrac{(5 + 5k) \cdot 2}{4 \cdot 2} + \dfrac{1 + k}{8}$

$\quad\quad\quad\quad\quad\quad\quad\quad\quad\quad\quad$ LCD = 8

$\quad\quad\quad\quad\quad\quad = \dfrac{2(5 + 5k) + (1 + k)}{8}$

$\quad\quad\quad\quad\quad\quad = \dfrac{10 + 10k + 1 + k}{8}$

$\quad\quad\quad\quad\quad\quad = \dfrac{11 + 11k}{8}$

$\quad\quad\quad\quad\quad\quad = \dfrac{11(1 + k)}{8}$

23. $\dfrac{7}{3p^2} - \dfrac{2}{p} = \dfrac{7}{3p^2} - \dfrac{2 \cdot 3p}{p \cdot 3p} \quad$ LCD $= 3p^2$

$\quad\quad\quad\quad\quad = \dfrac{7 - 6p}{3p^2}$

27. $\dfrac{4}{x - 5} + \dfrac{6}{5 - x}$

The two denominators, x − 5 and 5 − x, are negatives of each other, so either one may be used as the common denominator. We will work the exercise both ways and compare the answers.

$\dfrac{4}{x - 5} + \dfrac{6}{5 - x} = \dfrac{4}{x - 5} + \dfrac{6(-1)}{(5 - x)(-1)}$

$\quad\quad\quad\quad\quad\quad\quad\quad\quad\quad$ LCD = x + 5

$\quad\quad\quad\quad\quad\quad = \dfrac{4 + 6(-1)}{x - 5}$

$\quad\quad\quad\quad\quad\quad = \dfrac{4 - 6}{x - 5} = \dfrac{-2}{x - 5}$

$\dfrac{4}{x - 5} + \dfrac{6}{5 - x} = \dfrac{4(-1)}{(x - 5)(-1)} + \dfrac{6}{5 - x}$

$\quad\quad\quad\quad\quad\quad\quad\quad\quad\quad$ LCD = 5 − x

$\quad\quad\quad\quad\quad\quad = \dfrac{-4}{5 - x} + \dfrac{6}{5 - x}$

$\quad\quad\quad\quad\quad\quad = \dfrac{2}{5 - x}$

The two answers are equivlent, since

$\dfrac{-2}{x - 5} \cdot \dfrac{-1}{-1} = \dfrac{2}{5 - x}.$

31. $\dfrac{6}{c - 2} - \dfrac{8}{c + 2}$

$= \dfrac{6(c + 2)}{(c - 2)(c + 2)} - \dfrac{8(c - 2)}{(c + 2)(c - 2)}$

$\quad\quad\quad\quad\quad$ LCD = (c − 2)(x + 2)

$= \dfrac{6(c + 2) - 8(c - 2)}{(c - 2)(c + 2)}$

$= \dfrac{6c + 12 - 8c + 16}{(c - 2)(c + 2)}$

$= \dfrac{-2c + 28}{(c - 2)(c + 2)}$

$= \dfrac{-2(c - 14)}{(c - 2)(c + 2)}$

35. $\dfrac{-2}{x^2 - 4} + \dfrac{7}{4x + 8}$

$= \dfrac{-2}{(x + 2)(x - 2)} + \dfrac{7}{4(x + 2)}$

$\quad\quad\quad\quad\quad$ LCD = 4(x + 2)(x − 2)

$= \dfrac{-2(4)}{(x + 2)(x - 2)(4)}$

$\quad + \dfrac{7(x - 2)}{4(x + 2)(x - 2)}$

$= \dfrac{-2(4) + 7(x - 2)}{4(x + 2)(x - 2)}$

$= \dfrac{-8 + 7x - 14}{4(x + 2)(x - 2)}$

$= \dfrac{7x - 22}{4(x + 2)(x - 2)}$

39. $\dfrac{8}{m - 2} + \dfrac{3}{5m} + \dfrac{7}{5m(m - 2)}$

$= \dfrac{8 \cdot 5m}{(m - 2) \cdot 5m} + \dfrac{3(m - 2)}{5m(m - 2)} + \dfrac{7}{5m(m - 2)}$

$\quad\quad\quad\quad\quad$ LCD = 5m(m − 2)

$= \dfrac{8 \cdot 5m + 3(m - 2) + 7}{5m(m - 2)}$

$= \dfrac{40m + 3m - 6 + 7}{5m(m - 2)}$

$= \dfrac{43m + 1}{5m(m - 2)}$

Chapter 5 Rational Expressions

43. $\dfrac{\frac{5}{6}+\frac{7}{6}}{\frac{2}{3}-\frac{1}{3}} = \dfrac{\frac{12}{6}}{\frac{1}{3}}$

$= \dfrac{2}{\frac{1}{3}} = 2 \cdot 3 = 6$

Section 5.5 Complex Fractions

5.5 Margin Exercises

1. (a) $\dfrac{6+\frac{1}{x}}{5-\frac{2}{x}} = \dfrac{\frac{6x}{x}+\frac{1}{x}}{\frac{5x}{x}-\frac{2}{x}}$

$= \dfrac{\frac{6x+1}{x}}{\frac{5x-2}{x}}$

$= \dfrac{6x+1}{x} \div \dfrac{5x-2}{x}$

$= \dfrac{6x+1}{x} \cdot \dfrac{x}{5x-2}$ *Multiply by reciprocal*

$= \dfrac{6x+1}{5x-2}$

(b) $\dfrac{9-\frac{4}{p}}{\frac{2}{p}+1} = \dfrac{\frac{9p}{p}-\frac{4}{p}}{\frac{2}{p}+\frac{p}{p}}$

$= \dfrac{\frac{9p-4}{p}}{\frac{2+p}{p}}$

$= \dfrac{9p-4}{p} \div \dfrac{2+p}{p}$

$= \dfrac{9p-4}{p} \cdot \dfrac{p}{2+p}$ *Multiply by reciprocal*

$= \dfrac{9p-4}{2+p}$

2. (a) $\dfrac{\frac{rs^2}{t}}{\frac{r^2s}{t^2}} = \dfrac{rs^2}{t} \div \dfrac{r^2s}{t^2} = \dfrac{rs^2}{t} \cdot \dfrac{t^2}{r^2s}$

$= \dfrac{st}{r}$

(b) $\dfrac{\frac{m^2n^3}{p}}{\frac{m^4n}{p^2}} = \dfrac{m^2n^3}{p} \div \dfrac{m^4n}{p^2}$

$= \dfrac{m^2n^3}{p} \cdot \dfrac{p^2}{m^4n}$

$= \dfrac{n^2p}{m^2}$

3. $\dfrac{\frac{2}{x-1}+\frac{1}{x+1}}{\frac{3}{x-1}-\frac{4}{x+1}}$

$= \dfrac{\dfrac{2(x+1)}{(x-1)(x+1)}+\dfrac{1(x-1)}{(x+1)(x-1)}}{\dfrac{3(x+1)}{(x-1)(x+1)}-\dfrac{4(x-1)}{(x+1)(x-1)}}$

$= \dfrac{\dfrac{2x+2}{(x-1)(x+1)}+\dfrac{x-1}{(x+1)(x-1)}}{\dfrac{3x+3}{(x-1)(x+1)}-\dfrac{4x-4}{(x+1)(x-1)}}$

$= \dfrac{\dfrac{(2x+2)+(x-1)}{(x+1)(x-1)}}{\dfrac{(3x+3)-(4x-4)}{(x-1)(x+1)}}$

$= \dfrac{\dfrac{2x+2+x-1}{(x+1)(x-1)}}{\dfrac{3x+3-4x+4}{(x-1)(x+1)}}$

$= \dfrac{\dfrac{3x+1}{(x+1)(x-1)}}{\dfrac{-x+7}{(x-1)(x+1)}}$

$= \dfrac{3x+1}{(x+1)(x-1)} \cdot \dfrac{(x-1)(x+1)}{-x+7}$

$= \dfrac{3x+1}{-x+7}$

5.5 Section Exercises

In Exercises 3-27, either Method 1 or Method 2 can be used to simplify each complex fraction. Only one method will be shown for each exercise.

4. (a) $\dfrac{2 - \frac{6}{a}}{3 + \frac{4}{a}}$ LCD = a

$= \dfrac{a\left(2 - \frac{6}{a}\right)}{a\left(3 + \frac{4}{a}\right)}$ Multiply by a/a

$= \dfrac{2a - \frac{6}{a} \cdot a}{3a + \frac{4}{a} \cdot a}$

$= \dfrac{2a - 6}{3a + 4}$

(b) $\dfrac{\frac{5}{p} - 6}{\frac{2p+1}{p}}$ LCD = p

$= \dfrac{p\left(\frac{5}{p} - 6\right)}{p\left(\frac{2p+1}{p}\right)}$ Multiply by p/p

$= \dfrac{\frac{5}{p} \cdot p - 6p}{\frac{2p+1}{p} \cdot p}$

$= \dfrac{5 - 6p}{2p + 1}$

(c) $\dfrac{\frac{-4}{3+x}}{\frac{5}{2-x}}$ LCD = $(3+x)(2-x)$

$= \dfrac{\frac{-4}{3+x}}{\frac{5}{2-x}} \cdot \dfrac{(3+x)(2-x)}{(3+x)(2-x)}$

$= \dfrac{-4(2-x)}{5(3+x)}$

$= \dfrac{-4(2-x)}{5(3+x)} \cdot \dfrac{-1}{-1}$

$= \dfrac{4(x-2)}{5(3+x)}$

3. $\dfrac{-\frac{4}{3}}{\frac{2}{9}}$

To use Method 1, divide the numerator of the complex fraction by the denominator.

$\dfrac{-\frac{4}{3}}{\frac{2}{9}} = -\dfrac{4}{3} \cdot \dfrac{9}{2}$

$= -\dfrac{36}{6} = -6$

7. $\dfrac{\frac{x}{y^2}}{\frac{x^2}{y}} = \dfrac{x}{y^2} \cdot \dfrac{y}{x^2}$

$= \dfrac{xy}{y^2 x^2} = \dfrac{1}{xy}$

11. $\dfrac{\frac{m+2}{3}}{\frac{m-4}{m}}$

To use Method 2, multiply the numerator and denominator of the complex fraction by the least common denominator, $3m$.

$\dfrac{\frac{m+2}{3}}{\frac{m-4}{m}} = \dfrac{\frac{m+2}{3} \cdot 3m}{\frac{m-4}{m} \cdot 3m}$

$= \dfrac{m(m+2)}{3(m-4)}$

188 Chapter 5 Rational Expressions

15. $\dfrac{\frac{1}{x} + x}{\frac{x^2+1}{8}} = \dfrac{8x\left(\frac{1}{x} + x\right)}{8x\left(\frac{x^2+1}{8}\right)}$ *Use Method 2; LCD = 8x*

$= \dfrac{8 + 8x^2}{x(x^2+1)}$ *Multiply*

$= \dfrac{8(1 + x^2)}{x(x^2+1)}$ *Factor*

$= \dfrac{8}{x}$ *Lowest terms*

19. $\dfrac{\frac{1}{2} + \frac{1}{p}}{\frac{2}{3} + \frac{1}{p}} = \dfrac{\left(\frac{1}{2} + \frac{1}{p}\right) \cdot 6p}{\left(\frac{2}{3} + \frac{1}{p}\right) \cdot 6p}$ *Use Method 2; LCD = 6p*

$= \dfrac{\frac{1}{2}(6p) + \frac{1}{p}(6p)}{\frac{2}{3}(6p) + \frac{1}{p}(6p)}$

$= \dfrac{3p + 6}{4p + 6}$

$= \dfrac{3(p + 2)}{2(2p + 3)}$

23. $\dfrac{\frac{1}{k+1} - 1}{\frac{1}{k+1} + 1} = \dfrac{\frac{1}{k+1} - \frac{k+1}{k+1}}{\frac{1}{k+1} + \frac{k+1}{k+1}}$

                               Use Method 1

$= \dfrac{\frac{1 - (k+1)}{k+1}}{\frac{1 + (k+1)}{k+1}}$

$= \dfrac{\frac{1 - k - 1}{k+1}}{\frac{1 + k + 1}{k+1}}$

$= \dfrac{\frac{-k}{k+1}}{\frac{2 + k}{k+1}}$

$= \dfrac{-k}{k+1} \cdot \dfrac{k+1}{2+k}$

$= \dfrac{-k}{2+k}$

27. $2 - \dfrac{2}{2 + \frac{2}{2+2}} = 2 - \dfrac{2}{2 + \frac{1}{2}}$

$= 2 - \dfrac{2}{\frac{5}{2}}$

$= 2 - 2 \cdot \dfrac{2}{5}$

$= 2 - \dfrac{4}{5}$

$= \dfrac{10}{5} - \dfrac{4}{5} = \dfrac{6}{5}$

31. $5(2q + 1) - 2 = 8q$

$10q + 5 - 2 = 8q$

$10q + 3 = 8q$

$3 = -2q$

$-\dfrac{3}{2} = q$

Section 5.6 Equations Involving Rational Expressions

5.6 Margin Exercises

1. **(a)** $\dfrac{x}{5} + 3 = \dfrac{3}{5}$

The least common denominator is 5, so multiply each side of the equation by 5 to remove the denominators.

$5\left(\dfrac{x}{5} + 3\right) = 5\left(\dfrac{3}{5}\right)$ *Multiply by the LCD, 5*

$5\left(\dfrac{x}{5}\right) + 5(3) = 5\left(\dfrac{3}{5}\right)$ *Distributive property*

$x + 15 = 3$

$x = -12$ *Subtract 15*

Check by substituting -12 for x in the original equation.

$$\frac{x}{5} + 3 = \frac{3}{5}$$

$$\frac{-12}{5} + 3 = \frac{3}{5} \quad ? \quad \text{Let } x = -12$$

$$\frac{-12}{5} + \frac{15}{5} = \frac{3}{5} \quad ?$$

$$\frac{3}{5} = \frac{3}{5} \quad \text{True}$$

(b) $\quad \dfrac{x}{2} - \dfrac{x}{3} = \dfrac{5}{6}$

$$6\left(\frac{x}{2} - \frac{x}{3}\right) = 6\left(\frac{5}{6}\right) \quad \begin{array}{l}\text{Multiply by}\\ \text{the LCD, 6}\end{array}$$

$$6\left(\frac{x}{2}\right) - 6\left(\frac{x}{3}\right) = 6\left(\frac{5}{6}\right) \quad \begin{array}{l}\text{Distributive}\\ \text{property}\end{array}$$

$$3x - 2x = 5$$

$$x = 5 \quad \text{Combine terms}$$

Check by substituting 5 for x.

2. (a) $\quad \dfrac{k}{6} - \dfrac{k+1}{4} = -\dfrac{1}{2}$

$$12\left(\frac{k}{6}\right) - 12\left(\frac{k+1}{4}\right) = 12\left(-\frac{1}{2}\right)$$

$$\begin{array}{l}\text{Multiply by}\\ \text{the LCD, 12}\end{array}$$

$$2k - 3(k+1) = -6$$

$$2k - 3k - 3 = -6$$

$$-k - 3 = -6$$

$$-k = -3 \quad \text{Add 3}$$

$$k = 3 \quad \text{Divide by } -1$$

Check by substituting 3 for k.

(b) $\quad \dfrac{2m - 3}{5} - \dfrac{m}{3} = -\dfrac{6}{5}$

$$15\left(\frac{2m-3}{5}\right) - 15\left(\frac{m}{3}\right) = 15\left(-\frac{6}{5}\right)$$

$$\begin{array}{l}\text{Multiply by}\\ \text{the LCD, 15}\end{array}$$

$$3(2m - 3) - 5m = 3(-6)$$
$$6m - 9 - 5m = -18$$
$$m - 9 = -18$$
$$m = -9 \quad \text{Add 9}$$

Check by substituting -9 for m.

3. $\quad 1 - \dfrac{2}{x+1} = \dfrac{2x}{x+1}$

$$(x+1)\left(1 - \frac{2}{x+1}\right) = (x+1)\frac{2x}{x+1}$$

$$\begin{array}{l}\text{Multiply by the}\\ \text{LCD, } x+1\end{array}$$

$$(x+1) - 2 = 2x$$
$$x - 1 = 2x$$
$$-1 = x \quad \text{Subtract } x$$

Check by substituting -1 for x.

$$1 - \frac{2}{x+1} = \frac{2x}{x+1}$$

$$1 - \frac{2}{-1+1} = \frac{2(-1)}{-1+1} \quad ?$$

$$\text{Let } x = -1$$

$$1 - \frac{2}{0} = \frac{-2}{0} \quad ?$$

Fractions are undefined

The equation has no solution.

4. $\quad \dfrac{2m}{m^2 - 4} + \dfrac{1}{m - 2} = \dfrac{2}{m + 2}$

To check whether -6 is a solution, substitute -6 every place there is an m.

$$\frac{2(-6)}{(-6)^2 - 4} + \frac{1}{-6 - 2} = \frac{2}{-6 + 2} \quad ?$$

$$\frac{-12}{32} + \frac{1}{-8} = \frac{2}{-4} \quad ?$$

$$\frac{-12}{32} + \frac{1(-4)}{(-8)(-4)} = \frac{2}{-4} \quad ?$$

$$\frac{-12}{32} + \frac{-4}{32} = \frac{-2}{4} \quad ?$$

$$\frac{-16}{32} = -\frac{2}{4} \;?$$

$$-\frac{1}{2} = -\frac{1}{2} \quad \text{True}$$

The check shows that -6 is a solution.

5. (a) $\quad \dfrac{2p}{p^2 - 1} = \dfrac{2}{p + 1} - \dfrac{1}{p - 1}$

$\dfrac{2p}{(p + 1)(p - 1)} = \dfrac{2}{p + 1} - \dfrac{1}{p - 1} \quad$ *Factor*

Multiply by $(p + 1)(p - 1)$, the least common denominator.

$(p + 1)(p - 1)\left(\dfrac{2p}{(p + 1)(p - 1)}\right)$

$= (p + 1)(p - 1)\left(\dfrac{2}{p + 1} - \dfrac{1}{p - 1}\right)$

$2p = 2(p - 1) - (p + 1)$

$2p = 2p - 2 - p - 1$

$2p = p - 3$

$p = -3 \quad$ *Subtract p*

Check by substituting -3 for p.

(b) $\quad \dfrac{8r}{4r^2 - 1} = \dfrac{3}{2r + 1} + \dfrac{3}{2r - 1}$

$\dfrac{8r}{(2r + 1)(2r - 1)} = \dfrac{3}{2r + 1} + \dfrac{3}{2r - 1} \quad$ *Factor*

Multiply by the least common denominator, $(2r + 1)(2r - 1)$.

$(2r + 1)(2r - 1)\dfrac{8r}{(2r + 1)(2r - 1)}$

$= (2r + 1)(2r - 1)\dfrac{3}{2r + 1}$

$+ (2r + 1)(2r - 1)\dfrac{3}{2r - 1}$

$8r = 3(2r - 1) + 3(2r + 1)$

$8r = 6r - 3 + 6r + 3$

$8r = 12r$

$0 = 4r \quad$ *Subtract 8r*

$0 = r \quad$ *Divide by 4*

Check by substituting 0 for r.

6. (a) $\quad \dfrac{4}{3m + 3} = \dfrac{m + 1}{m^2 + m}$

$\dfrac{4}{3(m + 1)} = \dfrac{(m + 1)}{m(m + 1)} \quad$ *Factor*

Multiply by $3m(m + 1)$, the LCD.

$3m(m + 1)\dfrac{4}{3(m + 1)}$

$= 3m(m + 1)\dfrac{(m + 1)}{m(m + 1)}$

$4m = 3(m + 1)$

$4m = 3m + 3$

$m = 3 \quad$ *Subtract 3m*

Check by substituting 3 for m.

(b) $\quad \dfrac{2}{p^2 - 2p} = \dfrac{3}{p^2 - p}$

$\dfrac{2}{p(p - 2)} = \dfrac{3}{p(p - 1)} \quad$ *Factor*

Multiply by $p(p - 2)(p - 1)$, the LCD.

$p(p - 2)(p - 1)\dfrac{2}{p(p - 2)}$

$= p(p - 2)(p - 1)\dfrac{3}{p(p - 1)}$

$2(p - 1) = 3(p - 2)$

$2p - 2 = 3p - 6$

$2p + 4 = 3p \quad$ *Add 6*

$4 = p \quad$ *Subtract 2p*

Check by substituting 4 for p.

7. (a) $\quad \dfrac{1}{x - 2} + \dfrac{1}{5} = \dfrac{2}{5(x^2 - 4)}$

Factor the denominators.

$\dfrac{1}{x - 2} + \dfrac{1}{5} = \dfrac{2}{5(x - 2)(x + 2)}$

Multiply by $5(x - 2)(x + 2)$, the least common denominator.

$$5(x-2)(x+2)\left(\frac{1}{x-2}+\frac{1}{5}\right)$$
$$=5(x-2)(x+2)\frac{2}{5(x-2)(x+2)}$$
$$5(x+2)+(x+2)(x-2)=2$$
$$5x+10+x^2-4=2$$
$$x^2+5x+6=2$$
$$x^2+5x+4=0$$
$$(x+4)(x+1)=0$$
$$x+4=0 \quad \text{or} \quad x+1=0$$
$$x=-4 \quad \text{or} \quad x=-1$$

Check -4 by substituting -4 for x. Then check -1 by substituting -1 for x.

(b) $\quad \dfrac{6}{5a+10} - \dfrac{1}{a-5} = \dfrac{4}{a^2-3a-10}$

Factor each denominator.

$$\frac{6}{5(a+2)} - \frac{1}{a-5} = \frac{4}{(a-5)(a+2)}$$

Multiply by the LCD, $5(a+2)(a-5)$.

$$6(a-5) - 5(a+2) = 4(5)$$
$$6a - 30 - 5a - 10 = 20$$
$$a - 40 = 20$$
$$a = 60$$

Check by substituting 60 for a. You may wish to use a calculator.

8. (a) $\quad z = \dfrac{x}{x+y}$ for y

Multiply by the least common denominator, $x+y$.

$$z(x+y) = \frac{x}{x+y}(x+y)$$
$$zx + zy = x$$
$$zy = x - zx \quad \text{Subtract } xz$$
$$y = \frac{x-zx}{z} \quad \text{Divide by } z$$

(b) $\quad a = \dfrac{v-w}{t}$ for v

$at = \dfrac{v-w}{t} \cdot t \quad$ Multiply by the LCD, t

$at = v - w$

$at + w = v \quad$ Add w

5.6 Section Exercises

In Exercises 3–23 and 31–39, all proposed solutions should be checked by substituting in the original equation. It is essential to check whether a proposed solution will make any denominator in the original equation equal zero. Checks will be shown here for only a few of the exercises.

3. $\dfrac{5}{m} - \dfrac{3}{m} = 8$

Multiply each side by the LCD, m.

$$m\left(\frac{5}{m} - \frac{3}{m}\right) = m \cdot 8$$

Use the distributive property to remove parentheses; then solve.

$$m\left(\frac{5}{m}\right) - m\left(\frac{3}{m}\right) = 8m$$
$$5 - 3 = 8m$$
$$2 = 8m$$
$$m = \frac{2}{8} = \frac{1}{4}$$

Check $\quad \dfrac{5}{m} - \dfrac{3}{m} = 8$

$\dfrac{5}{\frac{1}{4}} - \dfrac{3}{\frac{1}{4}} = 8 \ ? \quad$ Let $m = 1/4$

$\dfrac{5 \cdot 4}{\frac{1}{4} \cdot 4} - \dfrac{3 \cdot 4}{\frac{1}{4} \cdot 4} = 8 \ ?$

$20 - 12 = 8 \ ?$

$8 = 8 \quad$ True

Thus, the solution is $1/4$.

7. $\dfrac{p}{3} - \dfrac{p}{6} = 4$

$6\left(\dfrac{p}{3} - \dfrac{p}{6}\right) = 6 \cdot 4$ *Multiply by LCD, 6*

$6\left(\dfrac{p}{3}\right) - 6\left(\dfrac{p}{6}\right) = 24$ *Distributive property*

$2p - p = 24$

$p = 24$

A check will verify that 24 is the solution.

11. $\dfrac{4m}{7} + m = 11$

$7\left(\dfrac{4m}{7} + m\right) = 7 \cdot 11$ *Multiply by LCD, 7*

$7\left(\dfrac{4m}{7}\right) + 7 \cdot m = 77$

$4m + 7m = 77$

$11m = 77$

$m = 7$

A check will verify that 7 is the solution.

15. $\dfrac{3p + 6}{8} = \dfrac{3p - 3}{16}$

$16\left(\dfrac{3p + 6}{8}\right) = 16\left(\dfrac{3p - 3}{16}\right)$ *Multiply by LCD, 16*

$2(3p + 6) = 3p - 3$

$6p + 12 = 3p - 3$

$3p + 12 = -3$

$3p = -15$

$p = -5$

A check will verify that -5 is the solution.

19. $\dfrac{k}{k - 4} - 5 = \dfrac{4}{k - 4}$

$(k - 4)\left(\dfrac{k}{k - 4} - 5\right) = (k - 4)\left(\dfrac{4}{k - 4}\right)$

   Multiply by LCD, $k - 4$

$(k - 4)\left(\dfrac{k}{k - 4}\right) - 5(k - 4) = (k - 4)\left(\dfrac{4}{k - 4}\right)$

$k - 5(k - 4) = 4$

$k - 5k + 20 = 4$

$-4k + 20 = 4$

$-4k = -16$

$k = 4$

The proposed solution is 4. However, 4 cannot be a solution because it makes the denominator $k - 4$ equal 0. Therefore, the given equation has no solution.

23. $\dfrac{t}{6} + \dfrac{4}{3} = \dfrac{t - 2}{3}$

$6\left(\dfrac{t}{6} + \dfrac{4}{3}\right) = 6\left(\dfrac{t - 2}{3}\right)$ *Multiply by LCD, 6*

$6\left(\dfrac{t}{6}\right) + 6\left(\dfrac{4}{3}\right) = 2(t - 2)$

$t + 8 = 2t - 4$

$8 = t - 4$

$12 = t$

A check will verify that 12 is the solution.

27. $\dfrac{1}{x - 4} = \dfrac{3}{2x}$

x cannot take a value which would make $x - 4$ or $2x$ equal to 0.

$x - 4 = 0$ $2x = 0$

$x = 4$ $x = 0$

Thus, 0 and 4 cannot be solutions of the given equation.

31. $\dfrac{y}{3y+3} = \dfrac{2y-3}{y+1} - \dfrac{2y}{3y+3}$

$\dfrac{y}{3(y+1)} = \dfrac{2y-3}{y+1} - \dfrac{2y}{3(y+1)}$

$3(y-1)\left(\dfrac{y}{3(y+1)}\right)$

$= 3(y-1)\left[\dfrac{2y-3}{y+1} - \dfrac{2y}{3(y+1)}\right]$

Multiply by LCD, $3(y+1)$

$y = 3(y-1)\left(\dfrac{2y-3}{y+1}\right)$

$\quad - 3(y-1)\left(\dfrac{2y}{3(y+1)}\right)$

$y = 3(2y-3) - 2y$

$y = 6y - 9 - 2y$

$y = 4y - 9$

$-3y = -9$

$y = 3$

Check

$\dfrac{y}{3y+3} = \dfrac{2y-3}{y+1} - \dfrac{2y}{3y+3}$

$\dfrac{3}{3(3)+3} = \dfrac{2(3)-3}{3+1} - \dfrac{2(3)}{3(3)+3}$?

Let $y = 3$

$\dfrac{3}{9+3} = \dfrac{6-3}{4} - \dfrac{6}{9+3}$?

$\dfrac{3}{12} = \dfrac{3}{4} - \dfrac{6}{12}$?

$\dfrac{1}{4} = \dfrac{3}{4} - \dfrac{2}{4}$?

$\dfrac{1}{4} = \dfrac{1}{4}$ *True*

Thus, the solution is 3.

35. $\dfrac{-2}{z+5} + \dfrac{3}{z-5} = \dfrac{20}{z^2-25}$

$\dfrac{-2}{z+5} + \dfrac{3}{z-5} = \dfrac{20}{(z+5)(z-5)}$

$(z+5)(x-5)\left(\dfrac{-2}{z+5} + \dfrac{3}{z-5}\right)$

$= (z+5)(z-5)$

$\quad \cdot \left(\dfrac{20}{(z+5)(z-5)}\right)$

Multiply by LCD, $(z+5)(x-5)$

$(z+5)(z-5)\left(\dfrac{-2}{z+5}\right) + (z+5)(z-5)\left(\dfrac{3}{z-5}\right) = 20$

$-2(z-5) + 3(z+5) = 20$

$-2z + 10 + 3z + 15 = 20$

$z + 25 = 20$

$z = -5$

The proposed solution, -5, cannot be a solution because it would make the denominators $z+5$ and z^2-25 equal 0 and the corresponding fractions undefined. Since -5 cannot be a solution, the equation has no solution.

39. $\dfrac{5x}{14x+3} = \dfrac{1}{x}$

Multiply each side by the LCD, $x(14x+3)$.

$x(14x+3)\left(\dfrac{5x}{14x+3}\right) = x(14x+3)\left(\dfrac{1}{x}\right)$

$x \cdot 5x = (14x+3)(1)$

$5x^2 = 14x + 3$

Solve the quadratic equation by factoring.

$5x^2 - 14x + 3 = 0$

$(5x+1)(x-3) = 0$

$5x + 1 = 0$ or $x - 3 = 0$

$5x = -1$

$x = -\dfrac{1}{5}$ or $x = 3$

Checking will verify that $-1/5$ and 3 are both solutions of the given equation.

194 Chapter 5 Rational Expressions

43. $m = \dfrac{kF}{a}$ for F

We need to isolate F on one side of the equation.

$$m \cdot a = \left(\dfrac{kF}{a}\right)(a) \quad \text{Multiply by } a$$

$$ma = kF$$

$$\dfrac{ma}{k} = \dfrac{kF}{k} \quad \text{Divide by } k$$

$$\dfrac{ma}{k} = F$$

47. $I = \dfrac{E}{R + r}$ for R

We need to isolate R on one side of the equation.

$$(R + r)I = (R + r)\left(\dfrac{E}{R + r}\right)$$
$$\qquad\qquad\qquad\qquad \text{Multiply by } R + r$$

$$RI + rI = E \quad \text{Distributive property}$$

$$RI = E - rI \quad \text{Subtract } rI$$

$$R = \dfrac{E - rI}{I} \quad \text{Divide by } I$$

51. $d = \dfrac{2S}{n(a + L)}$ for a

We need to isolate a on one side of the equation.

$$n(a + L)d = n(a + L)\left(\dfrac{2S}{n(a + L)}\right)$$

$$nd(a + L) = 2S$$

$$and + ndL = 2S$$

$$and = 2S - ndL$$

$$a = \dfrac{2S - ndL}{nd}$$

55. Since Joshua can do a job in x hours, he does 1/x of a job per hour.

Summary Exercises on Rational Expressions

3. $\dfrac{1}{x^2 + x - 2} \div \dfrac{4x^2}{2x - 2}$

To divide the first rational expression by the second, multiply the first expression by the reciprocal of the second.

$$\dfrac{1}{x^2 + x - 2} \div \dfrac{4x^2}{2x - 2}$$

$$= \dfrac{1}{x^2 + x - 2} \cdot \dfrac{2x - 2}{4x^2}$$

$$= \dfrac{1}{(x + 2)(x - 1)} \cdot \dfrac{2(x - 1)}{2 \cdot 2x^2}$$

$$= \dfrac{1}{2x^2(x + 2)}$$

7. $\dfrac{x - 4}{5} = \dfrac{x + 3}{6}$

To solve this equation, multiply each side by the LCD, 30.

$$30\left(\dfrac{x - 4}{5}\right) = 30\left(\dfrac{x + 3}{6}\right)$$

$$6(x - 4) = 5(x + 3)$$

$$6x - 24 = 5x + 15$$

$$x - 24 = 15$$

$$x = 39$$

11. $\dfrac{3}{t - 1} + \dfrac{1}{t} = \dfrac{7}{2}$

To solve this equation, multiply both sides by the LCD, $2t(t + 1)$.

$$2t(t-1)\left(\frac{3}{t-1}+\frac{1}{t}\right) = 2t(t-1)\left(\frac{7}{2}\right)$$

$$2t(t-1)\left(\frac{3}{t-1}\right) + 2t(t-1)\left(\frac{1}{t}\right) = 7t(t-1)$$

$$2t(3) + 2(t-1) = 7t(t-1)$$

$$6t + 2t - 2 = 7t^2 - 7t$$

$$8t - 2 = 7t^2 - 7t$$

$$0 = 7t^2 - 15t + 2$$

$$0 = (7t-1)(t-2)$$

$$7t - 1 = 0 \quad \text{or} \quad t - 2 = 0$$

$$7t = 1$$

$$t = \frac{1}{7} \quad \text{or} \quad t = 2$$

Checks will verify that 1/7 and 2 are both solutions of the given equation.

15. $\dfrac{1}{m^2 + 5m + 6} + \dfrac{2}{m^2 + 4m + 3}$

To add these rational expressions, we need a common denominator. To find the LCD, factor each denominator.

$$m^2 + 5m + 6 = (m+2)(m+3)$$
$$m^2 + 4m + 3 = (m+1)(m+3)$$

The least common denominator is $(m+1)(m+2)(m+3)$.

$$\frac{1}{m^2 + 5m + 6} + \frac{2}{m^2 + 4m + 3}$$

$$= \frac{1}{(m+2)(m+3)} + \frac{2}{(m+1)(m+3)}$$

$$= \frac{1(m+1)}{(m+1)(m+2)(m+3)}$$

$$+ \frac{2(m+2)}{(m+1)(m+2)(m+3)}$$

$$= \frac{(m+1) + (2m+4)}{(m+1)(m+2)(m+3)}$$

$$= \frac{3m+5}{(m+1)(m+2)(m+3)}$$

Section 5.7 Applications of Rational Expressions

5.7 Margin Exercises

1. **(a)** Let x = the number added to the numerator and subtracted from the denominator of 5/8,

    $$\frac{5+x}{8-x} = \text{reciprocal of 5/8}.$$

    Write this as an equation.

    $$\frac{5+x}{8-x} = \frac{8}{5}$$

    The least common denominator is $5(8-x)$; multiply by $5(8-x)$.

    $$5(8-x)\frac{5+x}{8-x} = 5(8-x)\frac{8}{5}$$

    $$5(5+x) = 8(8-x)$$

    $$25 + 5x = 64 - 8x$$

    $$13x = 39$$

    $$x = 3$$

    The number is 3.

    (b) Let x = the numerator of the original fraction,

    x + 1 = the denominator.

    The fraction is $\dfrac{x}{x+1}$.

    Add 6 to the numerator and subtract 6 from the denominator of this fraction to get

    $$\frac{x+6}{x+1-6} = \frac{x+6}{x-5}.$$

    The result is 15/4, or

    $$\frac{x+6}{x-5} = \frac{15}{4}.$$

Multiply by the LCD, $4(x - 5)$.

$$4(x - 5)\frac{x + 6}{x - 5} = 4(x - 5)\left(\frac{15}{4}\right)$$
$$4(x + 6) = 15(x - 5)$$
$$4x + 24 = 15x - 75$$
$$24 = 11x - 75$$
$$99 = 11x$$
$$9 = x$$

The original fraction is

$$\frac{9}{9 + 1} = \frac{9}{10}.$$

2. (a) $r = \frac{d}{t}$

   $r = \frac{500}{2.984}$

   $= 167.560$

   His average speed was 167.560 miles per hour.

   (b) $t = \frac{d}{r}$

   $t = \frac{500}{144.8}$

   $= 3.453$

   Rick Mears drove for 3.453 hours.

   (c) $d = rt$

   $d = 164 \cdot 2$

   $= 328$

   The distance between Warsaw and Rome is 328 miles.

3. (a) Let t = the number of hours until the distance between Lupe and Maria is 55 miles.

	Rate	Time	Distance
Lupe	10	t	10t
Maria	12	t	12t

$$12t + 10t = 55$$
$$22t = 55$$
$$t = \frac{55}{22} = 2.5 \text{ or } 2\frac{1}{2}$$

They will be 55 miles apart in 2 1/2 hours.

(b) Let t = the number of hours until the distance between the boats is 35 miles.

	Rate	Time	Distance
Slower boat	18	t	18t
Faster boat	25	t	25t

$$25t - 18t = 35$$
$$7t = 35$$
$$t = 5$$

They will be 35 miles apart in 5 hours.

4. (a) Let x = the speed of the boat with no current.

Complete a chart.

	d	r	t
With current	60	x + 4	$\frac{60}{x + 4}$
Against current	20	x - 4	$\frac{20}{x - 4}$

Since the times of the trips are equal,

$$\frac{60}{x + 4} = \frac{20}{x - 4}.$$

Multiply by the least common denominator, $(x + 4)(x - 4)$, to get

$$60(x - 4) = 20(x + 4)$$
$$60x - 240 = 20x + 80$$
$$40x - 240 = 80$$
$$40x = 320$$
$$x = 8.$$

The speed of the boat with no current is 8 miles per hour.

(b) Let x = the speed of the plane in still air.

Fill in the chart.

	d	r	t
With wind	450	$x + 15$	$\frac{450}{x + 15}$
Against wind	375	$x - 15$	$\frac{375}{x - 15}$

The times are the same, so

$$\frac{450}{x + 15} = \frac{375}{x - 15}.$$

Multiply by $(x + 15)(x - 15)$, the LCD, to obtain

$$450(x - 15) = 375(x + 15).$$

Divide by 25 (to make the numbers smaller).

$$18(x - 15) = 15(x + 15)$$
$$18x - 270 = 15x + 225$$
$$3x = 495$$
$$x = 165$$

The plane's speed is 165 miles per hour.

5. (a)

	Rate	Time working together	Fractional part of the job done when working together
Michael	$\frac{1}{8}$	x	$\frac{1}{8}x$
Lindsay	$\frac{1}{6}$	x	$\frac{1}{6}x$

$$\frac{1}{8}x + \frac{1}{6}x = 1$$
$$48\left(\frac{1}{8}x + \frac{1}{6}x\right) = 48(1)$$
$$48\left(\frac{1}{8}x\right) + 48\left(\frac{1}{6}x\right) = 48(1)$$
$$6x + 8x = 48$$
$$14x = 48$$
$$x = \frac{48}{14} = \frac{24}{7} = 3\frac{3}{7}$$

Working together, Michael and Lindsay can paint the room in 3 3/7 hours.

(b)

	Rate	Time working together	Fractional part of the job done when working together
Roberto	$\frac{1}{2}$	x	$\frac{1}{2}x$
Marco	$\frac{1}{3}$	x	$\frac{1}{3}x$

$$\frac{1}{2}x + \frac{1}{3}x = 1$$
$$6\left(\frac{1}{2}x + \frac{1}{3}x\right) = 6(1)$$
$$6\left(\frac{1}{2}x\right) + 6\left(\frac{1}{3}x\right) = 6$$
$$3x + 2x = 6$$
$$5x = 6$$
$$x = \frac{6}{5} = 1\frac{1}{5}$$

Working together, Roberto and Marco can tune up the Bronco in 1 1/5 hours.

6. (a) $z = kt$ z varies directly as t

$11 = 4k$ Substitute $z = 11$ and $t = 4$

$\dfrac{11}{4} = k$

Since $z = kt$ and $k = 11/4$,

$z = \dfrac{11}{4}t$ Substitute $k = 11/4$ in $z = kt$

$z = \dfrac{11}{4}(32)$ Substitute $t = 32$

$z = 88.$

(b) $C = kr$ Circumference varies directly as radius

$43.96 = 7k$ Substitute $r = 7$ and $C = 43.96$

$6.28 = k$

Since $C = kr$ and $k = 6.28$,

$C = 6.28r$

$C = 6.28(11)$ Substitute $r = 11$

$C = 69.08.$

The circumference is 69.08 centimeters when the radius is 11 centimeters.

7. (a) $z = \dfrac{k}{t}$ z varies inversely as t

$8 = \dfrac{k}{2}$ Substitute $z = 8$ and $t = 2$

$16 = k$

Since $z = k/t$ and $k = 16$,

$z = \dfrac{16}{t}$ Substitute $k = 16$ in $z = k/t$

$z = \dfrac{16}{32}$ Substitute $t = 32$

$z = \dfrac{1}{2}.$

(b) $c = \dfrac{k}{r}$ Current varies inversely as resistance

$80 = \dfrac{k}{10}$ Substitute $c = 80$ and $r = 10$

$800 = k$

Since $c = k/r$ and $k = 800$,

$c = \dfrac{800}{r}$

$c = \dfrac{800}{16}$ Substitute $r = 16$

$c = 50.$

If the resistance is 16 ohms, the current is 50 amperes.

5.7 Section Exercises

3. Let x = numerator of the original fraction;

$x - 4$ = denominantor of the original fraction.

$\dfrac{x + 3}{(x - 4) + 3} = \dfrac{3}{2}$

$\dfrac{x + 3}{x - 1} = \dfrac{3}{2}$

$2(x - 1)\left(\dfrac{x + 3}{x - 1}\right) = 2(x - 1)\left(\dfrac{3}{2}\right)$

$2(x + 3) = 3(x - 1)$

$2x + 6 = 3x - 3$

$6 = x - 3$

$9 = x$

The original fraction was

$\dfrac{9}{9 - 4} = \dfrac{9}{5}.$

5.7 Section Exercises

7. Let x = the amount invested in Mexico by the United States;

   $\frac{1}{10}x$ = the amount invested in Mexico by Germany.

   We will express amounts in billions of dollars. The total amount invested by two countries is $2 billion, so

   $$x + \frac{1}{10}x = 2.$$

   Solve this equation.

   $$10\left(x + \frac{1}{10}x\right) = 10(2)$$
   $$10x + x = 20$$
   $$11x = 20$$
   $$x = \frac{20}{11} \approx 1.818$$
   $$\frac{1}{10}x \approx .1818$$

   Germany invests approximately $.18 billion and the United States invests approximately $1.82 billion.

11. time = $\frac{\text{distance}}{\text{rate}}$

    We know the times for Amanda and Kenneth are the same, so

    $$\frac{D}{R} = \frac{d}{r}.$$

15. Let x = the time required for the trip at the lower speed.

	d	r	t
Dallas to Indianapolis	180x	180	x
Indianapolis to Dallas	150(x + 1)	150	x + 1

    The distances are equal, so

    $$180x = 150(x + 1).$$

    Solve this equation.

    $$180x = 150x + 150$$
    $$30x = 150$$
    $$x = 5$$

    The distance between the two cities is 180 · 5 = 900 miles.

19. If it takes Bill 10 hours to do a job, his rate is 1/10 of a job per hour.

23. Let x = the number of hours to pump the water using both pumps.

	Rate	Time working together	Fractional part of the job done when working together
Pump 1	$\frac{1}{10}$	x	$\frac{1}{10}x$
Pump 2	$\frac{1}{12}$	x	$\frac{1}{12}x$

    $$\frac{1}{10}x + \frac{1}{12}x = 1$$
    $$60\left(\frac{1}{10}x + \frac{1}{12}x\right) = 60 \cdot 1$$
    $$60\left(\frac{1}{10}x\right) + 60\left(\frac{1}{12}x\right) = 60$$
    $$6x + 5x = 60$$
    $$11x = 60$$
    $$x = \frac{60}{11} \approx 5.45$$

    It would take 60/11 or 5.45 hours to pump out the basement if both pumps were used.

27. If y varies directly as x, then y increases as x *increases*.

31. Let s = speed;
t = time.

The speed varies inversely with time, so there is a constant k such that s = k/t. Find the value of k by replacing s with 30 and t with 1/2.

$$s = \frac{k}{t}$$

$$30 = \frac{k}{\frac{1}{2}}$$

$$30 = 2k$$

$$15 = k$$

Since s = k/t and k = 15,

$$s = \frac{15}{t}.$$

Now find s when t = 3/4.

$$s = \frac{15}{\frac{3}{4}}$$

$$s = \frac{15}{1} \cdot \frac{4}{3}$$

$$s = \frac{60}{3} = 20$$

A speed of 20 miles per hour is needed to go the same distance in 3/4 hour.

35. Let P = pressure;
V = volume.

The variation equation is

$$P = \frac{k}{V}.$$

$$10 = \frac{k}{3} \quad \text{Let } P = 10, V = 3$$

$$30 = k$$

Thus,

$$P = \frac{30}{V}.$$

Find P when V = 1.5.

$$P = \frac{30}{1.5} = \frac{300}{15} = 20$$

The pressure is 20 pounds per square feet.

39. (a) $6x - 2 = y$
$6(-2) - 2 = y$ Let x = -2
$-12 - 2 = y$
$-14 = y$

(b) $6x - 2 = y$
$6(4) - 2 = y$ Let x = 4
$24 - 2 = y$
$22 = y$

Chapter 5 Review Exercises

1. $\frac{4}{x - 3}$

To find the values for which this expression is undefined, set the denominator equal to zero and solve for x.

$$x - 3 = 0$$
$$x = 3$$

Because x = 3 will make the denominator zero, the given expression is undefined for 3.

Chapter 5 Review Exercises   201

2. $\dfrac{y + 3}{2y}$

Set the denominator equal to zero and solve for y.

$$2y = 0$$
$$y = 0$$

The given expression is undefined for 0.

3. $\dfrac{m - 2}{m^2 - 2m - 3}$

Set the denominator equal to zero and solve for m.

$$m^2 - 2m - 3 = 0$$
$$(m - 3)(m + 1) = 0$$
$$m - 3 = 0 \quad \text{or} \quad m + 1 = 0$$
$$m = 3 \quad \text{or} \quad m = -1$$

The given expression is undefined for -1 and 3.

4. $\dfrac{2k + 1}{3k^2 + 17k + 10}$

Set the denominator equal to zero and solve for k.

$$3k^2 + 17k + 10 = 0$$
$$(3k + 2)(k + 5) = 0$$
$$3k + 2 = 0 \quad \text{or} \quad k + 5 = 0$$
$$3k = -2$$
$$k = -\dfrac{2}{3} \quad \text{or} \quad k = -5$$

The given expression is undefined for -5 and $-2/3$.

5. (a) $\dfrac{x^2}{x - 5} = \dfrac{(-2)^2}{-2 - 5}$ Let $x = -2$

$$= \dfrac{4}{-7} = -\dfrac{4}{7}$$

(b) $\dfrac{x^2}{x - 5} = \dfrac{4^2}{4 - 5}$ Let $x = 4$

$$= \dfrac{16}{-1} = -16$$

6. (a) $\dfrac{4x - 3}{5x + 2} = \dfrac{4(-2) - 3}{5(-2) + 2}$ Let $x = -2$

$$= \dfrac{-8 - 3}{-10 + 2} = \dfrac{-11}{-8} = \dfrac{11}{8}$$

(b) $\dfrac{4x - 3}{5x + 2} = \dfrac{4(4) - 3}{5(4) + 2}$ Let $x = 4$

$$= \dfrac{16 - 3}{20 + 2} = \dfrac{13}{22}$$

7. (a) $\dfrac{3x}{x^2 - 4} = \dfrac{3(-2)}{(-2)^2 - 4}$ Let $x = -2$

$$= \dfrac{-6}{4 - 4} = \dfrac{-6}{0}$$

Substituting -2 for x makes the denominator zero, so the given expression is undefined when $x = -2$.

(b) $\dfrac{3x}{x^2 - 4} = \dfrac{3(4)}{4^2 - 4}$ Let $x = 4$

$$= \dfrac{12}{16 - 4} = \dfrac{12}{12} = 1$$

8. (a) $\dfrac{x - 1}{x + 2} = \dfrac{-2 - 1}{-2 + 2}$ Let $x = -2$

$$= \dfrac{-3}{0}$$

Substituting -2 for x makes the denominator zero, so the given expression is undefined when $x = -2$.

(b) $\dfrac{x - 1}{x + 2} = \dfrac{4 - 1}{4 + 2}$ Let $x = 4$

$$= \dfrac{3}{6} = \dfrac{1}{2}$$

9. $\dfrac{5a^3b^3}{15a^4b^2} = \dfrac{b \cdot 5a^3b^2}{3a \cdot 5a^3b} = \dfrac{b}{3a}$

10. $\dfrac{m-4}{4-m} = \dfrac{-1(4-m)}{4-m} = -1$

11. $\dfrac{4x^2-9}{6-4x} = \dfrac{(2x+3)(2x-3)}{-2(2x-3)}$
    $= \dfrac{2x+3}{-2} = \dfrac{-1(2x+3)}{2}$
    $= \dfrac{-(2x+3)}{2}$

12. $\dfrac{4p^2+8pq-5q^2}{10p^2-3pq-q^2} = \dfrac{(2p-q)(2p+5q)}{(5p+q)(2p-q)}$
    $= \dfrac{2p+5q}{5p+q}$

13. $\dfrac{18p^3}{6} \cdot \dfrac{24}{p^4} = \dfrac{432p^3}{6 \cdot p^4} = \dfrac{72}{p}$

14. $\dfrac{8x^2}{12x^5} \cdot \dfrac{6x^4}{2x} = \dfrac{48x^6}{24x^6} = 2$

15. $\dfrac{9m^2}{(3m)^4} \div \dfrac{6m^5}{36m} = \dfrac{9m^2}{(3m)^4} \cdot \dfrac{36m}{6m^5}$
    $= \dfrac{9m^2}{81m^4} \cdot \dfrac{36m}{6m^5}$
    $= \dfrac{6m^3}{9m^9} = \dfrac{2}{3m^6}$

16. $\dfrac{x-3}{4} \cdot \dfrac{5}{2x-6} = \dfrac{x-3}{4} \cdot \dfrac{5}{2(x-3)} = \dfrac{5}{8}$

17. $\dfrac{3q+3}{5-6q} \div \dfrac{4q+4}{2(5-6q)}$
    $= \dfrac{3q+3}{5-6q} \cdot \dfrac{2(5-6q)}{4q+4}$
    $= \dfrac{3(q+1)}{5-6q} \cdot \dfrac{2(5-6q)}{4(q+1)}$
    $= \dfrac{6}{4} = \dfrac{3}{2}$

18. $\dfrac{2r+3}{r-4} \cdot \dfrac{r^2-16}{6r+9}$
    $= \dfrac{2r+3}{r-4} \cdot \dfrac{(r+4)(r-4)}{3(2r+3)}$
    $= \dfrac{r+4}{3}$

19. $\dfrac{6a^2+7a-3}{2a^2-a-6} \div \dfrac{a+5}{a-2}$
    $= \dfrac{6a^2+7a-3}{2a^2-a-6} \cdot \dfrac{a-2}{a+5}$
    $= \dfrac{(3a-1)(2a+3)}{(2a+3)(a-2)} \cdot \dfrac{a-2}{a+5}$
    $= \dfrac{3a-1}{a+5}$

20. $\dfrac{y^2-6y+8}{y^2+3y-18} \div \dfrac{y-4}{y+6}$
    $= \dfrac{y^2-6y+8}{y^2+3y-18} \cdot \dfrac{y+6}{y-4}$
    $= \dfrac{(y-4)(y-2)}{(y+6)(y-3)} \cdot \dfrac{y+6}{y-4}$
    $= \dfrac{y-2}{y-3}$

21. $\dfrac{2p^2+13p+20}{p^2+p-12} \cdot \dfrac{p^2+2p-15}{2p^2+7p+5}$
    $= \dfrac{(2p+5)(p+4)}{(p+4)(p-3)} \cdot \dfrac{(p+5)(p-3)}{(2p+5)(p+1)}$
    $= \dfrac{p+5}{p+1}$

22. $\dfrac{3z^2+5z-2}{9z^2-1} \cdot \dfrac{9z^2+6z+1}{z^2+5z+6}$
    $= \dfrac{(3z-1)(z+2)}{(3z-1)(3z+1)} \cdot \dfrac{(3z+1)^2}{(z+3)(z+2)}$
    $= \dfrac{3z+1}{z+3}$

23. $\dfrac{1}{8}, \dfrac{5}{12}, \dfrac{7}{32}$

   Factor the denominators.
   $$8 = 2^3$$
   $$12 = 2^2 \cdot 3$$
   $$32 = 2^5$$
   $$\text{LCD} = 2^5 \cdot 3 = 96$$

24. $\dfrac{4}{9y}, \dfrac{7}{12y^2}, \dfrac{5}{27y^4}$

   Factor the denominators.
   $$9y = 3^2 y$$
   $$12y^2 = 2^2 \cdot 3 \cdot y^2$$
   $$27y^4 = 3^3 \cdot y^4$$
   $$\text{LCD} = 2^2 \cdot 3^3 \cdot y^4 = 108y^4$$

25. $\dfrac{1}{m^2 + 2m}, \dfrac{4}{m^2 + 7m + 10}$

   Factor the denominators.
   $$m^2 + 2m = m(m + 2)$$
   $$m^2 + 7m + 10 = (m + 2)(m + 5)$$
   $$\text{LCD} = m(m + 2)(m + 5)$$

26. $\dfrac{3}{x^2 + 4x + 3}, \dfrac{5}{x^2 + 5x + 4}$

   Factor the denominators.
   $$x^2 + 4x + 3 = (x + 3)(x + 1)$$
   $$x^2 + 5x + 4 = (x + 1)(x + 4)$$
   $$\text{LCD} = (x + 3)(x + 1)(x + 4)$$

27. $\dfrac{5}{8} = \dfrac{}{56}$

   $$\dfrac{5}{8} = \dfrac{5}{8} \cdot \dfrac{7}{7} = \dfrac{35}{56}$$

28. $\dfrac{10}{k} = \dfrac{}{4k}$

   $$\dfrac{10}{k} = \dfrac{10}{k} \cdot \dfrac{4}{4} = \dfrac{40}{4k}$$

29. $\dfrac{3}{2a^3} = \dfrac{}{10a^4}$

   $$\dfrac{3}{2a^3} = \dfrac{3}{2a^3} \cdot \dfrac{5a}{5a} = \dfrac{15a}{10a^4}$$

30. $\dfrac{9}{x - 3} = \dfrac{}{18 - 6x} = \dfrac{}{6(3 - x)}$

   $$\dfrac{9}{x - 3} = \dfrac{9}{x - 3} \cdot \dfrac{-6}{-6}$$
   $$= \dfrac{-54}{-6x + 18}$$
   $$= \dfrac{-54}{18 - 6x}$$

31. $\dfrac{-3y}{2y - 10} = \dfrac{}{50 - 10y} = \dfrac{}{-5(2y - 10)}$

   $$\dfrac{-3y}{2y - 10} = \dfrac{-3y}{2y - 10} \cdot \dfrac{-5}{-5}$$
   $$= \dfrac{15y}{-10y + 50}$$
   $$= \dfrac{15y}{50 - 10y}$$

32. $\dfrac{4b}{b^2 + 2b - 3} = \dfrac{}{(b + 3)(b - 1)(b + 2)}$

   $$\dfrac{4b}{b^2 + 2b - 3} = \dfrac{4b}{(b + 3)(b - 1)}$$
   $$= \dfrac{4}{(b + 3)(b - 1)} \cdot \dfrac{b + 2}{b + 2}$$
   $$= \dfrac{4b(b + 2)}{(b + 3)(b - 1)(b + 2)}$$

33. $\dfrac{10}{x} + \dfrac{5}{x} = \dfrac{10 + 5}{x} = \dfrac{15}{x}$

34. $\dfrac{6}{3p} - \dfrac{12}{3p} = \dfrac{6 - 12}{3p} = \dfrac{-6}{3p} = \dfrac{-2}{p}$

35. $\dfrac{9}{k} - \dfrac{5}{k-5} = \dfrac{9(k-5)}{k(k-5)} - \dfrac{5k}{(k-5)k}$ $\quad LCD = k(k-5)$

$= \dfrac{9(k-5) - 5k}{k(k-5)}$

$= \dfrac{9k - 45 - 5k}{k(k-5)}$

$= \dfrac{4k - 45}{k(k-5)}$

36. $\dfrac{4}{y} + \dfrac{7}{7+y} = \dfrac{4(7+y)}{y(7+y)} + \dfrac{7 \cdot y}{(7+y)y}$ $\quad LCD = y(7+y)$

$= \dfrac{28 + 4y + 7y}{y(7+y)}$

$= \dfrac{28 + 11y}{y(7+y)}$

37. $\dfrac{m}{3} - \dfrac{2 + 5m}{6} = \dfrac{m \cdot 2}{3 \cdot 2} - \dfrac{2 + 5m}{6}$ $\quad LCD = 6$

$= \dfrac{2m - (2 + 5m)}{6}$

$= \dfrac{2m - 2 - 5m}{6}$

$= \dfrac{-2 - 3m}{6}$

38. $\dfrac{12}{x^2} - \dfrac{3}{4x} = \dfrac{12 \cdot 4}{x^2 \cdot 4} - \dfrac{3 \cdot x}{4x \cdot x}$ $\quad LCD = 4x^2$

$= \dfrac{48 - 3x}{4x^2}$

$= \dfrac{3(16 - x)}{4x^2}$

39. $\dfrac{5}{a - 2b} + \dfrac{2}{a + 2b}$

$= \dfrac{5(a + 2b)}{(a - 2b)(a + 2b)} + \dfrac{2(a - 2b)}{(a + 2b)(a - 2b)}$
$\quad LCD = (a - 2b)(a + 2b)$

$= \dfrac{5(a + 2b) + 2(a - 2b)}{(a - 2b)(a + 2b)}$

$= \dfrac{5a + 10b + 2a - 4b}{(a - 2b)(a + 2b)}$

$= \dfrac{7a + 6b}{(a - 2b)(a + 2b)}$

40. $\dfrac{4}{k^2 - 9} - \dfrac{k + 3}{3k - 9}$

$= \dfrac{4}{(k + 3)(k - 3)} - \dfrac{k + 3}{3(k - 3)}$
$\quad LCD = 3(k + 3)(k - 3)$

$= \dfrac{4 \cdot 3}{(k + 3)(k - 3) \cdot 3} - \dfrac{(k + 3)(k + 3)}{3(k - 3)(k + 3)}$

$= \dfrac{12 - (k + 3)(k + 3)}{3(k + 3)(k - 3)}$

$= \dfrac{12 - (k^2 + 6k + 9)}{3(k + 3)(k - 3)}$

$= \dfrac{12 - k^2 - 6k - 9}{3(k + 3)(k - 3)}$

$= \dfrac{-k^2 - 6k + 3}{3(k + 3)(k - 3)}$

41. $\dfrac{8}{z^2 + 6z} - \dfrac{3}{z^2 + 4z - 12}$

$= \dfrac{8}{z(z + 6)} - \dfrac{3}{(z + 6)(z - 2)}$
$\quad LCD = z(z + 6)(z - 2)$

$= \dfrac{8(z - 2)}{z(z + 6)(z - 2)} - \dfrac{3 \cdot z}{(z + 6)(z - 2) \cdot z}$

$= \dfrac{8(z - 2) - 3z}{z(z + 6)(z - 2)}$

$= \dfrac{8z - 16 - 3z}{z(z + 6)(z - 2)}$

$= \dfrac{5z - 16}{z(z + 6)(z - 2)}$

42. $\dfrac{11}{2p - p^2} - \dfrac{2}{p^2 - 5p + 6}$

$= \dfrac{11}{p(2 - p)} - \dfrac{2}{(p - 3)(p - 2)}$

$\quad LCD = p(p - 3)(p - 2)$

$= \dfrac{11(-1)(p - 3)}{p(2 - p)(-1)(p - 3)}$

$ - \dfrac{2p}{(p - 3)(p - 2)p}$

$= \dfrac{-11(p - 3) - 2p}{p(p - 2)(p - 3)}$

$= \dfrac{-11p + 33 - 2p}{p(p - 2)(p - 3)}$

$= \dfrac{-13p + 33}{p(p - 2)(p - 3)}$

43. $\dfrac{\frac{a^4}{b^2}}{\frac{a^3}{b}} = \dfrac{a^4}{b^2} \cdot \dfrac{b}{a^3}$

$= \dfrac{a^4 b}{a^3 b^2} = \dfrac{a}{b}$

44. $\dfrac{\frac{y - 3}{y}}{\frac{y + 3}{4y}} = \dfrac{y - 3}{y} \cdot \dfrac{4y}{y + 3}$

$= \dfrac{4(y - 3)}{y + 3}$

45. $\dfrac{\frac{3m + 2}{m}}{\frac{2m - 5}{6m}} = \dfrac{3m + 2}{m} \cdot \dfrac{6m}{2m - 5}$

$= \dfrac{6(3m + 2)}{2m - 5}$

46. $\dfrac{\frac{1}{p} - \frac{1}{q}}{\frac{1}{q - p}} = \dfrac{\left(\frac{1}{p} - \frac{1}{q}\right)pq(q - p)}{\left(\frac{1}{q - p}\right)pq(q - p)}$

  Multiply by the LCD, $pq(q - p)$

$= \dfrac{\frac{1}{p}[pq(q - p)] - \frac{1}{q}[pq(q - p)]}{pq}$

$= \dfrac{q(q - p) - p(q - p)}{pq}$

$= \dfrac{q^2 - pq - pq + p^2}{pq}$

$= \dfrac{q^2 - 2pq + p^2}{pq}$

$= \dfrac{(q - p)^2}{pq}$

47. $\dfrac{x + \frac{1}{w}}{x - \frac{1}{w}}$

$= \dfrac{\left(x + \frac{1}{w}\right) \cdot w}{\left(x - \frac{1}{w}\right) \cdot w}$ *Multiply by the LCD, w*

$= \dfrac{xw + \left(\frac{1}{w}\right)w}{xw - \left(\frac{1}{w}\right)w}$

$= \dfrac{xw + 1}{xw - 1}$

48. $\dfrac{\frac{1}{r + t} - 1}{\frac{1}{r + t} + 1}$

$= \dfrac{\left(\frac{1}{r + t} - 1\right)(r + t)}{\left(\frac{1}{r + t} + 1\right)(r + t)}$

  Multiply by the LCD, $r + t$

$= \dfrac{\frac{1}{r + t}(r + t) - 1(r + t)}{\frac{1}{r + t}(r + t) + 1(r + t)}$

$= \dfrac{1 - r - t}{1 + r + t}$

49. $\dfrac{k}{5} - \dfrac{2}{3} = \dfrac{1}{2}$

Multiply each side by the LCD, $2 \cdot 3 \cdot 5 = 30$.

$$30\left(\frac{k}{5} - \frac{2}{3}\right) = 30 \cdot \frac{1}{2}$$

$$30\left(\frac{k}{5}\right) - 30\left(\frac{2}{3}\right) = 15$$

$$6k - 20 = 15$$

$$6k = 35$$

$$k = \frac{35}{6}$$

A check will verify that 35/6 is the solution.

50. $\frac{4-z}{z} + \frac{3}{2} = \frac{-4}{z}$

Multiply each side by the LCD, 2z.

$$2z\left(\frac{4-z}{z} + \frac{3}{2}\right) = 2z\left(-\frac{4}{z}\right)$$

$$2z\left(\frac{4-z}{z}\right) + 2z\left(\frac{3}{2}\right) = -8$$

$$2(4 - z) + 3z = -8$$

$$8 - 2z + 3z = -8$$

$$8 + z = -8$$

$$z = -16$$

A check will verify that −16 is the solution.

51. $\frac{x}{2} - \frac{x-3}{7} = -1$

Multiply each side by the LCD, 14.

$$14\left(\frac{x}{2} - \frac{x-3}{7}\right) = 14(-1)$$

$$14\left(\frac{x}{2}\right) - 14\left(\frac{x-3}{7}\right) = -14$$

$$7x - 2(x - 3) = -14$$

$$7x - 2x + 6 = -14$$

$$5x + 6 = -14$$

$$5x = -20$$

$$x = -4$$

A check will verify that −4 is the solution.

52. $\frac{3y - 1}{y - 2} = \frac{5}{y - 2} + 1$

$$(y - 2)\left(\frac{3y - 1}{y - 2}\right) = (y - 2)\left(\frac{5}{y - 2} + 1\right)$$

Multiply by LCD, y − 2

$$(y - 2)\left(\frac{3y - 1}{y - 2}\right) = (y - 2)\left(\frac{5}{y - 2}\right)$$
$$+ (y + 2)(1)$$

Distributive property

$$3y - 1 = 5 + y - 2$$

$$3y - 1 = 3 + y$$

$$2y - 1 = 3$$

$$2y = 4$$

$$y = 2$$

There is no solution because y = 2 makes the original denominators equal to zero.

53. $\frac{3}{m - 2} + \frac{1}{m - 1} = \frac{7}{m^2 - 3m + 2}$

$$\frac{3}{m - 2} + \frac{1}{m - 1} = \frac{7}{(m - 2)(m - 1)}$$

$$(m - 2)(m - 1)\left(\frac{3}{m - 2} + \frac{1}{m - 1}\right)$$
$$= (m - 2)(m - 1) \cdot \frac{7}{(m - 2)(m - 1)}$$

Multiply by LCD, (m − 2)(m − 1)

$$3(m - 1) + 1(m - 2) = 7$$

$$3m - 3 + m - 2 = 7$$

$$4m - 5 = 7$$

$$4m = 12$$

$$m = 3$$

A check will verify that 3 is the solution.

54. $m = \frac{Ry}{t}$ for t

$$t \cdot m = t\left(\frac{Ry}{t}\right)$$

$$tm = Ry$$

$$t = \frac{Ry}{m}$$

55. $$x = \frac{3y - 5}{4} \text{ for } y$$
$$4x = 4\left(\frac{3y - 5}{4}\right)$$
$$4x = 3y - 5$$
$$4x + 5 = 3y$$
$$\frac{4x + 5}{3} = y$$

56. $$p^2 = \frac{4}{3m - q} \text{ for } m$$
$$(3m - q)p^2 = (3m - q)\left(\frac{4}{3m - q}\right)$$
$$3mp^2 - p^2q = 4$$
$$3mp^2 = 4 + p^2q$$
$$m = \frac{4 + p^2q}{3p^2}$$

57. Let x = the unknown number.
$$\tfrac{2}{3}x - \tfrac{1}{2}x = 2$$
$$6\left(\tfrac{2}{3}x - \tfrac{1}{2}x\right) = 6 \cdot 2$$
$$6\left(\tfrac{2}{3}x\right) - 6\left(\tfrac{1}{2}x\right) = 12$$
$$4x - 3x = 12$$
$$x = 12$$

The unknown number is 12.

58. Let x = the commission for selling a large car;
$\tfrac{2}{3}x$ = the commission for selling a small car.
$$x + \tfrac{2}{3}x = 300$$
$$3\left(x + \tfrac{2}{3}x\right) = 3 \cdot 300$$
$$3x + 3\left(\tfrac{2}{3}x\right) = 900$$
$$3x + 2x = 900$$
$$5x = 900$$
$$x = 180$$

The commission for a large car is $180, and the commission for a small car is $\tfrac{2}{3}(\$180) = \$120$.

59. $$t = \frac{d}{r}$$
$$= \frac{500}{74.59} \approx 6.7$$

The race took Ray Harroun about 6.7 hours.

60. Let x = the number of hours it takes them to do the job working together.

	Rate	Time working together	Fractional part of the job done when working together
Man	$\tfrac{1}{5}$	x	$\tfrac{1}{5}x$
Daughter	$\tfrac{1}{8}$	x	$\tfrac{1}{8}x$

Working together, they do 1 whole job, so
$$\tfrac{1}{5}x + \tfrac{1}{8}x = 1.$$

Solve this equation by multiplying both sides by the LCD, 40.
$$40\left(\tfrac{1}{5}x + \tfrac{1}{8}x\right) = 40(1)$$
$$8x + 5x = 40$$
$$13x = 40$$
$$x = \tfrac{40}{13} \approx 3.1$$

Working together, it takes them 40/13 or about 3.1 hours.

61. Let x = the time needed by the head gardener to mow the lawns;

$2x$ = the time needed by the assistant to mow the lawns.

	Rate	Time working together	Fractional part of the job done when working together
Head gardener	$\frac{1}{x}$	$1\frac{1}{3} = \frac{4}{3}$	$\frac{1}{x} \cdot \frac{4}{3} = \frac{4}{3x}$
Assistant	$\frac{1}{2x}$	$1\frac{1}{3} = \frac{4}{3}$	$\frac{1}{2x} \cdot \frac{4}{3} = \frac{2}{3x}$

$$\frac{4}{3x} + \frac{2}{3x} = 1$$

$$\frac{6}{3x} = 1$$

$$3x\left(\frac{6}{3x}\right) = 3x(1)$$

$$6 = 3x$$

$$2 = x$$

It takes the head gardener 2 hours to mow the lawns.

62. Let A = area of circle;
    r = radius of circle.

The area varies directly as the radius, so

$$A = kr.$$

Find k by replacing A with 78.5 and r with 5.

$$78.5 = k(5)$$
$$15.7 = k$$

Since $A = kr$ and $k = 15.7$,

$$A = 15.7r.$$

Now find r when $A = 100$.

$$100 = 15.7r$$
$$r = \frac{100}{15.7} \approx 6.37$$

The radius is about 6.37 inches.

63. Let h = height of parallelogram;
    b = length of base of parallelogram.

The height varies inversely as the base, so

$$h = \frac{k}{b}.$$

Find k by replacing h with 8 and b with 12.

$$8 = \frac{k}{12}$$
$$96 = k$$

Thus,

$$h = \frac{96}{b}.$$

Now find b when $h = 24$.

$$24 = \frac{96}{b}$$
$$24b = 96$$
$$b = \frac{96}{24} = 4$$

The height of the parallelogram is 4 centimeters.

64. $\dfrac{\dfrac{5}{x-y}+2}{3-\dfrac{2}{x+y}}$

To simplify this complex fraction, multiply numerator and denominator by the LCD of all the fractions, $(x-y)(x+y)$.

$= \dfrac{\left(\dfrac{5}{x-y}+2\right)(x-y)(x+y)}{\left(3-\dfrac{2}{x+y}\right)(x-y)(x+y)}$

$= \dfrac{\left(\dfrac{5}{x-y}\right)(x-y)(x+y)+2(x-y)(x+y)}{3(x-y)(x+y)-\left(\dfrac{2}{x+y}\right)(x-y)(x+y)}$

$= \dfrac{5(x+y)+2(x-y)(x+y)}{3(x-y)(x+y)-2(x-y)}$

$= \dfrac{(x+y)[5+2(x-y)]}{(x-y)[3(x+y)-2]}$

$= \dfrac{(x+y)(5+2x-2y)}{(x-y)(3x+3y-2)}$

65. $\dfrac{4}{m-1}-\dfrac{3}{x+1}$

To perform the indicated subtraction, use $(m-1)(m+1)$ as the LCD.

$\dfrac{4}{m-1}-\dfrac{3}{m+1}$

$= \dfrac{4(m+1)}{(m-1)(m+1)}-\dfrac{3(m-1)}{(m+1)(m-1)}$

$= \dfrac{4(m+1)-3(m-1)}{(m-1)(m+1)}$

$= \dfrac{4m+4-3m+3}{(m-1)(m+1)}$

$= \dfrac{m+7}{(m-1)(m+1)}$

66. $\dfrac{8p^5}{5} \div \dfrac{2p^3}{10}$

To perform the indicated division, multiply the first rational expression by the reciprocal of the second.

$\dfrac{8p^5}{5} \div \dfrac{2p^3}{10} = \dfrac{8p^5}{5} \cdot \dfrac{10}{2p^3}$

$= \dfrac{80p^5}{10p^3}$

$= 8p^2$

67. $\dfrac{r-3}{8} \div \dfrac{3r-9}{4} = \dfrac{r-3}{8} \cdot \dfrac{4}{3r-9}$

$= \dfrac{r-3}{8} \cdot \dfrac{4}{3(r-3)}$

$= \dfrac{4}{24} = \dfrac{1}{6}$

68. $\dfrac{\dfrac{5}{x}-1}{\dfrac{5-x}{3x}} = \dfrac{\left(\dfrac{5}{x}-1\right)3x}{\left(\dfrac{5-x}{3x}\right)3x}$ Multiply by LCD, $3x$

$= \dfrac{\dfrac{5}{x}(3x)-1(3x)}{5-x}$

$= \dfrac{15-3x}{5-x}$

$= \dfrac{3(5-x)}{5-x} = 3$

69. $\dfrac{4}{z^2-2z+1}-\dfrac{3}{z^2-1}$

$= \dfrac{4}{(z-1)^2}-\dfrac{3}{(z+1)(z-1)}$
$\qquad\qquad$ LCD $= (z+1)(z-1)^2$

$= \dfrac{4(z+1)}{(z-1)^2(z+1)}$

$\quad - \dfrac{3(z-1)}{(z+1)(z-1)(z-1)}$

$= \dfrac{4(z+1)-3(z-1)}{(z+1)(z-1)^2}$

$$= \frac{4z + 4 - 3z + 3}{(z + 1)(z - 1)^2}$$

$$= \frac{z + 7}{(z + 1)(z - 1)^2}$$

70. $$F = \frac{k}{d - D} \quad \text{for } d$$

$$F(d - D) = \left(\frac{k}{d - D}\right)(d - D)$$

$$Fd - FD = k$$

$$Fd = k + FD$$

$$d = \frac{k + FD}{F}$$

$$\text{or } d = \frac{k}{F} + \frac{FD}{F} = \frac{k}{F} + D$$

71. $$\frac{2}{z} - \frac{z}{z + 3} = \frac{1}{z + 3}$$

Multiply each side of the equation by the LCD, $z(z + 3)$.

$$z(z + 3)\left(\frac{2}{z} - \frac{z}{z + 3}\right) = z(z + 3)\left(\frac{1}{z + 3}\right)$$

$$z(z + 3)\left(\frac{2}{z}\right) - z(z + 3)\left(\frac{z}{z + 3}\right) = z$$

$$2(z + 3) - z^2 = z$$

$$2z + 6 - z^2 = z$$

$$0 = z^2 - z - 6$$

$$0 = (z - 3)(z + 2)$$

$$z - 3 = 0 \quad \text{or} \quad z + 2 = 0$$

$$z = 3 \quad \text{or} \quad z = -2$$

Checks will verify that -2 and 3 are both solutions.

72. Let x = number of people speaking Spanish;

$\frac{1}{10}x$ = number of people speaking French.

$$x + \frac{1}{10}x = 19.1$$

$$10\left(x + \frac{1}{10}x\right) = 10(19.1)$$

$$10x + 10\left(\frac{1}{10}x\right) = 191$$

$$10x + x = 191$$

$$11x = 191$$

$$x \approx 17.4$$

About 17.4 million people speak Spanish.

73.

	Rate	Time working together	Fractional part of the job done when working together
Pipe 1	$\frac{1}{6}$	x	$\frac{1}{6}x$
Pipe 2	$\frac{1}{9}$	x	$\frac{1}{9}x$

Working together, the two pipes fill the pool 3/4 full, so

$$\frac{1}{6}x + \frac{1}{9}x = \frac{3}{4}.$$

To clear fractions, multiply both sides by the LCD, 36.

$$36\left(\frac{1}{6}x + \frac{1}{9}x\right) = 36\left(\frac{3}{4}\right)$$

$$36\left(\frac{1}{6}x\right) + 36\left(\frac{1}{9}x\right) = 27$$

$$6x + 4x = 27$$

$$10x = 27$$

$$x = \frac{27}{10} = 2\frac{7}{10}$$

The two pipes would need to work 27/10 or 2 7/10 hours.

74. Let x = the speed of the plane in still air.

   Use the formula d = rt to fill in the table.

	d	r	t
With wind	400	x + 50	$\frac{400}{x + 50}$
Against wind	200	x − 50	$\frac{200}{x − 50}$

   The times are the same; therefore the equation is

   $$\frac{400}{x + 50} = \frac{200}{x − 50}.$$

   $400(x − 50) = 200(x + 50)$ *Cross products*

   $400x − 20{,}000 = 200x + 10{,}000$

   $200x = 30{,}000$

   $x = \frac{30{,}000}{200} = 150$

   The speed of the plane in still air is 150 kilometers per hour.

75. Let x = sales for Ford;

   $\frac{1}{2}x$ = sales for Phillip Morris.

   $x + \frac{1}{2}x = 151$

   $2\left(x + \frac{1}{2}x\right) = 2 \cdot 151$

   $2x + 2\left(\frac{1}{2}x\right) = 302$

   $2x + x = 302$

   $3x = 302$

   $x \approx 101$

   The sales for Ford were about $101 billion.

Chapter 5 Test

1. $\dfrac{3x − 1}{x^2 − 2x − 8}$

   Set the denominator equal to zero and solve for x.

   $x^2 − 2x − 8 = 0$

   $(x + 2)(x − 4) = 0$

   $x + 2 = 0$ or $x − 4 = 0$

   $x = -2$ or $x = 4$

   The expression is undefined for −2 and 4.

2. $\dfrac{4m^2 − 2m}{6m − 3} = \dfrac{2m(2m − 1)}{3(2m − 1)} = \dfrac{2m}{3}$

3. $\dfrac{6a^2 + a − 2}{2a^2 − 3a + 1} = \dfrac{(3a + 2)(2a − 1)}{(2a − 1)(a − 1)}$

   $= \dfrac{3a + 2}{a − 1}$

4. $\dfrac{x^6 y}{x^3} \cdot \dfrac{y^2}{x^2 y^3} = \dfrac{x^6 y^3}{x^5 y^3} = x$

5. $\dfrac{5(d − 2)}{9} \div \dfrac{3(d − 2)}{5}$

   $= \dfrac{5(d − 2)}{9} \cdot \dfrac{5}{3(d − 2)}$

   $= \dfrac{5 \cdot 5}{9 \cdot 3} = \dfrac{25}{27}$

6. $\dfrac{6k^2 − k − 2}{8k^2 + 10k + 3} \cdot \dfrac{4k^2 + 7k + 3}{3k^2 + 5k + 2}$

   $= \dfrac{(3k − 2)(2k + 1)}{(4k + 3)(2k + 1)} \cdot \dfrac{(4k + 3)(k + 1)}{(3k + 2)(k + 1)}$

   $= \dfrac{3k − 2}{3k + 2}$

7. $\dfrac{4a^2 + 9a + 2}{3a^2 + 11a + 10} \div \dfrac{4a^2 + 17a + 4}{3a^2 + 2a - 5}$

$= \dfrac{4a^2 + 9a + 2}{3a^2 + 11a + 10} \cdot \dfrac{3a^2 + 2a - 5}{4a^2 + 17a + 4}$

$= \dfrac{(4a + 1)(a + 2)}{(3a + 5)(a + 2)} \cdot \dfrac{(3a + 5)(a - 1)}{(4a + 1)(a + 4)}$

$= \dfrac{a - 1}{a + 4}$

8. $\dfrac{15}{4p} = \dfrac{}{64p^3}$

$\dfrac{15}{4p} = \dfrac{15 \cdot 16p^2}{4p \cdot 16p^2} = \dfrac{240p^2}{64p^3}$

9. $\dfrac{3}{6m - 12} = \dfrac{}{42m - 84} = \dfrac{}{7(6m - 12)}$

$\dfrac{3}{6m - 12} = \dfrac{3 \cdot 7}{(6m - 12)7} = \dfrac{21}{42m - 84}$

10. $\dfrac{8}{c} - \dfrac{5}{c} = \dfrac{8 - 5}{c} = \dfrac{3}{c}$

11. $\dfrac{-4}{y + 2} + \dfrac{6}{5y + 10}$

$= \dfrac{-4}{y + 2} + \dfrac{6}{5(y + 2)} \quad LCD = 5(y + 2)$

$= \dfrac{-4 \cdot 5}{(y + 2) \cdot 5} + \dfrac{6}{5(y + 2)}$

$= \dfrac{-20 + 6}{5(y + 2)} = \dfrac{-14}{5(y + 2)}$

12. $\dfrac{3}{2m^2 - 9m - 5} - \dfrac{m + 1}{2m^2 - m - 1}$

$= \dfrac{3}{(2m + 1)(m - 5)} - \dfrac{m + 1}{(2m + 1)(m - 1)}$

$\quad LCD = (2m + 1)(m - 5)(m - 1)$

$= \dfrac{3(m - 1)}{(2m + 1)(m - 5)(m - 1)}$

$\quad - \dfrac{(m + 1)(m - 5)}{(2m + 1)(m - 1)(m - 5)}$

$= \dfrac{3(m - 1) - (m + 1)(m - 5)}{(2m + 1)(m - 5)(m - 1)}$

$= \dfrac{(3m - 3) - (m^2 - 4m - 5)}{(2m + 1)(m - 5)(m - 1)}$

$= \dfrac{3m - 3 - m^2 + 4m + 5}{(2m + 1)(m - 5)(m - 1)}$

$= \dfrac{-m^2 + 7m + 2}{(2m + 1)(m - 5)(m - 1)}$

13. $\dfrac{\frac{2p}{k^2}}{\frac{3p^2}{k^3}} = \dfrac{2p}{k^2} \cdot \dfrac{k^3}{3p^2}$

$= \dfrac{2k^3 p}{3k^2 p^2} = \dfrac{2k}{3p}$

14. $\dfrac{\frac{1}{x + 3} - 1}{1 + \frac{1}{x + 3}}$

$= \dfrac{(x + 3)\left(\frac{1}{x + 3} - 1\right)}{(x + 3)\left(1 + \frac{1}{x + 3}\right)} \quad \text{Multiply by } LCD, \ x + 3$

$= \dfrac{(x + 3)\left(\frac{1}{x + 3}\right) - (x + 3)(1)}{(x + 3)(1) + (x + 3)\left(\frac{1}{x + 3}\right)}$

$= \dfrac{1 - (x + 3)}{(x + 3) + 1}$

$= \dfrac{1 - x - 3}{x + 4}$

$= \dfrac{-2 - x}{x + 4}$

15. x cannot have a value that would make a denominator equal zero. Find these values.

$x + 1 = 0 \qquad x - 4 = 0$

$x = -1 \qquad x = 4$

x cannot be −1 or 4.

16. $\dfrac{4}{3y} + \dfrac{3}{5y} = -\dfrac{11}{30}$ LCD = $30y$

$30y\left(\dfrac{4}{3y} + \dfrac{3}{5y}\right) = 30y\left(-\dfrac{11}{30}\right)$

$30y\left(\dfrac{4}{3y}\right) + 30y\left(\dfrac{3}{5y}\right) = -11y$

$40 + 18 = -11y$

$58 = -11y$

$-\dfrac{58}{11} = y$

17. $\dfrac{2}{p^2 - 2p - 3} = \dfrac{3}{p - 3} + \dfrac{2}{p + 1}$

To find the LCD, factor the denominator on the left.

$\dfrac{2}{(p - 3)(p + 1)} = \dfrac{3}{p - 3} + \dfrac{2}{p + 1}$

Multiply each side by the LCD, $(p - 3)(p + 1)$.

$(p - 3)(p + 1)\left(\dfrac{2}{(p - 3)(p + 1)}\right)$

$= (p - 3)(p + 1)\left(\dfrac{3}{p - 3} + \dfrac{2}{p + 1}\right)$

$2 = (p - 3)(p + 1)\left(\dfrac{3}{p - 3}\right)$

$ + (p - 3)(p + 1)\left(\dfrac{2}{p + 1}\right)$

$2 = 3(p + 1) + 2(p - 3)$

$2 = 3p + 3 + 2p - 6$

$2 = 5p - 3$

$5 = 5p$

$1 = p$

A check will verify that 1 is the solution.

18. Let x = the unknown number.

$4x + \dfrac{1}{2x} = 3$

$2x\left(4x + \dfrac{1}{2x}\right) = 2x(3)$

$(2x)(4x) + 2x\left(\dfrac{1}{2x}\right) = 3(2x)$

$8x^2 + 1 = 6x$

$8x^2 - 6x + 1 = 0$

$(4x - 1)(2x - 1) = 0$

$4x - 1 = 0$ or $2x - 1 = 0$

$4x = 1$ $2x = 1$

$x = \dfrac{1}{4}$ or $x = \dfrac{1}{2}$

The number is 1/4 or 1/2.

19. Let x = the speed of the current.

	Rate	Time	Distance
Upstream	$7 - x$	$\dfrac{20}{7 - x}$	20
Downstream	$7 + x$	$\dfrac{50}{7 + x}$	50

The times are equal, so

$\dfrac{20}{7 - x} = \dfrac{50}{7 + x}$.

To solve this equation, begin by multiplying both sides by the LCD, $(7 - x)(7 + x)$.

$(7 - x)(7 + x)\left(\dfrac{20}{7 - x}\right)$

$= (7 - x)(7 + x)\left(\dfrac{50}{7 + x}\right)$

$20(7 + x) = 50(7 - x)$

$140 + 20x = 350 - 50x$

$140 + 70x = 350$

$70x = 210$

$x = 3$

The speed of the current is 3 miles per hour.

20. Let c = current;
    r = resistance.

    The current varies inversely as the resistance, so

    $$c = \frac{k}{r}.$$

    Find k by replacing c with 50 and r with 10.

    $$50 = \frac{k}{10}$$
    $$500 = k$$

    Thus,

    $$c = \frac{500}{r}.$$

    Now find c when r = 5.

    $$c = \frac{500}{5} = 100$$

    If the resistance is 5 ohms, the current is 100 amperes.

Cumulative Review: Chapters R–5

1. $$\frac{5}{8}k = 4$$
   $$8 \cdot \frac{5}{8}k = 8 \cdot 4$$
   $$5k = 32$$
   $$k = \frac{32}{5}$$

2. $$3(2y - 5) = 2 + 5y$$
   $$6y - 15 = 2 + 5y$$
   $$y - 15 = 2$$
   $$y = 17$$

3. $$A = \frac{1}{2}bh \text{ for } b$$
   $$2 \cdot A = 2 \cdot \frac{1}{2}bh$$
   $$2A = bh$$
   $$\frac{2A}{h} = b$$

4. $$\frac{2 + m}{2 - m} = \frac{3}{4}$$
   $$4(2 - m)\left(\frac{2 + m}{2 - m}\right) = 4(2 - m)\left(\frac{3}{4}\right)$$
   $$4(2 + m) = 3(2 - m)$$
   $$8 + 4m = 6 - 3m$$
   $$8 + 7m = 6$$
   $$7m = -2$$
   $$m = -\frac{2}{7}$$

5. $$5y \leq 6y + 8$$
   $$-y \leq 8$$
   $$y \geq -8$$

6. $$5m - 9 > 2m + 3$$
   $$3m - 9 > 3$$
   $$3m > 12$$
   $$m > 4$$

7. $$\left(\frac{3}{4}\right)^{-2} = \left(\frac{4}{3}\right)^{2} = \frac{4^2}{3^2} = \frac{16}{9}$$

8. $$\frac{7^{-1}}{7} = 7^{-1-1} = 7^{-2} = \frac{1}{7^2} = \frac{1}{49}$$

9. $\dfrac{(4^{-2})^3}{4^6 4^{-3}} = \dfrac{4^{(-2)(3)}}{4^{6+(-3)}} = \dfrac{4^{-6}}{4^3} = 4^{-6-3}$
$= 4^{-9} = \dfrac{1}{4^9}$

10. $\dfrac{(2x^3)^{-1}x}{2^3 x^5} = \dfrac{2^{-1}(x^3)^{-1}x}{2^3 x^5} = \dfrac{2^{-1}x^{-3}x}{2^3 x^5}$
$= \dfrac{2^{-1}x^{-2}}{2^3 x^5} = 2^{-1-3}x^{-2-5}$
$= 2^{-4}x^{-7} = \dfrac{1}{2^4 x^7}$

11. $\dfrac{(m^{-2})^3 m}{m^5 m^{-4}} = \dfrac{m^{-6}m}{m^5 m^{-4}} = \dfrac{m^{-6+1}}{m^{5+(-4)}} = \dfrac{m^{-5}}{m}$
$= m^{-5-1} = m^{-6} = \dfrac{1}{m^6}$

12. $\dfrac{2p^3 q^4}{8p^5 q^3} = \dfrac{p^{-2}q}{4} = \dfrac{q}{4p^2}$

13. $(2k^2 + 3k) - (k^2 + k - 1)$
$= 2k^2 + 3k - k^2 - k + 1$
$= k^2 + 2k + 1$

14. $8x^2 y^2 (9x^4 y^5) = 72x^6 y^7$

15. $(2a - b)^2 = (2a)^2 - 2(2a)(b) + (b)^2$
$= 4a^2 - 4ab + b^2$

16. $(y^2 + 3y + 5)(3y - 1)$

   Multiply vertically.

$$\begin{array}{r} y^2 + 3y + 5 \\ 3y - 1 \\ \hline -y^2 - 3y - 5 \\ 3y^3 + 9y^2 + 15y \\ \hline 3y^3 + 8y^2 + 12y - 5 \end{array}$$

17. $\dfrac{12p^3 + 2p^2 - 12p + 4}{2p - 2}$

$$\begin{array}{r} 6p^2 + 7p + 1 \\ 2p - 2 \overline{\smash{)}12p^3 + 2p^2 - 12p + 4} \\ \underline{12p^3 - 12p^2} \\ 14p^2 - 12p \\ \underline{14p^2 - 14p} \\ 2p + 4 \\ \underline{2p - 2} \\ 6 \end{array}$$

Answer: $6p^2 + 7p + 1 + \dfrac{6}{2p - 2}$
$= 6p^2 + 7p + 1 + \dfrac{2 \cdot 3}{2(p - 1)}$
$= 6p^2 + 7p + 1 + \dfrac{3}{p - 1}$

18. $8t^2 + 10tv + 3v^2$
$= 8t^2 + 6tv + 4tv + 3v^2$
$= 2t(4t + 3v) + v(4t + 3v)$
$= (4t + 3v)(2t + v)$

19. $8r^2 - 9rs + 12s^2$

To factor this polynomial by the grouping method, we must find two integers whose product is $(8)(12) = 96$ and whose sum is -9. There is no pair of integers that satisfies both of these conditions, so the polynomial is prime.

20. $6a^2 m + am - 2m$
$= m(6a^2 + a - 2)$
$= m(2a - 1)(3a + 2)$

21. $ r^2 = 2r + 15$
$r^2 - 2r - 15 = 0$
$(r - 5)(r + 3) = 0$
$r + 3 = 0 \quad \text{or} \quad r - 5 = 0$
$r = -3 \quad \text{or} \quad r = 5$

The solutions are -3 and 5.

216   Chapter 5   Rational Expressions

22. $8m^2 = 64m$
    $8m^2 - 64m = 0$
    $8m(m - 8) = 0$
    $8m = 0$ or $m - 8 = 0$
    $m = 0$ or $m = 8$

   The solutions are 0 and 8.

23. $(r - 5)(2r + 1)(3r - 2) = 0$
    $r - 5 = 0$ or $2r + 1 = 0$ or $3r - 2 = 0$
    $\qquad\qquad\qquad 2r = -1 \qquad 3r = 2$
    $r = 5$ or $\quad r = -\dfrac{1}{2}\quad$ or $\quad r = \dfrac{2}{3}$

   The solutions are $-1/2$, $2/3$, and 5.

24. Let x = the smaller number;
    $x + 4$ = the larger number.
    $\quad x(x + 4) = x - 2$
    $\quad x^2 + 4x = x - 2$
    $\quad x^2 + 3x + 2 = 0$
    $\quad (x + 2)(x + 1) = 0$
    $\quad x + 2 = 0$ or $x + 1 = 0$
    $\quad\quad x = -2$ or $\quad x = -1$

   The smaller number can be either -2 or -1.

25. Let x = the width of the rectangle
    $2x - 2$ = the length of the rectangle.
    Use the formula $A = LW$.
    $\quad x(2x - 2) = 60$
    $\quad\quad 2x^2 - 2x = 60$
    $\quad 2x^2 - 2x - 60 = 0$
    $\quad 2(x^2 - x - 30) = 0$
    $\quad 2(x - 6)(x + 5) = 0$
    $\quad x - 6 = 0$ or $x + 5 = 0$
    $\quad\quad x = 6$ or $\quad x = -5$

   Discard -5 because the width cannot be negative. The width of the rectangle is 6 meters.

26. $\dfrac{x^6 y^2}{y^5 x^9} \cdot \dfrac{x^2}{y^3} = \dfrac{x^8 y^2}{x^9 y^8} = x^{8-9} y^{2-8}$
    $\qquad\qquad\qquad = x^{-1} y^{-6} = \dfrac{1}{xy^6}$

27. $\dfrac{5}{q} - \dfrac{1}{q} = \dfrac{5 - 1}{q} = \dfrac{4}{q}$

28. $\dfrac{3}{7} + \dfrac{4}{r} = \dfrac{3 \cdot r}{7 \cdot r} + \dfrac{4 \cdot 7}{r \cdot 7}\quad$ LCD $= 7r$
    $\qquad = \dfrac{3r + 28}{7r}$

29. $\dfrac{4}{5q - 20} - \dfrac{1}{3q - 12}$
    $= \dfrac{4}{5(q - 4)} - \dfrac{1}{3(q - 4)}$
    $= \dfrac{4 \cdot 3}{5(q - 4) \cdot 3} - \dfrac{1 \cdot 5}{3(q - 4) \cdot 5}$
    $\qquad\qquad\qquad$ LCD $= 5 \cdot 3 \cdot (q - 4)$
    $\qquad\qquad\qquad\qquad = 15(q - 4)$
    $= \dfrac{12 - 5}{15(q - 4)} = \dfrac{7}{15(q - 4)}$

30. $\dfrac{2}{k^2 + k} - \dfrac{3}{k^2 - k}$
    $= \dfrac{2}{k(k + 1)} - \dfrac{3}{k(k - 1)}$
    $= \dfrac{2(k - 1)}{k(k + 1)(k - 1)} - \dfrac{3(k + 1)}{k(k - 1)(k + 1)}$
    $\qquad\qquad$ LCD $= k(k + 1)(k - 1)$
    $= \dfrac{2(k - 1) - 3(k + 1)}{k(k + 1)(k - 1)}$
    $= \dfrac{2k - 2 - 3k - 3}{k(k + 1)(k - 1)}$
    $= \dfrac{-k - 5}{k(k + 1)(k - 1)}$

31. $\dfrac{7z^2 + 49z + 70}{16z^2 + 72z - 40} \div \dfrac{3z + 6}{4z^2 - 1}$

$= \dfrac{7z^2 + 49z + 70}{16z^2 + 72z - 40} \cdot \dfrac{4z^2 - 1}{3z + 6}$

$= \dfrac{7(z^2 + 7z + 10)}{8(2z^2 + 9z - 5)} \cdot \dfrac{(2z + 1)(2z - 1)}{3(z + 2)}$

$= \dfrac{7(z + 5)(z + 2)}{8(2z - 1)(z + 5)} \cdot \dfrac{(2z + 1)(2z - 1)}{3(z + 2)}$

$= \dfrac{7(2z + 1)}{8 \cdot 3} = \dfrac{7(2z + 1)}{24}$

32. $\dfrac{\dfrac{4}{a} + \dfrac{5}{2a}}{\dfrac{7}{6a} - \dfrac{1}{5a}}$

$= \dfrac{\left(\dfrac{4}{a} + \dfrac{5}{2a}\right) \cdot 30a}{\left(\dfrac{7}{6a} - \dfrac{1}{5a}\right) \cdot 30a}$ *Mutiply by the LCD, 30a*

$= \dfrac{\dfrac{4}{a}(30a) + \dfrac{5}{5a}(30a)}{\dfrac{7}{6a}(30a) - \dfrac{1}{5a}(30a)}$

$= \dfrac{4 \cdot 30 + 5 \cdot 15}{7 \cdot 5 - 1 \cdot 6}$

$= \dfrac{120 + 75}{35 - 6}$

$= \dfrac{195}{29}$

33. $\dfrac{r + 2}{5} = \dfrac{r - 3}{3}$

$15\left(\dfrac{r + 2}{5}\right) = 15\left(\dfrac{r - 3}{3}\right)$ *Multiply by LCD, 15*

$3(r + 2) = 5(r - 3)$

$3r + 6 = 5r - 15$

$6 = 2r - 15$

$21 = 2r$

$\dfrac{21}{2} = r$

The solution is 21/2.

34. $\dfrac{1}{x} = \dfrac{1}{x + 1} + \dfrac{1}{2}$

$2x(x + 1)\left(\dfrac{1}{x}\right) = 2x(x + 1)\left(\dfrac{1}{x + 1} + \dfrac{1}{2}\right)$

  Multiply by LCD, $2x(x + 1)$

$2(x + 1) = 2x(x + 1)\left(\dfrac{1}{x + 1}\right)$

$\qquad\qquad + 2x(x + 1)\left(\dfrac{1}{2}\right)$

$2(x + 1) = 2x + x(x + 1)$

$2x + 2 = 2x + x^2 + x$

$2x + 2 = x^2 + 3x$

$0 = x^2 + x - 2$

$0 = (x + 2)(x - 1)$

$x + 2 = 0 \quad \text{or} \quad x - 1 = 0$

$x = -2 \quad \text{or} \quad\quad x = 1$

Checks will verify that -2 and 1 are both solutions.

35.

	Rate	Time	Distance
To business	60	x	$60x$
Coming home	50	$x + \dfrac{1}{2}$	$50\left(x + \dfrac{1}{2}\right)$

$60x = 50\left(x + \dfrac{1}{2}\right)$

$60x = 50x + 50\left(\dfrac{1}{2}\right)$

$60x = 50x + 25$

$10x = 25$

$x = \dfrac{25}{10} = \dfrac{5}{2}$

Arlene's one way trip was $60(5/2) = 30 \cdot 5 = 150$ miles.

Chapter 5 Rational Expressions

36.

	Rate	Time working together	Fractional part of the job done when working together
Juanita	$\frac{1}{3}$	x	$\frac{1}{3}x$
Benito	$\frac{1}{2}$	c	$\frac{1}{2}x$

$$\frac{1}{3}x + \frac{1}{2}x = 1$$

$$6\left(\frac{1}{3}x + \frac{1}{2}x\right) = 6 \cdot 1$$

$$6\left(\frac{1}{3}x\right) + 6\left(\frac{1}{2}x\right) = 6$$

$$2x + 3x = 6$$

$$5x = 6$$

$$x = \frac{6}{5}$$

If Juanita and Benito worked together, it would take them 6/5 or 1 1/5 hours to weed the yard.

CHAPTER 6 GRAPHING LINEAR EQUATIONS

Section 6.1 Linear Equations in Two Variables

6.1 Margin Exercises

1. Place the numbers in each ordered pair in alphabetical order.

   (a) $x = 5$ and $y = 7$ is written as the ordered pair $(5, 7)$.

   (b) $y = 6$ and $x = -1$ is written as the ordered pair $(-1, 6)$.

   (c) $q = 4$ and $p = -3$ is written as the ordered pair $(-3, 4)$.

   (d) $r = 3$ and $s = 12$ is written as the ordered pair $(3, 12)$.

2. $5x + 2y = 20$

   (a) $(0, 10)$

   $$5x + 2y = 20$$
   $$5(0) + 2(10) = 20 \ ? \quad Let \ x = 0,$$
   $$y = 10$$
   $$0 + 20 = 20 \ ?$$
   $$20 = 20 \quad True$$

   Yes, $(0, 10)$ is a solution.

   (b) $(2, -5)$

   $$5x + 2y = 20$$
   $$5(2) + 2(-5) = 20 \ ? \quad Let \ x = 2,$$
   $$y = -5$$
   $$10 + (-10) = 20 \ ?$$
   $$0 = 20 \quad False$$

   No, $(2, -5)$ is not a solution.

   (c) $(3, 2)$

   $$5x + 2y = 20$$
   $$5(3) + 2(2) = 20 \ ? \quad Let \ x = 3,$$
   $$y = 2$$
   $$15 + 4 = 20 \ ?$$
   $$19 = 20 \quad False$$

   No, $(3, 2)$ is not a solution.

   (d) $(-4, 20)$

   $$5x + 2y = 20$$
   $$5(-4) + 2(20) = 20 \ ? \quad Let \ x = -4,$$
   $$y = 20$$
   $$-20 + 40 = 20 \ ?$$
   $$20 = 20 \quad True$$

   Yes, $(-4, 20)$ is a solution.

3. $y = 2x - 9$

   (a) In the ordered pair $(5, \ )$, $x = 5$. Find the corresponding value of y by replacing x with 5 in the given equation.

   $$y = 2x - 9$$
   $$y = 2(5) - 9 \quad Let \ x = 5$$
   $$y = 10 - 9$$
   $$y = 1$$

   This gives the ordered pair $(5, 1)$.

   (b) In the ordered pair $(2, \ )$, $x = 2$.

   $$y = 2x - 9$$
   $$y = 2(2) - 9 \quad Let \ x = 2$$
   $$y = 4 - 9$$
   $$y = -5$$

   This gives the ordered pair $(2, -5)$.

(c) In the ordered pair (, 7), y = 7. Find the corresponding value of x by replacing y with 7 in the given equation.

$$y = 2x - 9$$
$$7 = 2x - 9 \quad \text{Let } y = 7$$
$$16 = 2x \quad \text{Add } 9$$
$$8 = x \quad \text{Divide by } 2$$

This gives the ordered pair (8, 7).

(d) In the ordered pair (, −13), y = −13.

$$y = 2x - 9$$
$$-13 = 2x - 9 \quad \text{Let } y = -13$$
$$-4 = 2x \quad \text{Add } 9$$
$$-2 = x \quad \text{Divide by } 2$$

This gives the ordered pair (−2, −13).

4. $2x - 3y = 12$

x	0		3	
y		0		−3

To complete the first ordered pair, let x = 0.

$$2x - 3y = 12$$
$$2(0) - 3y = 12 \quad \text{Let } x = 0$$
$$-3y = 12$$
$$y = -4$$

This gives the ordered pair (0, −4).

To complete the second ordered pair, let y = 0.

$$2x - 3y = 12$$
$$2x - 3(0) = 12 \quad \text{Let } y = 0$$
$$2x = 12$$
$$x = 6$$

This gives the ordered pair (6, 0).

To complete the third ordered pair, let x = 3.

$$2x - 3y = 12$$
$$2(3) - 3y = 12 \quad \text{Let } x = 3$$
$$6 - 3y = 12$$
$$-3y = 6$$
$$y = -2$$

This gives the ordered pair (3, −2).

To complete the last ordered pair, let y = −3.

$$2x - 3y = 12$$
$$2x - 3(-3) = 12 \quad \text{Let } y = -3$$
$$2x + 9 = 12$$
$$2x = 3$$
$$x = \frac{3}{2}$$

This gives the ordered pair (3/2, −3). The completed table of ordered pairs is as follows.

x	0	6	3	3/2
y	−4	0	−2	−3

5. **(a)** x = 3

x is 3 for every value of y.

x	3	3	3
y	2	−4	0

(b) y = −4

y is −4 for every value of x.

x	2	6	−5
y	−4	−4	−4

6. A is in quadrant II, B is in quadrant IV, C is in quadrant I, and D is in quadrant II. E is on the y-axis, so it is not in any quadrant.

7. To plot the ordered pairs, start at the origin in each case.

   (a) (3, 5)
   Go 3 units to the right along the x-axis; then go up 5 units, parallel to the y-axis.

   (b) (-2, 6)
   Go 2 units to the left along the x-axis; then go up 6 units.

   (c) (-4, 0)
   Go 4 units to the left. This point is on the x-axis since the y-coordinate is 0.

   (d) (-5, -2)
   Go 5 units to the left; then go down 2 units.

   (e) (5, -2)
   Go 5 units to the right; then go down 2 units.

   (f) (0, -6)
   Go down 6 units. This point is on the y-axis since the y-coordinate is 0.

   Refer to the coordinate system included with the answer to this margin exercise in the textbook.

8. **(a)** (4,)
   $$y = 25x + 250$$
   $$y = 25(4) + 250 \quad \text{Let } x = 4$$
   $$y = 100 + 250$$
   $$y = 350$$
   (4, 350)

(b) (5,)
$$y = 25x + 250$$
$$y = 25(5) + 250 \quad \text{Let } x = 5$$
$$y = 125 + 250$$
$$y = 375$$
(5, 375)

(c) (6,)
$$y = 25x + 250$$
$$y = 25(6) + 250 \quad \text{Let } x = 6$$
$$y = 150 + 250$$
$$y = 400$$
(6, 400)

(d) (7,)
$$y = 25x + 250$$
$$y = 25(7) + 250 \quad \text{Let } x = 7$$
$$y = 175 + 250$$
$$y = 425$$
(7, 425)

(e) (8,)
$$y = 25x + 250$$
$$y = 25(8) + 250 \quad \text{Let } x = 8$$
$$y = 200 + 250$$
$$y = 450$$
(8, 450)

(f) (9,)
$$y = 25x + 250$$
$$y = 25(9) + 250 \quad \text{Let } x = 9$$
$$y = 225 + 250$$
$$y = 475$$
(9, 475)

6.1 Section Exercises

3. The points whose graph has coordinates (−4, 2) is in quadrant *II*.

7. $x + y = 9$; (0, 9)

To determine whether (0, 9) is a solution of the given equation, substitute 0 for x and 9 for y.

$$x + y = 9$$
$$0 + 9 = 9 \quad ? \quad \text{Let } x = 0, y = 9$$
$$9 = 9 \quad \text{True}$$

The result is true, so (0, 9) is a solution of the given solution of the equation $x + y = 9$.

11. $4x - 3y = 6$; (2, 1)

Substitute 2 for x and 1 for y.

$$4x - 3y = 6$$
$$4(2) - 3(1) = 6 \quad ? \quad \text{Let } x = 2, y = 1$$
$$8 - 3 = 6 \quad ?$$
$$5 = 6 \quad \text{False}$$

The result is false, so (2, 1) is not a solution of $4x - 3y = 6$.

15. $x = -6$; (5, −6)

Since y does not appear in the equation, we just substitute 5 for x.

$$x = -6$$
$$5 = -6 \quad \text{Let } x = 5; \text{ false}$$

The result is false, so (5, −6) is not a solution of $x = -6$.

19. $y = 2x + 7$; (2,)

In this ordered pair, x = 2. Find the corresponding value of y by replacing x with 2 in the given equation.

$$y = 2x + 7$$
$$y = 2(2) + 7 \quad \text{Let } x = 2$$
$$y = 4 + 7$$
$$y = 11$$

The ordered pair is (2, 11).

23. $y = 2x + 7$; (, −3)

In this ordered pair, y = −3. Find the corresponding value of x by replacing y with −3 in the given equation.

$$y = 2x + 7$$
$$-3 = 2x + 7 \quad \text{Let } y = -3$$
$$-10 = 2x$$
$$-5 = x$$

The ordered pair is (−5, −3).

27. $y = -4x - 4$; (, 16)

$$y = -4x - 4$$
$$16 = -4x - 4 \quad \text{Let } y = 16$$
$$20 = -4x$$
$$-5 = x$$

The ordered pair is (−5, 16).

35. $y = -8 - 2x$ (2,) (0,) (−3,)

Substitute each of the given values for x and find the corresponding value for y.

$y = -8 - 2x$
$y = -8 - 2(2)$ Let $x = 2$
$y = -8 - 4$
$y = -12$

$y = -8 - 2x$
$y = -8 - 2(0)$ Let $x = 0$
$y = -8 - 0$
$y = -8$

$y = -8 - 2x$
$y = -8 - 2(-3)$ Let $x = -3$
$y = -8 + 6$
$y = -2$

The ordered pairs are $(2, -12)$, $(0, -8)$, and $(-3, -2)$.

39. $2p + 3q = 12$

p	0		
q		0	8

$2p + 3q = 12$
$2(0) + 3q = 12$ Let $p = 0$
$0 + 3q = 12$
$3q = 12$
$q = 4$

$2p + 3q = 12$
$2p + 3(0) = 12$ Let $q = 0$
$2p + 0 = 12$
$2p = 12$
$p = 6$

$2p + 3q = 12$
$2p + 3(8) = 12$ Let $q = 8$
$2p + 24 = 12$
$2p = -12$
$p = -6$

The completed table of ordered pairs is as follows.

p	0	6	-6
q	4	0	8

43. $(3, -2), (3, 0), (3, 4), (3, 5)$

To plot the ordered pair $(3, -2)$, start at the origin, go 3 units to the right along the x-axis, and then go down 2 units on a line parallel to the y-axis. To plot the ordered pair $(3, 0)$, start at the origin and go 3 units to the right along the x-axis. (Since the y-coordinate is 0, we do not go any units up or down; the point is located on the x-axis.)

To plot the ordered pair $(3, 4)$, start at the origin, go 3 units to the right, and then go up 4 units.

To plot the ordered pair $(3, 5)$, start at the origin, go 3 units to the right, and then go up 5 units.

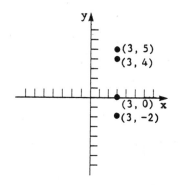

Based on this exercise, it would seem reasonable to conclude that if several ordered pairs have the same *x* value, they all lie on the same *vertical* line.

47. $y = -6$

x	8	4	-2
y			

The given equation is $y = -6$. No matter which value of x is chosen, the value of y will always be -6. Each ordered pair can be completed by placing -6 in the second position.

x	8	4	-2
y	-6	-6	-6

51. Point A was plotted by starting at the origin, going 2 units to the right along the x-axis, and then going up 4 units on a line parallel to the y-axis. The ordered pair for this point is (2, 4).

55. Point E was plotted by starting at the origin and going 3 units to the right along the x-axis. The ordered pair for this point is (3, 0).

For Exercises 59, 63, and 67, the ordered pairs are plotted on the graph following the solution for Exercise 67.

59. To plot (6, 2), start at the origin, go 6 units to the right, and then go up 2 units.

63. To plot (-4/5, -1), start at the origin, go 4/5 unit to the left, and then go down 1 unit.

67. To plot (4, 0), start at the origin and go 4 units to the left along the x-axis. The point lies on the x-axis.

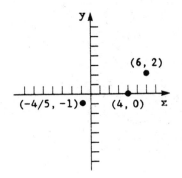

71. The point with coordinates (x, y) is in quadrant IV if x is *positive* and y is *negative*.

75. $3x - 4y = 12$

x	y
0	
	0
-4	
	-4

Substitute the given values to complete the ordered pairs.

$$3x - 4y = 12$$
$$3(0) - 4y = 12 \quad \text{Let } x = 0$$
$$0 - 4y = 12$$
$$-4y = 12$$
$$y = -3$$

$$3x - 4y = 12$$
$$3x - 4(0) = 12 \quad \text{Let } y = 0$$
$$3x - 0 = 12$$
$$3x = 12$$
$$x = 4$$

$$3x - 4y = 12$$
$$3(-4) - 4y = 12 \quad \text{Let } x = -4$$
$$-12 - 4y = 12$$
$$-4y = 24$$
$$y = -6$$

$$3x - 4y = 12$$
$$3x - 4(-4) = 12 \quad \text{Let } y = -4$$
$$3x + 16 = 12$$
$$3x = -4$$
$$x = -\frac{4}{3}$$

The completed table is as follows.

x	y
0	-3
4	0
-4	-6
$-\frac{4}{3}$	-4

Plot the points (0, -3), (4, 0), (-4, -6), and (-4/3, -4) on the given coordinate system. See the graph in the answer section of the textbook.

79. Plot the points (0, 275), (1, 300), (2, 400), (3, 500), (4, 425), (5, 610), and (6, 725) on the given axes. See the graph in the answer section of the textbook.

83.
$$9 - y = -4$$
$$9 - y - 9 = -4 - 9 \quad \text{Subtract } 9$$
$$-y = -13$$
$$\frac{-y}{-1} = \frac{-13}{-1} \quad \text{Divide by } -1$$
$$y = 13$$

Section 6.2 Graphing Linear Equations in Two Variables

6.2 Margin Exercises

1. **(a)** (-3,)
$$x + 2y = 7$$
$$-3 + 2y = 7 \quad \text{Let } x = -3$$
$$2y = 10$$
$$y = 5$$

The ordered pair is (-3, 5).

(b) (3,)
$$x + 2y = 7$$
$$3 + 2y = 7 \quad \text{Let } x = 3$$
$$2y = 4$$
$$y = 2$$

The ordered pair is (3, 2).

(c) (-1,)
$$x + 2y = 7$$
$$-1 + 2y = 7 \quad \text{Let } x = -1$$
$$2y = 8$$
$$y = 4$$

The ordered pair is (-1, 4).

(d) (7,)
$$x + 2y = 7$$
$$7 + 2y = 7 \quad \text{Let } x = 7$$
$$2y = 0$$
$$y = 0$$

The ordered pair is (7, 0).

226 Chapter 6 Graphing Linear Equations

2. $x + y = 6$

x	y
0	
	0
2	

Find the first missing value in the table by substituting 0 for x in the equation and solving for y.

$x + y = 6$
$0 + y = 6$ Let $x = 0$
$y = 6$

The first ordered pair is $(0, 6)$. Find the second missing value.

$x + y = 6$
$x + 0 = 6$ Let $y = 0$
$x = 6$

The second ordered pair is $(6, 0)$. Find the third missing value.

$x + y = 6$
$2 + y = 6$ Let $x = 2$
$y = 4$

The last ordered pair is $(2, 4)$. The completed table of ordered pairs follows.

x	y
0	6
6	0
2	4

To graph the line, plot the corresponding points and draw a line through them. Refer to the graph included with the answer to this margin exercise in the textbook.

3. Ordered pairs may vary; examples are given. Refer to the graphs included with the answers to this margin exercise in the textbook.

$2x = 3y + 6$
$2(0) = 3y + 6$ Let $x = 0$
$0 = 3y + 6$
$-3y = 6$
$y = -2$

This gives the ordered pair $(0, -2)$.

$2x = 3y + 6$
$2x = 3(0) + 6$ Let $y = 0$
$2x = 0 + 6$
$2x = 6$
$x = 3$

This gives the ordered pair $(3, 0)$.

$2x = 3y + 6$
$2(6) = 3y + 6$ Let $x = 6$
$12 = 3y + 6$
$-3y = -6$
$y = 2$

This gives the ordered pair $(6, 2)$. Graph the equation by plotting these three points and drawing a line through them.

4. $5x + 2y = 10$

Find the x-intercept by letting $y = 0$; find the y-intercept by letting $x = 0$.

$5x + 2y = 10$
$5x + 2(0) = 10$ Let $y = 0$
$5x + 0 = 10$
$5x = 10$
$x = 2$

x-intercept: $(2, 0)$

$$5x + 2y = 10$$
$$5(0) + 2y = 10 \quad \text{Let } x = 0$$
$$0 + 2y = 10$$
$$2y = 10$$
$$y = 5$$

y-intercept: (0, 5)

5. Ordered pairs may vary. Examples are given. Refer to the graphs included with the answers to this margin exercise in the textbook.

   (a) $2x = y$

   Three ordered pairs that can be used are shown in the following table.

x	y
0	0
1	2
-1	-2

   Graph the equation by plotting these points and drawing a line through them.

   (b) $x = -4y$

   Three ordered pairs that can be used are shown in the following table.

x	y
0	0
-4	1
4	-1

   Graph the equation by plotting these points and drawing a line through them.

6. Refer to the graphs included with the answers to this margin exercise in the textbook.

(a) $y = -5$

No matter what value we choose for x, y is always -5, as shown in the table of ordered pairs.

x	y
0	-5
-5	-5
3	-5

The graph is a horizontal line through (0, -5).

(b) $x = 2$

x is always 2 regardless of the value of y, as shown in the table of ordered pairs.

x	y
2	0
2	-5
2	5

The graph is a vertical line through (2, 0).

6.2 Section Exercises

3. To find the y-intercept for the graph of a linear equation, we let $x = 0$.

For Exercises 7 and 11, see the graphs in the answer section of the textbook.

7. $y = -x + 5$

   (0,), (, 0), (2,)

   If $x = 0$,   If $y = 0$,
   $y = -0 + 5$ $0 = -x + 5$
   $y = 5$. $x = 5$.

If $x = 2$,
$y = -2 + 5$
$y = 3$.

This gives us three ordered pairs: (0, 5), (5, 0), and (2, 3). Plot the corresponding points and draw a line through them.

11. $3x = -y - 6$

   (0,   ), (, 0), $(-\frac{1}{3},$)

   If $x = 0$,   If $y = 0$,
   $3(0) = -y - 6$ $3x = -0 - 6$
   $0 = -y - 6$ $3x = -6$
   $y = -6$. $x = -2$.

   If $x = -\frac{1}{3}$,
   $3(-\frac{1}{3}) = -y - 6$
   $-1 = -y - 6$
   $y - 1 = -6$
   $y = -5$.

   This gives us three ordered pairs: (0, -6), (-2, 0), and (-1/3, -5). Plot the corresponding points and draw a line through them.

15. $x + 6y = 0$

   To find the x-intercept, let $y = 0$.

   $x + 6y = 0$
   $x + 6(0) = 0$
   $x + 0 = 0$
   $x = 0$

   Letting $x = 0$ also gives the ordered pair (0, 0). The x-intercept is (0, 0); the y-intercept is also (0, 0).

For Exercises 19-35, see the graphs in answer section of the textbook.

19. $x = y + 2$

   Begin by finding the intercepts.

   $x = y + 2$ $x = y + 2$
   $x = 0 + 2$ $0 = y + 2$
     Let $y = 0$ Let $x = 0$
   $x = 2$ $-2 = y$

   The x-intercept is (2, 0) and the y-intercept is (0, -2). To find a third point, choose $y = 1$.

   $x = y + 2$
   $x = 1 + 2$ Let $y = 1$
   $x = 3$

   This gives the ordered pair (3, 1). Plot (2, 0), (0, -2), and (3, 1) and draw a line through them.

23. $2x + y = 6$

   Find the intercepts.

   $2x + y = 6$ $2x + y = 6$
   $2x + 0 = 6$ $2(0) + y = 6$
     Let $y = 0$ Let $x = 0$
   $2x = 6$ $0 + y = 6$
   $x = 3$ $y = 6$

   The x-intercept is (3, 0) and the y-intercept is (0, 6). To find a third point, choose $x = 1$.

   $2x + y = 6$
   $2(1) + y = 6$ Let $x = 1$
   $2 + y = 6$
   $y = 4$

This gives the point (1, 4). Plot (3, 0), (0, 6), and (1, 4) and draw a line through them.

27. $3x + 7y = 14$

   Find the intercepts.

   $3x + 7y = 14$ $3x + 7y = 14$
   $3x + 7(0) = 14$ $3(0) + 7y = 14$
       Let $y = 0$ Let $x = 0$
        $3x = 14$ $0 + 7y = 14$
        $x = \frac{14}{3}$ $y = 2$

   The x-intercept is (14/3, 0) and the y-intercept is (0, 2). To find a third point, choose $x = 2$.

   $3x + 7y = 14$
   $3(2) + 7y = 14$ Let $x = 2$
   $6 + 7y = 14$
   $7y = 8$
   $y = \frac{8}{7}$

   This gives the ordered pair (2, 8/7). Plot (14/3, 0), (0, 2), and (2, 8/7). Writing 14/3 as the mixed number 4 2/3 and 8/7 as 1 1/7 will be helpful for plotting. Draw a line through these three points.

31. $y = -6x$

   Find three points on the line.

   If $x = 0$, $y = -6(0) = 0$.
   If $x = 1$, $y = -6(1) = -6$.
   If $x = -1$, $y = -6(-1) = 6$.

   Plot (0, 0), (1, -6), and (-1, 6) and draw a line through these points.

35. $x = -2$

   For any value of y, the value of x will always be -2. Three ordered pairs are (-2, 3), (-2, 0), and (-2, -4). Plot these points and draw a line through them. The graph will be a vertical line.

39. $\frac{2 - 7}{-3 - 5} = \frac{-5}{-8} = \frac{5}{8}$

Section 6.3 The Slope of a Line

6.3 Margin Exercises

1. (a) For $y_2 = 4$, $y_1 = -1$, $x_2 = 3$, and $x_1 = 4$,

   $\frac{y_2 - y_1}{x_2 - x_1} = \frac{4 - (-1)}{3 - 4}$
   $= \frac{4 + 1}{3 - 4}$
   $= \frac{5}{-1}$
   $= -5.$

   (b) For $x_1 = 3$, $x_2 = -5$, $y_1 = 7$, and $y_2 = -9$,

   $\frac{y_2 - y_1}{x_2 - x_1} = \frac{-9 - 7}{-5 - 3}$
   $= \frac{-16}{-8}$
   $= 2.$

   (c) For $x_1 = 2$, $x_2 = 7$, $y_1 = 4$, and $y_2 = 9$,

   $\frac{y_2 - y_1}{x_2 - x_1} = \frac{9 - 4}{7 - 2}$
   $= \frac{5}{5}$
   $= 1.$

2. **(a)** Through $(6, -2)$ and $(5, 4)$

Let $(6, -2) = (x_1, y_1)$ and $(5, 4) = (x_2, y_2)$.

$$m = \frac{y_2 - y_1}{x_2 - x_1} = \frac{4 - (-2)}{5 - 6} = \frac{6}{-1} = -6$$

(b) Through $(-3, 5)$ and $(-4, 7)$

$$m = \frac{y_2 - y_1}{x_2 - x_1} = \frac{7 - 5}{-4 - (-3)} = \frac{2}{-1} = -2$$

(c) Through $(6, -8)$ and $(-2, 4)$

$$m = \frac{y_2 - y_1}{x_2 - x_1} = \frac{4 - (-8)}{-2 - 6} = \frac{12}{-8} = -\frac{3}{2}$$

3. **(a)** Through $(2, 5)$ and $(-1, 5)$

$$m = \frac{5 - 5}{-1 - 2} = \frac{0}{-3} = 0$$

(b) Through $(3, 1)$ and $(3, -4)$

$$m = \frac{-4 - 1}{3 - 3} = \frac{-5}{0} \quad \text{Undefined}$$

(c) With equation $y = -1$

All lines with equation $y = k$ are horizontal and have a slope of 0.

(d) With equation $x - 4 = 0$

$x - 4 = 0$ means $x = 4$. All lines with equation $x = k$ are vertical and have undefined slope.

4. **(a)** $y = -\frac{7}{2}x + 1$

The slope is the coefficient of x, $-7/2$.

(b) $3x + 2y = 9$

Solve for y: $3x + 2y = 9$

$$2y = -3x + 9$$
$$y = -\frac{3}{2}x + \frac{9}{2}$$

The slope is the coefficient of x, $-3/2$.

(c) $y + 4 = 0$

Solve for y: $y + 4 = 0$

$$y = -4$$

Equations of the form $y = k$ are horizontal and have a slope of 0.

(d) $x + 3 = 7$

$$x = 4$$

Equations of the form $x = k$ are vertical and have undefined slope.

5. Find the slope of each line by first solving each equation for y.

(a) $x + y = 6 \qquad x + y = 1$

$y = -x + 6 \qquad y = -x + 1$

Slope is -1. $\qquad$ Slope is -1.

The lines have the same slope, so they are parallel.

(b) $3x - y = 4 \qquad x + 3y = 9$

$-y = -3x + 4 \qquad 3y = -x + 9$

$y = 3x - 4 \qquad y = -\frac{1}{3}x + 3$

Slope is 3. $\qquad$ Slope is $-1/3$.

The product of the slopes is $3(-1/3) = -1$, so the lines are perpendicular.

(c) $2x - y = 5 \qquad 2x + y = 3$

$\qquad -y = -2x + 5 \qquad y = -2x + 3$

$\qquad y = 2x - 5$

Slope is 2.   Slope is -2.

The slopes are not equal, and the product of the slopes is $2(-2) = -4$, not -1. The lines are neither parallel nor perpendicular.

(d) $3x - 7y = 35$

$\qquad -7y = -3x + 35$

$\qquad y = \frac{3}{7}x - 5$

Slope is $\frac{3}{7}$.

$7x - 3y = -6$

$\qquad -3y = -7x - 6$

$\qquad y = \frac{7}{3}x + 2$

Slope is $\frac{7}{3}$.

The slopes are not equal and the product of the slopes is $(3/7)(7/3) = 1$, not -1. The lines are neither parallel nor perpendicular.

6.3 Section Exercises

3. undefined slope

   Sketches will vary. The line must be vertical. One such line is shown in the following graph.

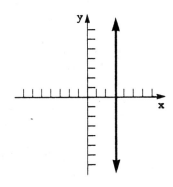

7. The graph of $y = 5x - 3$ is a line with a slope of 5. Since parallel lines have the same slope, the slope of a line parallel to $y = 5x - 3$ will also be 5.

11. The indicated points have coordinates $(-3, 5)$ and $(1, -2)$. Use the definition of slope with $(-3, 5) = (x_1, y_1)$ and $(1, -2) = (x_2, y_2)$.

    $\text{slope} = m = \frac{y_2 - y_1}{x_2 - x_1}$

    $= \frac{-2 - 5}{1 - (-3)}$

    $= \frac{-7}{4} = -\frac{7}{4}$

15. (a) Count up 6 units. The vertical change is 6.

    (b) Count 4 units to the right. The horizontal change is 4.

    (c) The quotient of the numbers found in parts (a) and (b) is

    $\frac{6}{4}$ or $\frac{3}{2}$.

    We call this number the slope of the line.

19. (−8, 0) and (0, −5)

Let (−8, 0) = (x_1, y_1) and (0, −5) = (x_2, y_2).

$$\text{slope} = m = \frac{y_2 - y_1}{x_2 - x_1}$$
$$= \frac{-5 - 0}{0 - (-8)}$$
$$= \frac{-5}{8} = -\frac{5}{8}$$

23. (−8, 6) and (−8, −1)

Let (−8, 6) = (x_1, y_1) and (−8, −1) = (x_2, y_2).

$$m = \frac{y_2 - y_1}{x_2 - x_1}$$
$$= \frac{-1 - 6}{-8 - (-8)}$$
$$= \frac{-7}{0},$$

which is undefined.

27. $y = 2x - 3$

Since the equation is already solved for y, the slope is given by the coefficient of x, which is 2. Thus, the slope is 2.

31. $y = 6 - 4x$
$y = -4x - 6$

Since the equation is already solved for y, the slope is given by the coefficient of x, so the slope is −4.

35. $y = 4$

This is the equation of a horizontal line. Its slope is 0. (This equation may be rewritten in the form $y = 0x + 4$, where the coefficient of x gives the slope.)

39. (a) Because the line rises from left to right, its slope is positive.

(b) Because the line intersects the y-axis *at* the origin, the y-coordinate of its y-intercept is zero.

43. $2x + 5y = 4$
$4x + 10y = 1$

Find the slope of each line by solving each equation for y.

$2x + 5y = 4$
$5y = -2x + 4$ Subtract 2x
$y = -\frac{2}{5}x + \frac{4}{5}$ Divide by 5

The slope of the first line is −2/5.

$4x + 10y = 1$
$10y = -4x + 1$ Subtract 4x
$y = -\frac{4}{10}x + \frac{1}{4}$ Divide by 10
$y = -\frac{2}{5}x + \frac{1}{4}$ Lowest terms

The slope of the second line is −2/5.
The slopes are equal, so the lines are parallel.

47. $3x - 2y = 6$
$2x + 3y = 3$

Find the slopes by solving the equation for y.

$3x - 2y = 6$
$-2y = -3x + 6$ *Subtract 3x*
$y = \frac{3}{2}x - 3$ *Divide by 2*

The slope of the first line is 3/2.

$2x + 3y = 3$
$3y = -2x + 3$ *Subtract 2x*
$y = -\frac{2}{3}x + 1$ *Divide by 3*

The slope of the second line is $-2/3$.

The product of the slopes is

$$\frac{3}{2}\left(-\frac{2}{3}\right) = -1,$$

so the lines are perpendicular.

51. $y - (-8) = 2(x - 4)$
$y + 8 = 2x - 8$ *Distributive property*
$y = 2x - 16$ *Subtract 8*

Section 6.4 Equations of a Line

6.4 Margin Exercises

1. **(a)** $m = 1/2$; $b = -4$
    Use the form $y = mx + b$ to get
    $$y = \frac{1}{2}x - 4.$$

    (b) $m = -1$; $b = 8$
    Use the form $y = mx + b$ to get
    $$y = -1x + 8 \quad \text{or} \quad y = -x + 8.$$

    (c) $m = 3$; $b = 0$
    Use the form $y = mx + b$ to get
    $$y = 3x + 0 \quad \text{or} \quad y = 3x.$$

    (d) $m = 0$; $b = 2$
    Use the form $y = mx + b$ to get
    $$y = (0)x + 2 \quad \text{or} \quad y = 2.$$

2. Refer to the graphs included with the answers to this margin exercise in the textbook.

    (a) Through $(-1, 2)$, with slope 3/2

    First, locate the point $(-1, 2)$. Write the slope as

    $$m = \frac{\text{difference in y-values}}{\text{difference in x-values}} = \frac{3}{2}.$$

    Locate another point by counting 3 units up and then 2 units to the right of $(-1, 2)$. Draw a line through this new point, $(1, 5)$, and the given point, $(-1, 2)$.

    (b) Through $(2, -3)$, with slope $-1/3$

    To graph the line, write the slope as

    $$m = \frac{\text{difference in y-values}}{\text{difference in x-values}} = \frac{-1}{3}.$$

    (This could also be $\frac{1}{-3}$.) Locate $(2, -3)$. Count 1 unit down and 3 units to the right. Draw a line through this point, $(5, -4)$, and $(2, -3)$.

3. **(a)** Through $(-1, 3)$, with slope -2

    Use the point-slope form with $x_1 = -1$, $y_1 = 3$, and $m = -2$.

$$y - y_1 = m(x - x_1)$$
$$y - 3 = -2(x + 1)$$
$$y - 3 = -2x - 2 \quad \text{Distributive property}$$
$$y = -2x + 1 \quad \text{Add 3}$$
$$2x + y = 1 \quad \text{Add } 2x; Ax + By = C \text{ form}$$

(b) Through $(5, 2)$, with slope $-1/3$

Use the point-slope form with $x_1 = 5$, $y_1 = 2$, and $m = -1/3$.

$$y - y_1 = m(x - x_1)$$
$$y - 2 = -\frac{1}{3}(x - 5)$$
$$3(y - 2) = -1(x - 5) \quad \text{Multiply by 3}$$
$$3y - 6 = -x + 5 \quad \text{Distributive property}$$
$$3y = -x + 11 \quad \text{Add 6}$$
$$x + 3y = 11 \quad \text{Add } x; Ax + By = C \text{ form}$$

4. (a) $(-3, 1)$ and $(2, 4)$

Find the slope.

$$m = \frac{4 - 1}{2 - (-3)} = \frac{3}{5}$$

Let $(-3, 1) = (x_1, y_1)$ and use the point-slope form.

$$y - y_1 = m(x - x_1)$$
$$y - 1 = \frac{3}{5}[x - (-3)]$$
$$5(y - 1) = 3(x + 3)$$
$$5y - 5 = 3x + 9$$
$$-14 = 3x - 5y$$
$$\text{or } 3x - 5y = -14$$

(b) $(2, 5)$ and $(-1, 6)$

Find the slope.

$$m = \frac{6 - 5}{-1 - 2} = \frac{1}{-3} = -\frac{1}{3}$$

Let $(2, 5) = (x_1, y_1)$ and use the point-slope form.

$$y - y_1 = m(x - x_1)$$
$$y - 5 = -\frac{1}{3}(x - 2)$$
$$3(y - 5) = -1(x - 2)$$
$$3y - 15 = -x + 2$$
$$3y = -x + 17$$
$$x + 3y = 17$$

6.4 Section Exercises

3. Since the line falls from left to right, the "rise" is negative. For this line, the "rise" is -3 and the "run" is 3, so the slope is

$$m = \frac{\text{rise}}{\text{run}} = \frac{-3}{3} = -1.$$

The y-intercept is $(0, 3)$, so $b = 3$. The $y = mx + b$ form of the equation is

$$y = -1x + 3$$
$$y = -x + 3.$$

7. $m = 0$, $b = 3$

Use the slope-intercept form.

$$y = mx + b$$
$$y = 0x + 3$$
$$y = 3.$$

6.4 Section Exercises

For Exercises 11-19, see the graphs in the answer section of the textbook.

11. $(-2, 3)$, $m = \frac{1}{2}$

First, locate the point $(-2, 3)$. Write the slope as

$$m = \frac{\text{difference in y-values}}{\text{difference in x-values}} = \frac{1}{2}.$$

Locate another point by counting 1 unit up and then 2 units to the right. Draw a line through this new point, $(0, 2)$, and the given point, $(-2, 3)$.

15. $(0, 2)$, $m = 3$

First, locate the point $(0, 2)$, which is the y-intercept of the line to be graphed. Write the slope as

$$m = \frac{\text{difference in y-values}}{\text{difference in x-values}} = \frac{3}{1}.$$

Locate another point by counting 3 units up and then 1 unit to the right. Draw a line through this new point, $(1, 5)$, and the given point $(0, 2)$.

19. $(3, -2)$, undefined slope.

First, locate the point $(3, -2)$. Because the slope is undefined, the line will be vertical. Draw a vertical line through $(3, -2)$.

23. $(4, 1)$, $m = 2$

The given point is $(4, 1)$, so $x_1 = 4$ and $y_1 = 2$. Also, $m = 2$. Substitute these values into the point-slope form.

$$y - y_1 = m(x - x_1)$$
$$y - 1 = 2(x - 4)$$
$$y - 1 = 2x - 8 \quad \textit{Distributive property}$$
$$y = 2x - 7 \quad \textit{Add 1}$$
$$y + 7 = 2x \quad \textit{Add 7}$$
$$7 = 2x - y \quad \textit{Subtract y}$$
$$\text{or} \quad 2x - y = 7 \quad \textit{Standard form}$$

27. $(-2, 5)$, $m = \frac{2}{3}$

Use the values $x_1 = -2$, $y_1 = 5$, and $m = 2/3$ in the point-slope form.

$$y - y_1 = m(x - x_1)$$
$$y - 5 = \frac{2}{3}[x - (-2)]$$
$$y - 5 = \frac{2}{3}(x + 2)$$
$$3(y - 5) = 3 \cdot \frac{2}{3}(x + 2)$$
$$\quad \textit{Multiply by 3 to clear fractions}$$
$$3(y - 5) = 2(x + 2)$$
$$3y - 15 = 2x + 4$$
$$\quad \textit{Distributive property}$$
$$3y - 19 = 2x \quad \textit{Subtract 4}$$
$$-19 = 2x - 3y$$
$$\quad \textit{Subtract 3y}$$
$$\text{or} \quad 2x - 3y = -19 \quad \textit{Standard form}$$

31. An equation of the line which passes through the origin and a second point whose x- and y-coordinates are equal is y = x. This equation may be written in other forms, such as x - y = 0 or y - x = 0.

35. (-1, -7) and (-8, -2)

Begin by finding the slope of the line, using the definition of slope.

$$m = \frac{-2 - (-7)}{-8 - (-1)} = \frac{5}{-7} = -\frac{5}{7}.$$

Now use either point for (x_1, y_1) and $m = -5/7$ in the point-slope form. If we use (-1, -7), we get the following result.

$$y - y_1 = m(x - x_1)$$
$$y - (-7) = -\frac{5}{7}[x - (-1)]$$
$$y + 7 = -\frac{5}{7}(x + 1)$$
$$7(y + 7) = 7\left(-\frac{5}{7}\right)(x + 1)$$
$$7(y + 7) = -5(x + 1)$$
$$7y + 49 = -5x - 5$$
$$7y = -5x - 54$$
$$5x + 7y = -54$$

The same result would be found by using (-8, -2) for (x_1, y_1).

39. $\left(\frac{1}{2}, \frac{3}{2}\right)$, $\left(-\frac{1}{4}, \frac{5}{4}\right)$

Using the definition to find the slope.

$$m = \frac{\frac{5}{4} - \frac{3}{2}}{-\frac{1}{4} - \frac{1}{2}}$$

$$= \frac{\frac{5}{4} - \frac{6}{4}}{-\frac{1}{4} - \frac{2}{4}}$$

$$= \frac{-\frac{1}{4}}{-\frac{3}{4}} = \left(-\frac{1}{4}\right)\left(-\frac{4}{3}\right)$$

$$= \frac{1}{3}$$

Now use m = 1/3 and one of the two points, say (1/2, 3/2), in the point-slope form.

$$y - y_1 = m(x - x_1)$$
$$y - \frac{3}{2} = \frac{1}{3}\left(x - \frac{1}{2}\right)$$
$$6\left(y - \frac{3}{2}\right) = 6 \cdot \frac{1}{3}\left(x - \frac{1}{2}\right)$$
$$\qquad\qquad\text{Multiply by 6}$$
$$6\left(y - \frac{3}{2}\right) = 2\left(x - \frac{1}{2}\right)$$
$$6y - 9 = 2x - 1$$
$$\qquad\qquad\text{Distributive property}$$
$$6y - 8 = 2x \quad \text{Add 1}$$
$$-8 = 2x - 6y$$
$$\qquad\qquad\text{Subtract 6y}$$
$$-4 = x - 3y$$
$$\qquad\qquad\text{Divide by 2}$$
or $\quad x - 3y = -4 \quad$ Standard form

The same result would be found by using (-1/4, 5/4) for (x_1, y_1).

43. (a) The information given in the table corresponds to the ordered pairs (1, 4800) and (5, 24,800).

(b) First, find the slope.

$$m = \frac{24,800 - 4800}{5 - 1}$$

$$= \frac{20,000}{4}$$

$$= 5000$$

To find b, let $x = 1$, $y = 4800$, and $m = 5000$ in the sales equation.

$$y = mx + b$$
$$4800 = 5000(1) + b$$
$$4800 = 5000 + b$$
$$-200 = b$$

Thus, the sales equation is

$$y = 5000x - 200.$$

(c) In this situation, the slope, m, represents the change in sales per year.

47. The expression for F in terms of C obtained in Exercise 46 is

$$F = \frac{9}{5}C + 32.$$

To obtain an expression for C in terms of F, solve this equation for C.

$$F - 32 = \frac{9}{5}C \quad \text{Subtract 32}$$

$$\frac{5}{9}(F - 32) = \frac{5}{9} \cdot \frac{9}{5}C$$

Multiply by 5/9, the reciprocal of 9/59/5

$$\frac{5}{9}(F - 32) = C$$

An alternative form of this equation may be obtained as follows.

$$F = \frac{9}{5}C + 32$$

$$5F = 5\left(\frac{9}{5}C + 32\right)$$

Multiply by 5 to clear fractions

$$5F = 9C + 160$$

Distributive property

$$5F - 160 = 9C \quad \text{Subtract 160}$$

$$\frac{5F - 160}{9} = C$$

or $\frac{5}{9}F - \frac{160}{9} = C$

51.
$$5 - 3x < -10$$
$$5 - 3x - 5 < -10 - 5 \quad \text{Subtract 5}$$
$$-3x < -15$$
$$\frac{-3x}{-3} > \frac{-15}{-3} \quad \text{Divide by -3;}$$
reverse the symbol
$$x > 5$$

To graph this solution on the number line, place an open circle at 5 (to show that 5 is not part of the graph) and draw an arrow extending to the right. See the graph in the answer section of the textbook.

Section 6.5 Graphing Linear Inequalities in Two Variables

6.5 Margin Exercises

For Exercises 1-6, refer to the graphs included with the answers to the margin exercises in the textbook.

1. (a) $x + 2y \geq 6$

   $2y \geq -x + 6$ Subtract x

   $y \geq -\frac{1}{2}x + 3$ Divide by 2

   The ordered pairs for which y is equal to $-\frac{1}{2}x + 3$ are on the line with equation

   $$y = -\frac{1}{2}x + 3.$$

   Graph this line with y-intercept (0, 3) and slope -1/2. The ordered pairs for which y is greater than $-\frac{1}{2}x + 3$ are above this line. Indicate the solution by shading the region above the line.

   (b) $3x + 4y \leq 12$

   $4y \leq -3x + 12$ Subtract $3x$

   $y \leq -\frac{3}{4}x + 3$ Divide by 4

   The ordered pairs for which y is equal to $-\frac{3}{4}x + 3$ are on the line with equation

   $$y = -\frac{3}{4}x + 3.$$

   Graph this line with y-intercept (0, 3) and slope -3/4. The ordered pairs for which y is less than $-\frac{3}{4}x + 3$ are below this line. Indicate this solution by shading the region below the line.

2. $4x - 5y \leq 20$

   $4(0) - 5(0) \leq 20$?   Let $x = 0$, $y = 0$

   $0 - 0 \leq 20$?

   $0 \leq 20$ True

   Since the last statement is true, shade the region that includes the test point (0, 0), that is, the region above the line.

3. $3x + 5y > 15$

   $3(1) + 5(2) > 15$?   Let $x = 1$, $y = 2$

   $3 + 10 > 15$?

   $13 > 15$ False

   Since the last statement is false, shade the region that does *not* include the test point (1, 2), that is, the region above the line.

4. $2x - y \geq -4$

   Start by graphing the equation
   $$2x - y = -4.$$
   The intercepts are (-2, 0) and (0, 4). A third point on the line may be found. Through these points, draw a solid line to show that the points on the line are solutions to the inequality
   $$2x - y \geq -4.$$

Choose a test point not on the line, such as (0, 0).

$$2x - y \geq -4$$
$$2(0) - 0 \geq -4 \quad ? \quad \text{Let } x = 0, y = 0$$
$$0 \geq -4 \quad \text{True}$$

Since this statement is true, shade the region that includes test point (0, 0), that is, the region below the line.

5. $y < 4$

First graph $y = 4$, a horizontal line through the point (0, 4). Use a dashed line because of the < symbol. Choose (0, 0) as a test point.

$$y < 4$$
$$0 < 4 \quad \text{Let } y = 0; \text{ true}$$

Since $0 < 4$ is a true statement, shade the region below the line that includes test point (0, 0), that is, the region below the line.

6. $x \geq -3y$

Graph $x = -3y$ as a solid line through (0, 0) and (-3, 1). Since (0, 0) cannot be the test point, use (1, 1), which is not on the line.

$$x \geq -3y$$
$$1 \geq -3(1) \quad ? \quad \text{Let } x = 1, y = 1$$
$$1 \geq -3 \quad \text{True}$$

Since $1 \geq -3$ is a true statement, shade the region that includes test point (1, 1), that is, the region above the line.

6.5 Section Exercises

3. $3x - 2y \geq 0$

   (a) (4, 1)

   Substitute 4 for x and 1 for y in the given inequality.

   $$3x - 2y \geq 0$$
   $$3(4) - 2(1) \geq 0 \quad ?$$
   $$12 - 2 \geq 0 \quad ?$$
   $$10 \geq 0 \quad \text{True}$$

   The true result shows that (4, 1) is a solution of the inequality.

   (b) (0, 0)

   Substitute 0 for x and 0 for y in the given inequality.

   $$3x - 2y \geq 0$$
   $$3(0) - 2(0) \geq 0 \quad ?$$
   $$0 \geq 0 \quad \text{True}$$

   The true result shows that (0, 0) is a solution of the given inequality.

For Exercises 7 and 11, see the graphs in the answer section of the textbook.

7. $x + 2y \geq 7$

   Use (0, 0) as a test point.

   $$x + 2y \geq 7$$
   $$0 + 2(0) \geq 7 \quad ? \quad \text{Let } x = 0, y = 0$$
   $$0 \geq 7 \quad \text{False}$$

   Because the last statement is false, we shade the region that does *not* include the test point (0, 0). This is the region above the line.

11. $x > 4$

   Use $(0, 0)$ as a test point.

   $x > 4$
   $0 > 4$ Let $x = 0$; *false*

   Because $0 > 4$ is false, shade the region *not* containing $(0, 0)$. This is the region to the right of the line.

15. Whenever the inequality symbol is $>$ or $<$, use a dashed line to show that the boundary is *not* part of the solution. Whenever the inequality symbol is $\geq$ or $\leq$, use a solid line to show that the boundary is part of the solution.

For Exercises 19–27, see the graphs in the answer section of the textbook.

19. $x + 2y < 4$

   Step 1 Graph the boundary of the region, the line with equation $x + 2y = 4$.

   If $y = 0$, $x = 4$, so the x-intercept is $(4, 0)$. If $x = 0$, $y = 2$, so the y-intercept is $(0, 2)$.
   Draw the line with these intercepts. Make the line dashed because of the $<$ sign.

   Step 2 Choose $(0, 0)$ as a test point.

   $x + 2y < 4$
   $0 + 2(0) < 4$?   Let $x = 0$, $y = 0$
   $0 < 4$ True

   Because $0 < 4$ is true, shade the region containing the origin. The dashed line shows that the boundary is not part of the graph.

23. $y \geq 2x + 1$

   The boundary is the line with equation $y = 2x + 1$. This line has slope 2 and y-intercept $(0, 1)$. It may be graphed by starting at $(0, 1)$ and going 2 units up and then 1 unit to the right to reach the point $(1, 3)$. Draw a solid line through $(0, 1)$ and $(1, 3)$.
   Using $(0, 0)$ as a test point will result in the inequality $0 \geq 1$, which is false. Shade the region *not* containing the origin, that is, the region above the line. The solid line shows that the boundary is part of the graph.

27. $y < 5$

   The boundary has the equation $y = 5$. This is the horizontal line through $(0, 5)$. Make this line dashed because of the $<$ sign.
   Using $(0, 0)$ as a test point will result in the inequality $0 < 5$, which is true. Shade the region containing the origin, that is, the region below the line. The dashed line shows that the boundary is not part of the graph.

31. For every point in quadrant IV, the x-coordinate is positive and the y-coordinate s negative. Since a negative number is never less than a positive number, the inequality y > x can never be true for a point in quadrant IV. Thus, the graph of y > x cannot lie in quadrant IV.

35. 2x + 3y
    -3x - 3y

Add in columns.

$$\begin{array}{r} 2x + 3y \\ -3x - 3y \\ \hline -x + 0 = -x \end{array}$$

Chapter 6 Review Exercises

1. y = 3x + 2   (-1,)   (0,)   (, 5)

    y = 3x + 2   y = 3x + 2
    y = 3(-1) + 2   y = 3(0) + 2
        Let x = -1   Let x = 0
    y = -3 + 2   y = 0 + 2
    y = -1   y = 2

    y = 3x + 2
    5 = 3x + 2   Let y = 5
    3 = 3x
    1 = x

    The ordered pairs are (-1, -1), (0, 2), and (1, 5).

2. 4x + 3y = 6   (0,)   (, 0)   (-2,)

    4x + 3y = 6   4x + 3y = 6
    4(0) + 3y = 6   4x + 3(0) = 6
        Let x = 0   Let y = 0
    3y = 6   4x = 6
    y = 2 $x = \frac{6}{4} = \frac{3}{2}$

    4x + 3y = 6
    4(-2) + 3y = 6   Let x = -2
    -8 + 3y = 6
    3y = 14
    $y = \frac{14}{3}$

    The ordered pairs are (0, 2), (3/2, 0), and (-2, 14/3).

3. x = 3y   (0,)   (8,)   (, -3)

    x = 3y   x = 3y
    0 = 3y   Let x = 0   8 = 3y   Let x = 8
    0 = y $\frac{8}{3} = y$

    x = 3y
    x = 3(-3)   Let y = -3
    x = -9

    The ordered pairs are (0, 0), (8, 8/3), and (-9, -3).

4. x - 7 = 0   (, -3)   (, 0)   (, 5)

    The given equation may be written x = 7. For any value of y, the value of x will always be 7. The ordered pairs are (7, -3), (7, 0), and (7, 5).

Chapter 6 Graphing Linear Equations

5. $x + y = 7$; $(2, 5)$

   Substitute 2 for x and 5 for y in the given equation.

   $x + y = 7$
   $2 + 5 = 7$?
   $7 = 7$ True

   Yes, (2, 5) is a solution of the given equation.

6. $2x + y = 5$; $(-1, 3)$

   Substitute -1 for x and 3 for y in the given equation.

   $2x + y = 5$
   $2(-1) + 3 = 5$?
   $-2 + 3 = 5$?
   $1 = 5$ False

   No, $(-1, 3)$ is not a solution of the given equation.

7. $3x - y = 4$; $\left(\frac{1}{3}, -3\right)$

   Substitute 1/3 for x and -3 for y in the given equation.

   $3x - y = 4$
   $3\left(\frac{1}{3}\right) - (-3) = 4$?
   $1 - (-3) = 4$?
   $4 = 4$ True

   Yes, (1/3, -3) is a solution of the given equation.

8. $5x - 3y = 16$; $\left(0, -\frac{16}{3}\right)$

   Substitute 0 for x and $-16/3$ for x in the given equation.

   $5x - 3y = 16$
   $5(0) - 3\left(-\frac{16}{3}\right) = 16$?
   $0 + 16 = 16$?
   $16 = 16$ True

   Yes, (0, $-16/3$) is a solution of the given equation.

   For Exercises 9–12, see the graphs in the answer section of the textbook.

9. To plot (2, 3), start at the origin, go 2 units to the right, and then go up 3 units.

10. To plot $(-4, 2)$, start at the origin, go 4 units to the left, and then go up 2 units.

11. To plot (3, 0), start at the origin and go 3 units to the right. The point lies on the x-axis.

12. To plot (0, -6), start at the origin and go down 6 units. The point lies on the y-axis.

13. The product of two numbers is positive whenever the two numbers have the same sign. If $xy > 0$, either $x > 0$ and $y > 0$, so that (x, y) lies in quadrant I, or $x < 0$ and $y < 0$, so that (x, y) lies in quadrant III.

14. The y-coordinate is 0 for any point on the x-axis. Thus, for any real value of k, the point (k, 0) lies on the x-axis.

15. The point (−2, 3) has a negative x-coordinate and a positive y-coordinate, so it lies in quadrant II.

16. The point (−1, −4) has a negative x-coordinate and a negative y-coordinate, so it lies in quadrant III.

17. $y = 2x + 5$

    To find the x-intercept, let $y = 0$.

    $$y = 2x + 5$$
    $$0 = 2x + 5$$
    $$-2x = 5$$
    $$x = -\frac{5}{2}$$

    The x-intercept is (−5/2, 0).

    To find the y-intercept, let $x = 0$.

    $$y = 2x + 5$$
    $$y = 2(0) + 5$$
    $$y = 5$$

    The y-intercept is (0, 5).

18. $2x + y = -7$

    To find the x-intercept, let $y = 0$.

    $$2x + y = -7$$
    $$2x + 0 = -7$$
    $$2x = -7$$
    $$x = -\frac{7}{2}$$

    The x-intercept is (−7/2, 0).

    To find the y-intercept, let $x = 0$.

    $$2x + y = -7$$
    $$2(0) + y = -7$$
    $$y = -7$$

    The y-intercept is (0, −7).

19. $3x + 2y = 8$

$3x + 2y = 8$	$3x + 2y = 8$
$3x + 2(0) = 8$	$3(0) + 2y = 8$
Let $y = 0$	Let $x = 0$
$3x = 8$	$2y = 8$
$x = \frac{8}{3}$	$y = 4$

    The x-intercept is (8/3, 0); the y-intercept is (0, 4).

For Exercises 20–22, see the graphs in the answer section of the textbook.

20. $y = -2x - 5$

    Begin by finding the intercepts.

$y = -2x - 5$	$y = -2x - 5$
$0 = -2x - 5$	$y = -2(0) - 5$
Let $y = 0$	Let $x = 0$
$2x = -5$	$y = 0 - 5$
$x = -\frac{5}{2}$	$y = -5$

    The x-intercept is (−5/2, 0) and the y-intercept is (0, −5). To find a third point, choose $x = -1$.

    $$y = -2x - 5$$
    $$y = -2(-1) - 5 \quad Let\ x = -1$$
    $$y = 2 - 5$$
    $$y = -3$$

    This gives the ordered pair (−1, −3). Plot (−5/2, 0), (0, −5), and (−1, −3) and draw a line through them.

21. $x + 2y = -4$

   Find the intercepts.

   If $y = 0$, $x = -4$, so the x-intercept is $(-4, 0)$.
   If $x = 0$, $y = -2$, so the y-intercept is $(0, -2)$.

   To find a third point, choose $x = 2$.

   $$x + 2y = -4$$
   $$2 + 2y = -4 \quad \text{Let } x = 2$$
   $$2y = -6$$
   $$y = -3$$

   This gives the ordered pair $(2, -3)$. Plot $(-4, 0)$, $(0, -2)$, and $(2, -3)$ and draw a line through them.

22. $x + y = 0$

   This line goes through the origin, so both intercepts are $(0, 0)$. Two other points on the line are $(2, -2)$ and $(-2, 2)$.

23. Through $(2, 3)$ and $(-4, 6)$

   Let $(2, 3) = (x_1, y_1)$ and $(-4, 6) = (x_2, y_2)$.

   $$\text{slope} = m = \frac{y_2 - y_1}{x_2 - x_1}$$
   $$= \frac{6 - 3}{-4 - 2}$$
   $$= \frac{3}{-6} = -\frac{1}{2}$$

24. Through $(0, 0)$ and $(-3, 2)$

   Let $(0, 0) = (x_1, y_1)$ and $(-3, 2) = (x_2, y_2)$.

   $$m = \frac{2 - 0}{-3 - 0}$$
   $$= \frac{2}{-3} = -\frac{2}{3}$$

25. Through $(0, 6)$ and $(1, 6)$

   Let $(0, 6) = (x_1, y_1)$ and $(1, 6) = (x_2, y_2)$.

   $$m = \frac{6 - 6}{1 - 0}$$
   $$= \frac{0}{1} = 0$$

26. Through $(2, 5)$ and $(2, 8)$

   Let $(2, 5) = (x_1, y_1)$ and $(2, 8) = (x_2, y_2)$.

   $$m = \frac{8 - 5}{2 - 2}$$
   $$= \frac{3}{0},$$

   which is undefined.

27. $y = 3x - 4$

   The equation is already solved for y, so the slope of the line is given by the coefficient of x. Thus, the slope is 3.

28. $y = \frac{2}{3}x + 1$

   The equation is already solved for y, so the slope is given by the coefficient of x. Thus, the slope is 2/3.

29. The indicated points have coordinates (0, -2) and (2, 1). Use the definition of slope with $(0, -2) = (x_1, y_1)$ and $(2, 1) = (x_2, y_2)$.

$$m = \frac{y_2 - y_1}{x_2 - x_1}$$
$$= \frac{1 - (-2)}{2 - 0}$$
$$= \frac{3}{2}$$

30. The indicated points have coordinates (0, 1) and (3, 0). Use the definition of slope with $(0, 1) = (x_1, y_1)$ and $(3, 0) = (x_2, y_2)$.

$$m = \frac{y_2 - y_1}{x_2 - x_1}$$
$$= \frac{0 - 1}{3 - 0}$$
$$= \frac{-1}{3} = -\frac{1}{3}$$

31. $y = 4$ is the equation of a horizontal line. Its slope is 0.

32. $x = 0$ is the equation of the y-axis, which is a vertical line. Its slope is undefined.

33. The equation $y = 4x + 6$ is written in $y = mx + b$ form, so the slope of this line is 4.

(a) Two lines are perpendicular if the product of their slopes is -1. Since

$$4\left(-\frac{1}{4}\right) = -1,$$

a line perpendicular to $y = 4x + 6$ will have a slope of $-1/4$.

(b) Nonvertical parallel lines always have equal slopes, so a line parallel to $y = 4x + 6$ will have a slope of 4.

34. If the product of two numbers is negative, the numbers must have opposite signs. Because the product of the slopes of two perpendicular lines is -1, one line must have a positive slope and the other a negative slope. Therefore, the signs of the slopes cannot be the same.

35. $3x + 2y = 6$
$6x + 4y = 8$

Find the slope of each line by solving each equation for y.

$$3x + 2y = 6$$
$$2y = -3x + 6 \quad \textit{Subtract 3x}$$
$$y = -\frac{3}{2}x + 3 \quad \textit{Divide by 2}$$

The slope of the first line is $-3/2$.

$$6x + 4y = 8$$
$$4y = -6x + 8 \quad \textit{Subtract 6x}$$
$$y = -\frac{6}{4}x + 2 \quad \textit{Divide by 4}$$
$$y = -\frac{3}{2}x + 2 \quad \textit{Lowest terms}$$

The slope of the second line is $-3/2$.

The slopes are equal, so the lines are parallel.

36. $x - 3y = 1$
    $3x + y = 4$

    Find the slopes by solving the equations for y.

    $$x - 3y = 1$$
    $$-3y = -x + 1$$
    $$y = \tfrac{1}{3}x - \tfrac{1}{3}$$

    The slope of the first line is 1/3.

    $$3x + y = 4$$
    $$y = -3x + 4$$

    The slope of the second line is -3.
    The product of the slopes is

    $$\tfrac{1}{3}(-3) = -1,$$

    so the lines are perpendicular.

37. $x - 2y = 8$
    $x + 2y = 8$

    Find the slopes by solving the equations for y.

    $$x - 2y = 8$$
    $$-2y = -x + 8$$
    $$y = \tfrac{1}{2}x - 4$$

    The slope of the first line is 1/2.

    $$x + 2y = 8$$
    $$2y = -x + 8$$
    $$y = -\tfrac{1}{2}x + 4$$

    The slopes are not equal and their product is

    $$\left(\tfrac{1}{2}\right)\left(-\tfrac{1}{2}\right) = -\tfrac{1}{4} \neq -1,$$

    so they are neither parallel nor perpendicular.

38. A line with undefined slope is vertical. Any line perpendicular to a vertical line is a horizontal line, which has a slope of 0.

39. $m = -1$; $b = \tfrac{2}{3}$

    First write the equation in slope-intercept form.

    $$y = mx + b$$
    $$y = -x + \tfrac{2}{3}$$

    Now rewrite this equation in the standard form $Ax + By = C$.

    $3y = 3\left(-x + \tfrac{2}{3}\right)$ Multiply by 3 to clear fractions

    $3y = -3x + 2$ Distributive property

    $3x + 3y = 2$ Standard form

40. The line in Exercise 30 has slope $-1/3$ and y-intercept $(0, 1)$, so we may write its equation in slope-intercept form as

    $$y = -\tfrac{1}{3}x + 1.$$

    Rewrite this equation in the standard form $Ax + By = C$.

    $3y = 3\left(-\tfrac{1}{3}x + 1\right)$ Multiply by 3

    $3y = -x + 3$ Distributive property

    $x + 3y = 3$ Standard form

41. Through $(4, -3)$; $m = 1$

Let $x_1 = 4$, $y_1 = 3$, and $m = 1$ in the point-slope form.

$$y - y_1 = m(x - x_1)$$
$$y - (-3) = 1(x - 4)$$
$$y + 3 = x - 4$$
$$y + 7 = x$$
$$7 = x - y$$
or $x - y = 7$

42. Through $(-1, 4)$; $m = \frac{2}{3}$

Use the point-slope form with $x_1 = -1$, $y_1 = 4$, and $m = 2/3$.

$$y - y_1 = m(x - x_1)$$
$$y - 4 = \frac{2}{3}[x - (-1)]$$
$$3(y - 4) = 2(x + 1) \quad \textit{Multiply by 3}$$
$$3y - 12 = 2x + 2$$
$$-14 = 2x - 3y$$
or $2x - 3y = -14 \quad \textit{Standard form}$

43. Through $(1, -1)$; $m = -\frac{3}{4}$

Use the point-slope form with $m = -3/4$, $x_1 = 1$, and $y_1 = -1$.

$$y - y_1 = m(x - x_1)$$
$$y - (-1) = -\frac{3}{4}(x - 1)$$
$$y + 1 = -\frac{3}{4}(x - 1)$$
$$4(y + 1) = -3(x - 1)$$
$$\quad\quad \textit{Multiply by 4}$$
$$4y + 4 = -3x + 3$$
$$3x + 4y = -1 \quad \textit{Standard form}$$

44. Through $(2, 1)$ and $(-2, 2)$

First find the slope of the line.

$$m = \frac{1 - 2}{2 - (-2)} = -\frac{1}{4}$$

Using $(2, 1)$ and $m = -1/4$, substitute into the point-slope form.

$$y - y_1 = m(x - x_1)$$
$$y - 1 = -\frac{1}{4}(x - 2)$$
$$4(y - 1) = -1(x - 2)$$
$$\quad\quad \textit{Multiply by 4}$$
$$4y - 4 = -x + 2$$
$$x + 4y = 6 \quad \textit{Standard form}$$

45. Through $(-4, 1)$ with slope 0

Every line with slope 0 is horizontal. Every point on the horizontal line through $(-4, 1)$ has a y-coordinate of 1, so the equation of this line is $y = 1$.

46. Through $\left(\frac{1}{3}, -\frac{3}{4}\right)$ with undefined slope

Every line with undefined slope is vertical. Every point on the vertical line through $(1/3, -3/4)$ has an x-coordinate of 1/3, so the equation of this line is $x = 1/3$.

For Exercises 47–52, see the graphs in the answer section of the textbook.

47. $x - y \geq 3$

In order to determine which region to shade, use $(0, 0)$ as a test point.

$x - y \geq 3$

$0 - 0 \geq 3$? Let $x = 0$, $y = 0$

$0 \geq 3$ False

Since $0 \geq 3$ is a false statement, shade the region that does *not* include the origin, that is, the region below the line.

48. $3x - y \leq 5$

Using (0, 0) as a test point will result in the inequality $0 \leq 5$, which is true. Shade the region that contains the origin, that is, the region above the line.

49. $x + 2y < 6$

Using (0, 0) as a test point will result in the inequality $0 < 6$, which is true. Shade the region that contains the origin, that is, the region below the line.

50. $3x + 5y > 9$

To graph the boundary, which is the line $3x + 5y = 9$, find its intercepts.

$3x + 5y = 9$ $\quad\quad$ $3x + 5y = 9$
$3x + 5(0) = 9$ $\quad\quad$ $3(0) + 5y = 9$
$\quad$ Let $y = 0$ $\quad\quad\quad$ Let $x = 0$
$3x = 9$ $\quad\quad\quad\quad$ $5y = 9$
$x = 3$ $\quad\quad\quad\quad$ $y = \frac{9}{5}$

The x-intercept is (3, 0) and the y-intercept is (0, 9/5). (A third point may be used as a check.) Draw a dashed line through these points.

In order to determine which side of the line should be shaded, use (0, 0) as a test point. Substituting 0 for x and 0 for y will result in the inequality $0 > 9$, which is false. Shade the region *not* containing the origin. This is the region above the line. The dashed line shows that the boundary is not part of the graph.

51. $2x - 3y > -6$

Use intercepts to graph the boundary, $2x - 3y = -6$.

If $y = 0$, $x = -3$, so the x-intercept is (-3, 0).

If $x = 0$, $y = 2$, so the y-intercept is (0, 2).

Draw a dashed line through (-3, 0) and (0, 2).

Using (0, 0) as a test point will result in the inequality $0 > -6$, which is true. Shade the region containing the origin. This is the region below the line. The dashed line shows that the boundary is not part of the graph.

52. $x - 2y \geq 0$

The equation of the boundary is $x - 2y = 0$. This line goes through the origin, so both intercepts are (0, 0). Two other points on this line are (2, 1) and (-2, -1). Draw a solid line through (0, 0), (2, 1), and (-2, -1).

Because (0, 0) lies on the boundary, we must choose another point as the test point. Using (0, 3) results in the inequality $-6 \geq 0$, which is false. Shade the region *not* containing the test point (0, 3).
This is the region below the line. The solid line shows that the boundary is part of the graph.

53. $11x - 3y = 4$

    x-intercept:

    $11x - 3(0) = 4$ *Let y = 0*
    $11x = 4$
    $x = \frac{4}{11}$

    The x-intercept is $(4/11, 0)$.

    y-intercept:

    $11(0) - 3y = 4$ *Let x = 0*
    $-3y = 4$
    $y = -\frac{4}{3}$

    The y-intercept is $(0, -4/3)$.

    To find the slope, solve the equation for y.

    $-3y = 4 - 11x$
    $y = \frac{11}{3}x - \frac{4}{3}$

    The slope is the coefficient of x, so $m = 11/3$.

54. Through $(4, -1)$ and $(-2, -3)$

    To find the intercepts, we must find the equation of the line.

    $m = \frac{-3 - (-1)}{-2 - 4}$
    $= \frac{-3 + 1}{-6}$
    $= \frac{-2}{-6} = \frac{1}{3}$

    Use the point-slope form with $x_1 = 4$, $y_1 = -1$, and $m = 1/3$.

    $y - y_1 = m(x - x_1)$
    $y - (-1) = \frac{1}{3}(x - 4)$
    $y + 1 = \frac{1}{3}(x - 4)$
    $3(y + 1) = 3 \cdot \frac{1}{3}(x - 4)$
    $3y + 3 = 1(x - 4)$
    $3y + 3 = x - 4$
    $7 = x - 3y$

    or $x - 3y = 7$

    Use this equation to find the intercepts.

    If $y = 0$, $x = 7$, so the x-intercept is $(7, 0)$.
    If $x = 0$, $y = -7/3$, so the y-intercept is $(0, -7/3)$.

    The slope of the line is $1/3$.

55. Through $(0, -1)$ and $(9, -5)$

    $m = \frac{-5 - (-1)}{9 - 0} = \frac{-4}{9} = -\frac{4}{9}$

    The line goes through $(0, -1)$, so the y-intercept is $(0, -1)$.
    We may write the equation of this line in slope-intercept form as

    $y = mx + b$
    $y = -\frac{4}{9}x - 1.$

To find the x-intercept, let $y = 0$.

$$0 = -\frac{4}{9}x - 1$$

$$\frac{4}{9}x = -1$$

$$\left(\frac{9}{4}\right)\left(\frac{4}{9}x\right) = \frac{9}{4}(-1)$$

$$x = -\frac{9}{4}$$

Thus, the x-intercept is $(-9/4, 0)$, the y-intercept is $(0, -1)$, and the slope is $-4/9$.

56. $8x = 6 - 3y$

x-intercept:

$$8x = 6 - 3y$$
$$8x = 6 - 3(0) \quad \text{Let } y = 0$$
$$8x = 6$$
$$x = \frac{6}{8} = \frac{3}{4}$$

The x-intercept is $(3/4, 0)$.

y-intercept:

$$8x = 6 - 3y$$
$$8(0) = 6 - 3y \quad \text{Let } x = 0$$
$$0 = 6 - 3y$$
$$3y = 6$$
$$y = 2$$

The y-intercept is $(0, 2)$.

To find the slope, solve the given equation for y.

$$8x = 6 - 3y$$
$$3y + 8x = 6$$
$$3y = -8x + 6$$
$$y = -\frac{8}{3}x + 2$$

The slope is $-8/3$.

57. Through $(5, 0)$ and $(5, -1)$

First find the slope of the line, using 0 and -1 as the y-values and 5 and 5 as the x-values.

$$m = \frac{0 - (-1)}{5 - 5}$$

$$= \frac{1}{0} \quad \text{Undefined}$$

Since the slope is undefined, the line is vertical and its equation has the form $x = k$. In this case, since the line passes through $(5, 0)$, the equation is $x = 5$.

58. $m = -\frac{1}{4}$, $b = -\frac{5}{4}$

Substitute the values $m = -1/4$ and $b = -5/4$ into the slope-intercept form.

$$y = mx + b$$
$$y = -\frac{1}{4}x - \frac{5}{4}$$
$$4y = -x - 5 \quad \text{Multiply by 4}$$
$$x + 4y = -5 \quad \text{Standard form}$$

59. Through $(8, 6)$; $m = -3$

Use the point-slope form with $(x_1, y_1) = (8, 6)$ and $m = -3$.

$$y - y_1 = m(x - x_1)$$
$$y - 6 = -3(x - 8)$$
$$y - 6 = -3x + 24$$
$$3x + y = 30 \quad \text{Standard form}$$

60. Through $(3, -5)$ and $(-4, -1)$

First, find the slope of the line.

$$m = \frac{-1 - (-5)}{-4 - 3} = -\frac{4}{7}$$

Now use either point and the slope in the point-slope form. If we use $(3, -5)$ we get the following.

$$y - y_1 = m(x - x_1)$$
$$y + 5 = -\frac{4}{7}(x - 3)$$
$$7(y + 5) = -4(x - 3) \quad \textit{Multiply by 7}$$
$$7y + 35 = -4x + 12$$
$$4x + 7y + 35 = 12$$
$$4x + 7y = -23 \quad \textit{Standard form}$$

For Exercises 61–66, see the graphs in the answer section of the textbook.

61. $x - 2y \leq 6$

This is a linear inequality, so its graph will be a shaded region. Graph the boundary, $x - 2y = 6$, as a solid line through the intercepts $(0, -3)$ and $(6, 0)$.
Using $(0, 0)$ as a test point results in the true statement $0 < 6$, so shade the region containing the origin. This is the region above the line. The solid line shows that the boundary is part of the graph.

62. $x + 3y = 0$

This is a linear equation, so its graph will be a line.

If $x = 0$, $y = 0$, so both intercepts are $(0, 0)$. Two other points on the line are $(3, -1)$ and $(-3, 1)$. Draw a line through $(0, 0)$, $(3, -1)$ and $(-3, 1)$.

63. $y < -4x$

This is a linear inequality, so its graph will be a shaded region. Graph the boundary, $y = -4x$, as a dashed line through $(0, 0)$, $(1, -4)$, and $(-1, 4)$.
Choose a test point that does not lie on the line. Using $(2, 2)$ results in the inequality $2 < -8$, which is false so shade the region *not* containing $(2, 2)$. This is the region below the line. The dashed line shows that the boundary is not part of the graph.

64. $y - 5 = 0$

This is a linear equation, so its graph will be a line. The equation may be rewritten as $y = 5$, which is of the form $y = k$. Draw a horizontal line through $(0, 5)$.

65. $2x - y = 3$

This is a linear equation, so its graph will be a line.

If $y = 0$, $x = 3/2$, so the x-intercept is $(3/2, 0)$.
If $x = 0$, $y = -3$, so the y-intercept is $(0, -3)$.

Draw a line through these two points. A third point may be used as a check.

66. $x \geq -4$

This is a linear inequality, so its graph is a shaded region. Graph the boundary, $x = -4$, as a solid vertical line through $(-4, 0)$.
Using $(0, 0)$ as a test point results in the true statement $0 \geq -4$, so shade the region containing the origin.
This is the region to the right of the line. The solid line shows that the boundary is part of the graph.

Chapter 6   Test

1. $y = 4x - 9$ $(0, \ )$ $(1, \ )$ $(\ , -3)$

    $y = 4x - 9$ $\qquad$ $y = 4x - 9$
    $y = 4(0) - 9$ $\qquad$ $y = 4(1) - 9$
    $\quad$ Let $x = 0$ $\qquad$ $\quad$ Let $x = 1$
    $y = 0 - 9$ $\qquad\quad$ $y = 4 - 9$
    $y = -9$ $\qquad\qquad\quad$ $y = -5$

    $y = 4x - 9$
    $-3 = 4x - 9$ Let $y = -3$
    $6 = 4x$
    $\frac{6}{4} = x$
    $\frac{3}{2} = x$

    The ordered pairs are $(0, -9)$, $(1, -5)$, and $(3/2, -3)$.

2. $3x + 5y = -30$ $(0, \ )$ $(\ , 0)$ $(\ , 3)$

    $3x + 5y = -30$ $\qquad$ $3x + 5y = -30$
    $3(0) + 5y = -30$ $\qquad$ $3x + 5(0) = -30$
    $\quad$ Let $x = 0$ $\qquad\qquad$ $\quad$ Let $y = 0$
    $5y = -30$ $\qquad\qquad\quad$ $3x = -30$
    $y = -6$ $\qquad\qquad\qquad$ $x = -10$

    $3x + 5y = -30$
    $3x + 5(3) = -30$ Let $y = 3$
    $3x + 15 = -30$
    $3x = -45$
    $x = -15$

    The ordered pairs are $(0, -6)$, $(-10, 0)$, and $(-15, 3)$.

3. $y + 12 = 0$ $(0, \ )$ $(-4, \ )$ $\left(\frac{5}{2}, \ \right)$

    The equation may also be written as $y = -12$. For every value of x, the value of y will always be -12. Complete the ordered pairs by placing -12 in the second position in each pair.
    The ordered pairs are $(0, -12)$, $(-4, -12)$, and $(5/2, -12)$.

4. To find the x-intercept, let $y = 0$ and solve for x.
    To find the y-intercept, let $x = 0$ and solve for y.

For Exercises 5–9, see the graphs in the answer section of the textbook.

5. $x - y = 4$

    If $y = 0$, $x = 4$, so the x-intercept is $(4, 0)$.
    If $x = 0$, $y = -4$, so the y-intercept is $(0, -4)$.

A third point, such as (2, -2), can be used as a check. Draw a line through (0, -4), (2, -2), and (4, 0).

6. $3x + y = 6$

   If $y = 0$, $x = 2$, so the x-intercept is (2, 0).

   If $x = 0$, $y = 6$, so the y-intercept is (0, 6).

   A third point, such as (1, 3), can be used as a check. Draw a line through (0, 6), (1, 3), and (2, 0).

7. $y - 2x = 0$

   If $x = 0$, $y = 0$, so the x-intercept is (0, 0).

   The y-intercept is also (0, 0).

   Two other points on the line are (1, 2) and (-1, -2). Draw a line through (-1, -2), (0, 0), and (1, 2).

8. $x + 3 = 0$
   $x = -3$

   This is the equation of the vertical line through (-3, 0). The x-intercept is (-3, 0). Because this line is parallel to the y-axis, it has no y-intercept.

9. $y = 1$

   This is the equation of a horizontal line through (0, 1). Because this line is parallel to the x-axis, it has no x-intercept. The y-intercept is (0, 1).

10. Through (-4, 6) and (-1, -2)

    Use the definition of slope with $(x_1, y_1) = (-4, 6)$ and $(x_2, y_2) = (-1, -2)$.

    $$m = \frac{y_2 - y_1}{x_2 - x_1}$$
    $$= \frac{-2 - 6}{-1 - (-4)}$$
    $$= \frac{-8}{3} = -\frac{8}{3}$$

11. $2x + y = 10$

    To find the slope, solve the given equation for y.

    $$2x + y = 10$$
    $$y = -2x + 10$$

    The equation is now written in $y = mx + b$ form, so the slope is given by the coefficient of x, which is -2.

12. $x + 12 = 0$
    $x = -12$

    This is the equation of a vertical line. Its slope is undefined.

13. The indicated points are (0, -4) and (2, 1). Use the definition of slope with $(x_1, y_1) = (0, -4)$ and $(x_2, y_2) = (2, 1)$.

    $$m = \frac{y_2 - y_1}{x_2 - x_1}$$
    $$= \frac{1 - (-4)}{2 - 0}$$
    $$= \frac{5}{2}$$

14. A line parallel to the graph of $y - 4 = 6$.

The given equation may be rewritten as $y = 10$. This is a horizontal line; its slope is 0. Any line parallel to this line will also have a slope of 0.

15. Through $(-1, 4)$; $m = 2$

Let $x_1 = -1$, $y_1 = 4$, and $m = 2$ in the point-slope form.

$$y - y_1 = m(x - x_1)$$
$$y - 4 = 2[(x - (-1)]$$
$$y - 4 = 2(x + 1)$$
$$y - 4 = 2x + 2$$
$$y - 6 = 2x$$
$$-6 = 2x - y$$
$$\text{or } 2x - y = -6$$

16. The line in Exercise 13

In Exercise 13, we found that the slope of the line is 5/2. From the graph, we see that the y-intercept is $(0, -4)$. Therefore, we can write the equation of this line in slope-intercept form as

$$y = mx + b$$
$$y = \frac{5}{2}x - 4.$$

Rewrite this equation in the standard form $Ax + By = C$.

$$2y = 2\left(\frac{5}{2}x - 4\right)$$
$$2y = 5x - 8$$
$$2y + 8 = 5x$$
$$8 = 5x - 2y$$
$$\text{or } 5x - 2y = 8$$

17. Through $(2, -6)$ and $(1, 3)$

First find the slope.

$$m = \frac{3 - (-6)}{1 - 2} = \frac{9}{-1} = -9$$

Use the point-slope form with $x_1 = 1$, $y_1 = 3$, and $m = -9$. (We could also use $(2, -6)$ as (x_1, y_1).)

$$y - y_1 = m(x - x_1)$$
$$y - 3 = -9(x - 1)$$
$$y - 3 = -9x + 9$$
$$y = -9x + 12$$
$$9x + y = 12$$

For Exercises 18-20, see the graphs in the answer section of the textbook.

18. $x + y \leq 3$

Graph the boundary, $x + y = 3$, as a solid line through the intercepts $(3, 0)$ and $(0, 3)$. Using $(0, 0)$ as a test point results in the true statement $0 \leq 3$, so shade the region containing the origin. This is the region below the line. The solid line shows that the boundary is part of the graph.

19. $3x - y > 0$

The boundary, $3x - y = 0$, goes through the origin, so both intercepts are $(0, 0)$. Two other points on this line are $(1, 3)$ and $(-1, -3)$. Draw the boundary as a dashed line. Choose a test point which is not on the boundary. Using $(3, 0)$ results in the true statement $9 > 0$, so shade the region containing $(3, 0)$.

This is the region below the line. The dashed line shows that the boundary is not part of the graph.

20. $x \leq -4$

    The boundary, $x = -4$, is the vertical line through $(-4, 0)$. Graph this boundary as a solid line. Shade the region to the left of the boundary. The solid line shows that the boundary is part of the graph.

Cumulative Review: Chapters R–6

1. $-5(8 - 2z) + 4(7 - z) = 7(8 + z) - 3$

    $-40 + 10z + 28 - 4z = 56 + 7z - 3$
                            Distributive property

    $6z - 12 = 7z + 53$
               Combine like terms

    $-z - 12 = 53$
           Subtract 7z

    $-z = 65$
        Add 12

    $z = -65$
        Divide by –1

2. $A = p + prt$ for t

    Rewrite the equation with t alone on one side of the equation.

    $A = p + prt$

    $A - p = prt$ *Subtract p*

    $\dfrac{A - p}{pr} = t$ *Divide by pr*

    or $t = \dfrac{A - p}{pr}$

3. $7x^2 + 8x + 1 = 0$

    To solve this quadratic equation, begin by factoring the polynomial.

    $(7x + 1)(x + 1) = 0$

    Use the zero-factor property to set each factor equal to 0.

    $7x + 1 = 0$ or $x + 1 = 0$

    Solve each of the linear equations.

    $7x = -1$ or $x = -1$

    $x = -\dfrac{1}{7}$ or $x = -1$

    The solutions of the original equation are -1 and $-1/7$.

4. $\dfrac{4}{x - 2} = 4$

    To clear fractions, multiply both sides of the equation by $x - 2$.

    $\dfrac{4}{x - 2}(x - 2) = 4(x - 2)$

    $4 = 4(x - 2)$

    $4 = 4x - 8$

    $12 = 4x$

    $3 = x$

    Check that 3 does not make the denominator of the fraction equal to 0. The solution is 3.

5. $\dfrac{2}{x - 1} = \dfrac{5}{x - 1} - \dfrac{3}{4}$

    Multiply both sides by the LCD, $4(x - 1)$.

$4(x - 1)\left(\frac{2}{x - 1}\right) = 4(x - 1)\left(\frac{5}{x - 1} - \frac{3}{4}\right)$

$4(x - 1)\left(\frac{2}{x - 1}\right) = 4(x - 1)\left(\frac{5}{x - 1}\right)$
$\phantom{4(x - 1)\left(\frac{2}{x - 1}\right) =} - 4(x - 1)\left(\frac{3}{4}\right)$
 Distributive property

$8 = 20 - 3(x - 1)$
$8 = 20 - 3x + 3$ Distributive property
$8 = 23 - 3x$
$3x + 8 = 23$
$3x = 15$
$x = 5$

Check that 5 does not make any denominator equal to 0. The solution is 5.

For Exercises 6–8, see the graphs in the answer section of the textbook.

6. $-2.5x < 6.5$

To solve the inequality, we must divide both sides by -2.5. Whenever both sides of an inequality are multiplied or divided by a negative number, the direction of the inequality symbol is reversed.

$-2.5x < 6.5$
$\frac{-2.5x}{-2.5} > \frac{6.5}{-2.5}$
$x > -2.6$

To graph this solution on the number line, put an open circle at -2.6 and draw an arrow extending to the right.

7. $4(x + 3) - 5x < 12$
$4x + 12 - 5x < 12$ Distributive property
$-x + 12 < 12$ Combine like term
$-x < 0$ Subtract 12
$x > 0$ Divide by -1; reverse the symbol

To graph this solution on the number line, put an open circle at 0 and draw an arrow extending to the right.

8. $\frac{2}{3}y - \frac{1}{6}y \leq -2$

$6\left(\frac{2}{3}y - \frac{1}{6}y\right) \leq 6(-2)$ Multiply by 6 to clear fractions

$6\left(\frac{2}{3}y\right) - 6\left(\frac{1}{6}y\right) \leq 6(-2)$ Distributive property

$4y - y \leq -12$
$3y \leq -12$
$y \leq -4$

To graph this solution on the number line, put a dot at -4 and draw an arrow extending to the left.

9. $(x^2y^{-3})(x^{-4}y^2)$
$= (x^2x^{-4})(y^{-3}y^2)$ Commutative and associative properties
$= x^{2+(-4)}y^{-3+2}$ Product rule
$= x^{-2}y^{-1}$
$= \frac{1}{x^2y}$ Definition of negative exponent

10. $\frac{x^{-6}y^3z^{-1}}{x^7y^{-4}z}$
$= x^{-6-7}y^{3-(-4)}z^{-1-1}$ Quotient rule
$= x^{-13}y^7z^{-2}$
$= \frac{y^7}{x^{13}z^2}$ Definition of negative exponent

11. $(2m^{-2}n^3)^{-3}$

$= 2^{-3}(m^{-2})^{-3}(n^3)^{-3}$ *Power rule (b)*

$= 2^{-3}m^6n^{-9}$ *Power rule (a)*

$= \dfrac{m^6}{2^3n^9}$ *Definition of negative exponent*

12. $2(3x^2 - 8x + 1) - 4(x^2 - 3x - 9)$

$= 6x^2 - 16x + 2 - 4x^2 + 12x + 36$
 *Distributive property*

$= (6x^2 - 4x^2) + (-16x + 12x)$

$+ (2 + 36)$ *Combine like terms*

$= 2x^2 - 4x + 38$

13. $(3x + 2y)(5x - y)$

$= 3x(5x) + 3x(-y) + 2y(5x) + 2y(-y)$
 FOIL

$= 15x^2 - 3xy + 10xy - 2y^2$

$= 15x^2 + 7xy - 2y^2$

14. $(x + 2y)(x^2 - 2xy + 4y^2)$

Multiply vertically.

$$\begin{array}{r} x^2 - 2xy + 4y^2 \\ x + 2y \\ \hline 2x^2y - 4xy^2 + 8y^3 \\ x^3 - 2x^2y + 4xy^2 \\ \hline x^3 + 8y^3 \end{array}$$

Thus,

$(x + 2y)(x^2 - 2xy + 4y^2) = x^3 + 8y^3$.

15. $\dfrac{m^3 - 3m^2 + 5m - 3}{m - 1}$

$$\begin{array}{r} m^2 - 2m + 3 \\ m-1 \overline{\smash{)}m^3 - 3m^2 + 5m - 3} \\ \underline{m^3 - m^2} \\ -2m^2 + 5m \\ \underline{-2m^2 + 2m} \\ 3m - 3 \\ \underline{3m - 3} \\ 0 \end{array}$$

The remainder is 0. The answer is the quotient, $m^2 - 2m + 3$.

16. $y^2 + 4yk - 12k^2$

Look for two expressions whose product is $-12k^2$ and whose sum is $4k$. These expressions are $6k$ and $-2k$. Thus,

$$y^2 + 4yk - 12k^2 = (y + 6k)(y - 2k).$$

17. $9x^4 - 25y^2$

This polynomial is the difference of two squares.

$9x^4 - 25y^2 = (3x^2)^2 - (5y)^2$

$ = (3x^2 + 5y)(3x^2 - 5y)$

18. $125x^4 - 400x^3y + 195x^2y^2$

Begin by factoring out the greatest common factor, $5x^2$.

$125x^4 - 400x^3y + 195x^2y^2$

$= 5x^2(25x^2 - 80xy + 39y^2)$

Now factor $25x^2 - 80xy + 39y^2$ by trial and error.

$25x^2 - 80xy + 39y^2$

$= (5x - 13y)(5x - 3y)$

The complete factored form is

$125x^4 - 400x^3y + 195x^2y^2$

$= 5x^2(5x - 13y)(5x - 3y)$.

19. $f^2 + 20f + 100$

This polynomial is a perfect square trinomial.

$f^2 + 20f + 100 = f^2 + 2(f)(10) + 10^2$

$ = (f + 10)^2$

20. $100x^2 + 49$

The sum of two squares cannot be factored unless the two terms have a common factor.
This polynomial is the sum of two squares in which there is no common factor. Therefore, $100x^2 + 49$ is prime.

21. $\dfrac{3}{2x + 6} + \dfrac{2x + 3}{2x + 6}$

These rational expressions have a common denominator, so we add their numerators and keep the common denominator.

$$\dfrac{3}{2x + 6} + \dfrac{2x + 3}{2x + 6} = \dfrac{2x + 6}{2x + 6}$$
$$= 1$$

22. $\dfrac{8}{x + 1} - \dfrac{2}{x + 3}$

To add these rational expressions, we need a common denominator. The LCD is $(x + 1)(x + 3)$. Rewrite each factor with $(x + 1)(x + 3)$ as its denominator.

$$\dfrac{8}{x + 1} - \dfrac{2}{x + 3}$$
$$= \dfrac{8}{x + 1} \cdot \dfrac{x + 3}{x + 3} - \dfrac{2}{x + 3} \cdot \dfrac{x + 1}{x + 1}$$
$$= \dfrac{8(x + 3)}{(x + 1)(x + 3)} - \dfrac{2(x + 1)}{(x + 1)(x + 3)}$$
$$= \dfrac{8x + 24}{(x + 1)(x + 3)} - \dfrac{2x + 2}{(x + 1)(x + 3)}$$

$$= \dfrac{(8x + 24) - (2x + 2)}{(x + 1)(x + 3)}$$
$$= \dfrac{8x + 24 - 2x - 2}{(x + 1)(x + 3)}$$
$$= \dfrac{6x + 22}{(x + 1)(x + 3)}$$
or $\dfrac{2(3x + 11)}{(x + 1)(x + 3)}$

23. $\dfrac{x^2 - 25}{3x + 6} \cdot \dfrac{4x + 8}{x^2 + 10x + 25}$

$$= \dfrac{(x + 5)(x - 5)}{3(x + 2)} \cdot \dfrac{4(x + 2)}{(x + 5)(x + 5)}$$
Factor numerators and denominators

$$= \dfrac{(x + 5)(x - 5)(4)(x + 2)}{3(x + 2)(x + 5)(x + 5)}$$
Multiply

$$= \dfrac{4(x - 5)}{3(x + 5)} \quad \text{Lowest terms}$$

24. $\dfrac{x^2 + 2x - 3}{x^2 - 5x + 4} \cdot \dfrac{x^2 - 3x - 4}{x^2 + 3x}$

$$= \dfrac{(x - 1)(x + 3)}{(x - 1)(x - 4)} \cdot \dfrac{(x + 1)(x - 4)}{x(x + 3)}$$
Factor numerators and denominators

$$= \dfrac{(x - 1)(x + 3)(x + 1)(x - 4)}{(x - 1)(x - 4)(x)(x + 3)}$$
Multiply

$$= \dfrac{x + 1}{x} \quad \text{Lowest terms}$$

25. $\dfrac{x^2 + 5x + 6}{3x} \div \dfrac{x^2 - 4}{x^2 + x - 6}$

$$= \dfrac{x^2 + 5x + 6}{3x} \cdot \dfrac{x^2 + x - 6}{x^2 - 4}$$
Multiply by reciprocal of second expression

$$= \dfrac{(x + 2)(x + 3)}{3x} \cdot \dfrac{(x + 3)(x - 2)}{(x + 2)(x - 2)}$$
Factor numerators and denominators

$$= \dfrac{(x + 2)(x + 3)(x + 3)(x - 2)}{3x(x + 2)(x - 2)}$$
Multiply

$$= \dfrac{(x + 3)^2}{3x}$$

26. $\dfrac{6x^4y^3z^2}{8xyz^4} \div \dfrac{3x^2}{16y^2}$

$= \dfrac{6x^4y^3z^2}{8xyz^4} \cdot \dfrac{16y^2}{3x^2}$ *Multiply by reciprocal*

$= \dfrac{96x^4y^5z^2}{24x^3yz^4}$ *Multiply*

$= \dfrac{4xy^4}{z^2}$ *Lowest terms*

27. $\dfrac{\frac{2}{3} - \frac{1}{4}}{\frac{1}{2} + \frac{1}{6}} = \dfrac{12\left(\frac{2}{3} - \frac{1}{4}\right)}{12\left(\frac{1}{2} + \frac{1}{6}\right)}$ *LCD = 12*

$= \dfrac{12\left(\frac{2}{3}\right) - 12\left(\frac{1}{4}\right)}{12\left(\frac{1}{2}\right) + 12\left(\frac{1}{6}\right)}$ *Distributive property*

$= \dfrac{8 - 3}{6 + 2} = \dfrac{5}{8}$

28. $\dfrac{\frac{12}{x+6}}{\frac{4}{2x+12}} = \dfrac{12}{x+6} \div \dfrac{4}{2x+12}$

$= \dfrac{12}{x+6} \cdot \dfrac{2x+12}{4}$ *Multiply by reciprocal*

$= \dfrac{12 \cdot 2(x+6)}{(x+6)4}$

$= \dfrac{24(x+6)}{4(x+6)}$

$= 6$

29. Through $(-4, 5)$ and $(2, -3)$

Use the definition of slope with $x_1 = -4$, $y_1 = 5$, $x_2 = 2$, and $y_2 = -3$.

$m = \dfrac{y_2 - y_1}{x_2 - x_1}$

$= \dfrac{-3 - 5}{2 - (-4)}$

$= \dfrac{-8}{6} = -\dfrac{4}{3}$

30. Horizontal, through $(4, 5)$

The slope of every horizontal line is 0.

31. Through $(4, -1)$; $m = -4$

Use the point-slope form with $x_1 = 4$, $y_1 = -1$, and $m = -4$.

$y - y_1 = m(x - x_1)$
$y - (-1) = -4(x - 4)$
$y + 1 = -4x + 16$
$y = -4x + 15$
$4x + y = 15$

32. Through $(0, 0)$ and $(1, 4)$

Find the slope.

$m = \dfrac{4 - 0}{1 - 0} = 4$

Because the slope is 4 and the y-intercept is 0, the equation of the line in slope-intercept form is

$y = 4x$.

Rewrite this equation in standard form as

$4x - y = 0$.

For Exercises 33–35, see the graphs in the answer section of the textbook.

33. $-3x + 4y = 12$

If $y = 0$, $x = -4$, so the x-intercept is $(-4, 0)$.

If $x = 0$, $y = 3$, so the y-intercept is $(0, 3)$.

Draw a line through these intercepts. A third point may be used as a check.

34. $y \leq 2x - 6$

    Graph the boundary, $y = 2x - 6$, as a solid line through the intercepts (3, 0) and (0, -6). A third point such as (1, -4) can be used as a check. Using (0, 0) as a test point results in the false inequality $0 \leq -6$, so shade the region *not* containing the origin. This is the region below the line. The solid line shows that the boundary is part of the graph.

35. $3x + 2y < 0$

    Graph the boundary, $3x + 2y = 0$, as a dashed line through (0, 0), (-2, 3), and (2, -3). Choose a test point not on the line. Using (1, 1) results in the false statement $5 < 0$, so shade the region *not* containing (1, 1). This is the region below the line. The dashed line shows that the boundary is not part of the graph.

36. Let x = the number of women attending the concert.

    Then x + 36 = the number of men attending the concert.

    The total number of men and women was 196, so

    $$x + (x + 36) = 196.$$

    Solve this equation.

    $$2x + 36 = 196$$
    $$2x = 160$$
    $$x = 80$$
    $$x + 36 = 80 + 36 = 116$$

    The concert audience included 116 men and 80 women.

37. The sum of the measures of the angles of any triangle is 180°, so

    $$(x + 15) + (6x + 10) + (x - 5) = 180.$$

    Solve this equation.

    $$8x + 20 = 180$$
    $$8x = 160$$
    $$x = 20$$

    Substitute 20 for x to find the measures of the angles.

    $$x - 5 = 20 - 5 = 15$$
    $$x + 15 = 20 + 15 = 35$$
    $$6x + 10 = 6(20) + 10 = 130$$

    The measures of the angles of the triangle are 15°, 35°, and 130°.

38. Let x = the length of the shorter leg;

    3x + 4 = the length of the hypotenuse;

    2x + 10 = the length of the longer leg.

    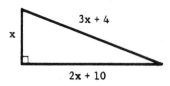

Use the Pythagorean formula with
$a = x$, $b = 3x + 4$, and $c = 2x + 10$.

$$a^2 + b^2 = c^2$$
$$x^2 + (2x + 10)^2 = (3x + 4)^2$$
$$x^2 + 4x^2 + 40x + 100 = 9x^2 + 24x + 16$$
Square the binomials
$$5x^2 + 40x + 100 = 9x^2 + 24x + 16$$
Combine like terms
$$0 = 4x^2 - 16x - 84$$
Standard form
$$0 = 4(x^2 - 4x - 21)$$
Factor out 4
$$0 = 4(x - 7)(x + 3)$$
Factor

$$x - 7 = 0 \quad \text{or} \quad x + 3 = 0$$
$$x = 7 \quad \text{or} \quad x = -3$$

Discard -7 because length cannot be negative. Thus, the length of the shorter leg is 7 inches.

39. Since x varies directly as y, there is a constant k such that $x = ky$. First find the value of k. We know that $x = 4$ when $y = 12$.

$$x = ky$$
$$4 = k \cdot 12 \quad \text{Let } x = 4, y = 12$$
$$k = \frac{4}{12} = \frac{1}{3}$$

Since $x = ky$ and $k = 1/3$,

$$x = \frac{1}{3}y.$$

Now find x when $y = 42$.

$$x = \frac{1}{3}(42) \quad \text{Let } y = 42$$
$$x = 14$$

40. Let x = the number of hours it will take the man and his wife to do the job, working together.

	Rate	Time working together	Part of job done working together
Man	$\frac{1}{3}$	x	$\frac{1}{3}x$
Wife	$\frac{1}{1.5}$	x	$\frac{1}{1.5}x$

Together, the man and his wife complete 1 whole job, so

$$\frac{1}{3}x + \frac{1}{1.5}x = 1.$$

Rewrite this equation by changing the decimal to a fraction.

$$\frac{1}{3}x + \frac{1}{3/2}x = 1$$
$$\frac{1}{3}x + \frac{2}{3}x = 1 \quad \begin{array}{l}\textit{Reciprocal}\\\textit{of 3/2 is 2/3}\end{array}$$

Multiply both sides of this equation by 3 to clear fractions.

$$3\left(\frac{1}{3}x + \frac{2}{3}x\right) = 3(1)$$
$$3\left(\frac{1}{3}x\right) + 3\left(\frac{2}{3}x\right) = 3$$
$$x + 2x = 3$$
$$3x = 3$$
$$x = 1$$

Working together, it will take the man and his wife 1 hour to do the job.

CHAPTER 7 LINEAR SYSTEMS

Section 7.1 Solving Systems of Linear Equations by Graphing

7.1 Margin Exercises

1. **(a)** (2, 5)

   $3x - 2y = -4$

   $5x + y = 15$

   To decide whether (2, 5) is a solution, substitute 2 for x and 5 for y in each equation.

   $3x - 2y = -4$
   $3(2) - 2(5) = -4$?
   $6 - 10 = -4$?
   $-4 = -4$ *True*

   $5x + y = 15$
   $5(2) + 5 = 15$?
   $10 + 5 = 15$?
   $15 = 15$ *True*

   Since (2, 5) satisfies both equations, it is a solution of the system.

   (b) (1, -2)

   $x - 3y = 7$

   $4x + y = 5$

   To decide whether (1, -2) is a solution, substitute 1 for x and -2 for y in each equation.

   $x - 3y = 7$
   $1 - 3(-2) = 7$?
   $1 + 6 = 7$?
   $7 = 7$ *True*

   $4x + y = 5$
   $4(1) + (-2) = 5$?
   $4 - 2 = 5$?
   $2 = 5$ *False*

   Since (1, -2) does not satisfy the second equation, it is not a solution of the system.

2. **(a)** $5x - 3y = 9$

   $x + 2y = 7$

   Graph each equation by finding several points to plot for each line.

   Graph $5x - 3y = 9$.   Graph $x + 2y = 7$.

x	y
9/5	0
0	-3
3	2

x	y
7	0
0	7/2
5	1

   (The line $5x - 3y = 9$ is already graphed in the textbook.)

   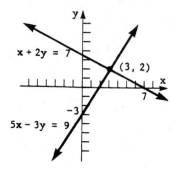

   As suggested by the figure, the solution is (3, 2), the point at which the graphs of the two lines intersect. Check by substituting 3 for x and 2 for y in both equations of the system.

(b) $x + y = 4$
$2x - y = -1$

Graph $x + y = 4$. Graph $2x - y = -1$.

x	y
4	0
0	4
3	1

x	y
$-\frac{1}{2}$	0
0	1
2	5

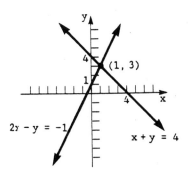

The solution is (1, 3), the point at which the graphs of the two lines intersect. Check by substituting 1 for x and 3 for y in both equations of the system.

3. (a) $3x - y = 4$
$6x - 2y = 12$

Graph $3x - y = 4$. Graph $6x - 2y = 12$.

x	y
$\frac{4}{3}$	0
0	-4
1	-1

x	y
2	0
0	-6
1	-3

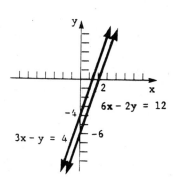

The two lines are parallel. Since they have no points in common, there is no solution.

(b) $-x + 3y = 2$
$2x - 6y = -4$

Graph $-x + 3y = 2$. Graph $2x - 6y = -4$.

x	y
-2	0
0	$\frac{2}{3}$
1	1

x	y
-2	0
0	$\frac{2}{3}$
1	1

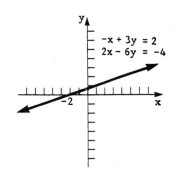

The graphs of these two equations are the same line. There are an infinite number of solutions.

7.1 Section Exercises

3. (6, 2)
$3x + y = 20$
$2x + 3y = 18$

To decide whether (6, 2) is a solution of the system, substitute 6 for x and 2 for y in each equation.

$3x + y = 20$
$3(6) + 2 = 20$?
$18 + 2 = 20$?
$20 = 20$ *True*

$$2x + 3y = 18$$
$$2(6) + 3(2) = 18 \quad ?$$
$$12 + 6 = 18 \quad ?$$
$$18 = 18 \quad ? \quad \textit{True}$$

Because (6, 2) satisfies both equations, it is a solution of the system.

$$3y = 2x + 30$$
$$3(-8) = 2(6) + 30 \quad ?$$
$$-24 = 12 + 30 \quad ?$$
$$-24 = 42 \quad \textit{False}$$

Because (6, −8) does not satisfy *both* equations, it is not a solution of the system.

7. (−1, −3)

$$3x + 5y = -18$$
$$4x + 2y = -10$$

Substitute −1 for x and −3 for y in each equation.

$$3x + 5y = -18$$
$$3(-1) + 5(-3) = -18 \quad ?$$
$$-3 - 15 = -18 \quad ?$$
$$-18 = -18 \quad \textit{True}$$

$$4x + 2y = -10$$
$$4(-1) + 2(-3) = -10 \quad ?$$
$$-4 - 6 = -10 \quad ?$$
$$-10 = -10 \quad \textit{True}$$

Because (−1, −3) satisfies both equations, it is a solution of the system.

11. (6, −8)

$$-2y = x + 10$$
$$3y = 2x + 30$$

Substitute 6 for x and −8 for y in each equation.

$$-2y = x + 10$$
$$-2(-8) = 6 + 10 \quad ?$$
$$16 = 16 \quad \textit{True}$$

15. $x - y = 2$
$x + y = 6$

To graph the equations, find the intercepts.

$x - y = 2$: Let $y = 0$; then $x = 2$.
Let $x = 0$; then $y = -2$.

Plot the intercepts, (2, 0) and (0, −2), and draw the line through them.

$x + y = 6$: Let $y = 0$; then $x = 6$.
Let $x = 0$; then $y = 6$.

Plot the intercepts, (6, 0) and (0, 6), and draw the line through them.

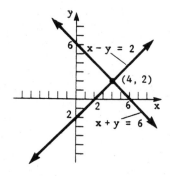

The lines intersect at the point (4, 2). Check this by substituting 4 for x and 2 for y in both equations. Because (4, 2) satisfies both equations, the solution of this system is (4, 2).

19. $x - 2y = 6$
    $x + 2y = 2$

To graph the equations, find the intercepts.

$x - 2y = 6$: Let $y = 0$; then $x = 6$.
Let $x = 0$; then $y = -3$.

Plot the intercepts (6, 0) and (0, -3), and draw the line through them.

$x + 2y = 2$: Let $y = 0$; then $x = 2$.
Let $x = 0$; then $y = 1$.

Plot the intercepts, (2, 0) and (0, 1), and draw the line through them.

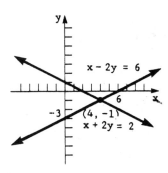

The lines intersect at the point (4, -1). This ordered pair is the solution of the system.

23. $2x - 3y = -6$
    $y = -3x + 2$

To graph the first line, find the intercepts.

$2x - 3y = -6$: Let $y = 0$; then $x = -3$.
Let $x = 0$; then $y = 2$.

Plot the intercepts, (-3, 0) and (0, 2), and draw the line through them.

To graph the second line, start by plotting the y-intercept, (0, 2). From this point, go 3 units down and 1 unit to the right (because the slope is -3) to reach the point (1, -1). Draw the line through (0, 2) and (1, -1).

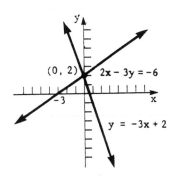

The lines intersect at (0, 2), the solution of the system.

27. A system of two linear equations cannot have exactly two solutions because two lines cannot intersect in exactly two points.

266 Chapter 7 Linear Systems

35. $-2x + y = -4$
$4x = 2y + 8$

To graph the lines, find the intercepts.

$-2x + y = -4$: Let $y = 0$; then $x = 2$.
Let $x = 0$; then $y = -4$.

Draw the line through $(2, 0)$ and $(0, -4)$.

$4x = 2y + 8$: Let $y = 0$; then $x = 2$.
Let $x = 0$; then $y = -4$.

Again, the intercepts are $(2, 0)$ and $(0, -4)$, so the graph is the same line.

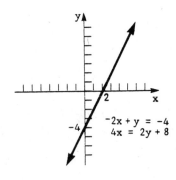

Because the two equations of the system produce the same line, the system has an infinite number of solutions.

39. $14x - 3y$
$\underline{2x + 3y}$
$16x + 0y = 16x$

43. The opposite of $6y$, which is $-6y$, must be added to $6y$ to get a sum of 0:

$6y + (-6y) = 0$.

Section 7.2 Solving Systems of Linear Equations by Addition

7.2 Margin Exercises

1. (a) $x + y = 5$
$4 + y = 5$ Let $x = 4$
$y = 5 - 4$
$y = 1$

(b) If $x = 4$, then $y = 1$. The solution is $(4, 1)$.

2. (a) $x + y = 8$
$\underline{x - y = 2}$
$2x = 10$
$x = 5$

To find y, substitute 5 for x in either of the original equations.

$x + y = 8$ *First equation*
$5 + y = 8$ *Let $x = 5$*
$y = 3$

The solution is $(5, 3)$. Check by substituting 5 for x and 3 for y in both equations of the system.

(b) $3x - y = 7$
$\underline{2x + y = 3}$
$5x = 10$
$x = 2$

To find y, substitute 2 for x in either of the original equations.

$2x + y = 3$ *Second equation*
$2(2) + y = 3$ *Let $x = 2$*
$4 + y = 3$
$y = -1$

The solution is (2, −1). Check by substituting 2 for x and −1 for y in both equations of the system.

3. (a) $2x - y = 2$
       $\underline{4x + y = 10}$
       $6x = 12$ Add
       $x = 2$ Divide by 2

   To find y, substitute 2 for x in either of the original equations.

   $4x + y = 10$ Second equation
   $4(2) + y = 10$ Let x = 2
   $8 + y = 10$
   $y = 2$

   The solution is (2, 2). Check by substituting 2 for x and 2 for y in both original equations.

   (b) $8x - 5y = 32$
        $\underline{4x + 5y = 4}$
        $12x = 36$ Add
        $x = 3$ Divide by 12

   To find y, substitute 3 for x in one of the original equations.

   $4x + 5y = 4$ Second equation
   $4(3) + 5y = 4$ Let x = 3
   $12 + 5y = 4$
   $5y = -8$
   $y = -\frac{8}{5}$

   The solution is (3, −8/5). Check by substituting 3 for x and −8/5 for y in both of the original equations.

4. (a) $x - 3y = -7$ (1)
        $3x + 2y = 23$ (2)

   Neither variable is eliminated if we add. First multiply equation (1) by −3 so the terms with variable x will drop out after adding.

   $-3(x - 3y) = -3(-7)$
   $-3x + 9y = 21$ (3)

   Now add equations (3) and (2).

   $-3x + 9y = 21$ (3)
   $\underline{3x + 2y = 23}$ (2)
   $11y = 44$ Add
   $y = 4$ Divide by 11

   Substitute 4 for y in equation (1) to find x.

   $x - 3y = -7$
   $x - 3(4) = -7$ Let y = 4
   $x - 12 = -7$
   $x = 5$

   The solution is (5, 4). Check that this ordered pair satisfies both of the original equations.

   (b) $8x + 2y = 2$ (1)
       $3x - y = 6$ (2)

   Neither variable is eliminated if we add. First multiply equation (2) by 2 so the terms with variable y will drop out after adding.

   $2(3x - y) = 2(6)$
   $6x - 2y = 12$ (3)

   Now add equations (1) and (3).

   $8x + 2y = 2$ (1)
   $\underline{6x - 2y = 12}$ (3)
   $14x = 14$ Add
   $x = 1$ Divide by 14

Substitute 1 for x in equation (2) to find y.

$$3x - y = 6$$
$$3(1) - y = 6 \quad \text{Let } x = 1$$
$$3 - y = 6$$
$$-y = 3$$
$$y = -3$$

The solution is (1, -3). Check that this ordered pair satisfies both of the original equations.

5. (a) $\quad 4x - 5y = -18 \quad (1)$
$\quad\quad\;\; 3x + 2y = -2 \quad\;\; (2)$

We will eliminate the y-terms. Multiply equation (1) by 2.

$$2(4x - 5y) = 2(-18)$$
$$8x - 10y = -36 \quad (3)$$

Multiply equation (2) by 5.

$$5(3x + 2y) = 5(-2)$$
$$15x + 10y = -10 \quad (4)$$

Now add equations (3) and (4).

$$8x - 10y = -36 \quad (3)$$
$$\underline{15x + 10y = -10} \quad (4)$$
$$23x \quad\quad\;\; = -46 \quad \text{Add}$$
$$x = -2 \quad \text{Divide by 23}$$

Substitute -2 for x in equation (2) to find y.

$$3x + 2y = -2$$
$$3(-2) + 2y = -2 \quad \text{Let } x = -2$$
$$-6 + 2y = -2$$
$$2y = 4$$
$$y = 2$$

The solution is (-2, 2). Check this in both of the original equations.

(b) $\quad 6x + 7y = 4 \quad\;\; (1)$
$\quad\quad\; 5x + 8y = -1 \quad (2)$

We will eliminate the x-terms. Multiply equation (1) by -5.

$$-5(6x + 7y) = -5(4)$$
$$-30x - 35y = -20 \quad (3)$$

Multiply equation (2) by 6.

$$6(5x + 8y) = 6(-1)$$
$$30x + 48y = -6 \quad (4)$$

Now add equations (3) and (4).

$$-30x - 35y = -20 \quad (3)$$
$$\underline{30x + 48y = -6} \quad\;\; (4)$$
$$13y = -26 \quad \text{Add}$$
$$y = -2 \quad \text{Divide by 13}$$

Substitute -2 for y in equation (1) to find x.

$$6x + 7y = 4$$
$$6x + 7(-2) = 4 \quad \text{Let } y = -2$$
$$6x - 14 = 4$$
$$6x = 18$$
$$x = 3$$

The solution is (3, -2). Check this in both of the original equations.

6. (a) $\quad 5x = 7 + 2y \quad (1)$
$\quad\quad\; 5y = 5 - 3x \quad (2)$

Rearrange the terms in both equations as follows so that like terms can be aligned.

$$5x - 2y = 7 \quad (3)$$
$$3x + 5y = 5 \quad (4)$$

Eliminate y. Multiply equation (3) by 5: $5(5x - 2y) = 5(7)$.
Multiply equation (4) by 2: $2(3x + 5y) = 2(5)$. Then add the resulting equations.

$$\begin{array}{r} 25x - 10y = 35 \\ \underline{6x + 10y = 10} \\ 31x = 45 \quad Add \end{array}$$

$$x = \frac{45}{31} \quad Divide\ by\ 31$$

Substituting 45/31 for x to get y in one of the original equations would be messy. Instead, solve for y by starting with equations (3) and (4) and eliminating x. Multiply equation (3) by -3: $-3(5x - 2y) = -3(7)$.
Multiply equation (4) by 5: $5(3x + 5y) = 5(5)$. Then add the resulting equations.

$$\begin{array}{r} -15x + 6y = -21 \\ \underline{15x + 25y = 25} \\ 31y = 4 \quad Add \end{array}$$

$$y = \frac{4}{31} \quad Divide\ by\ 31$$

The solution is (45/31, 4/31).

(b) $3y = 8 + 4x$ (1)
     $6x = 9 - 2y$ (2)

Rearrange the terms in both equations.

$$-4x + 3y = 8 \quad (3)$$
$$6x + 2y = 9 \quad (4)$$

Eliminate x. Multiply equation (3) by 3: $3(-4x + 3y) = 3(8)$.
Multiply equation (4) by 2: $2(6x + 2y) = 2(9)$. Notice that we

multiplied by numbers that will cause the absolute value of the coefficients of x to be the *least* common positive multiple of -4 and 6, that is, 12.

$$\begin{array}{r} -12x + 9y = 24 \\ \underline{12x + 4y = 18} \\ 13y = 42 \quad Add \end{array}$$

$$y = \frac{42}{13} \quad Divide\ by\ 13$$

Now eliminate y. Multiply equation (3) by 2: $2(-4x + 3y) = 2(8)$.
Multiply equation (4) by -3: $-3(6x + 2y) = -3(9)$.

$$\begin{array}{r} -8x + 6y = 16 \\ \underline{-18x - 6y = -27} \\ -26x = -11 \quad Add \end{array}$$

$$x = \frac{11}{26} \quad Divide\ by\ -26$$

The solution is (11/26, 42/13).

7. (a) $4x + 3y = 10$ (1)
     $2x + \frac{3}{2}y = 12$ (2)

Multiply equation (2) by -2, and then add the result to equation (1).

$$\begin{array}{r} 4x + 3y = 10 \\ \underline{-4x - 3y = -24} \\ 0 = -14 \quad False \end{array}$$

A false statement results. The graphs of these equations are parallel lines, so there is no solution.

(b) $-2x - 4y = -1$ (1)
     $5x + 10y = 15$ (2)

Multiply equation (1) by 5 and equation (2) by 2, and then add the results.

$$-10x - 20y = -5$$
$$\underline{10x + 20y = 30}$$
$$0 = 25 \quad \text{False}$$

Since a false statement results, the graphs of these equations are parallel lines. There is no solution.

8. **(a)** $\quad 6x + 3y = 9 \quad (1)$
$\quad\quad -8x - 4y = -12 \quad (2)$

To eliminate x, multiply equation (1) by 4 and equation (2) by 3; then add the resulting equations.

$$24x + 12y = 36$$
$$\underline{-24x - 12y = -36}$$
$$0 = 0 \quad \text{True}$$

This result means that every solution of one equation is also a solution of the other, so the system has an infinite number of solutions.

(b) $\quad 4x - 6y = 10 \quad (1)$
$\quad\quad -10x + 15y = -25 \quad (2)$

$\quad 20x - 30y = 50 \quad$ *Multiply (1) by 5*
$\underline{-20x + 30y = -50} \quad$ *Multiply (2) by 2*
$\quad\quad\quad\quad 0 = 0 \quad$ *True*

The system has an infinite number of solutions.

7.2 Section Exercises

In Exercises 3–19, check your answers by substituting into *both* of the original equations. The check will be shown only for Exercise 3.

3. $\quad x - y = -2 \quad (1)$
$\quad \underline{x + y = 10} \quad (2)$
$\quad 2x \quad\quad = 8 \quad$ *Add (1) and (2)*
$\quad\quad x = 4 \quad$ *Divide by 2*

This result gives the x-value of the solution. To find the y-value of the solution, substitute 4 for x in either equation. We will use equation (2).

$$x + y = 10$$
$$4 + y = 10 \quad \text{Let } x = 4$$
$$y = 6$$

The solution is (4, 6).

Check by substituting 4 for x and 6 for y in both equations of the original system.

Check

$x - y = -2$
$4 - 6 = -2 \quad ?$
$\quad -2 = -2 \quad$ *True*

$x + y = 10$
$4 + 6 = 10 \quad ?$
$\quad 10 = 10 \quad$ *True*

7. $\quad 2x + y = -5 \quad (1)$
$\quad \underline{x - y = 2} \quad (2)$
$\quad 3x \quad\quad = -3 \quad$ *Add (1) and (2)*
$\quad\quad x = -1 \quad$ *Divide by 3*

Substitute −1 for x in equation (1) to find the y-value of the solution.

$$2x + y = -5$$
$$2(-1) + y = -5 \quad \text{Let } x = -1$$
$$-2 + y = -5$$
$$y = -3$$

The solution is (−1, −3).

11. $6x - y = -1$ (1)
    $\underline{-6x + 5y = 17}$ (2)
    $4y = 16$ Add
    $y = 4$

Substitute 4 for y in equation (1) to find the x-value of the solution.

$$6x - y = -1$$
$$6x - 4 = -1 \quad \text{Let } y = 4$$
$$6x = 3$$
$$x = \frac{1}{2}$$

The solution is (1/2, 4).

15. $x + 3y = 19$ (1)
    $2x - y = 10$ (2)

If we simply add the equations, we will not eliminate either variable. To eliminate y, multiply equation (2) by 3 and add the result to equation (1).

$$x + 3y = 19$$
$$\underline{6x - 3y = 30}$$
$$7x = 49$$
$$x = 7$$

Substitute 7 for x in equation (1) to find the y-value of the solution.

$$x + 3y = 19$$
$$7 + 3y = 19$$
$$3y = 12$$
$$y = 4$$

The solution is (7, 4).

19. $5x - 3y = -20$ (1)
    $-3x + 6y = 12$ (2)

To eliminate y, multiply equation (1) by 2 and add the result to equation (2).

$$10x - 6y = -40$$
$$\underline{-3x + 6y = 12}$$
$$7x = -28$$
$$x = -4$$

To find the y-value of the solution, substitute −4 for x in equation (2).

$$-3x + 6y = 12$$
$$-3(-4) + 6y = 12$$
$$12 + 6y = 12$$
$$6y = 0$$
$$y = 0$$

The solution is (−4, 0).

23. If x = 0 and y = 0,

$$Ax + By = A \cdot 0 + B \cdot 0$$
$$= 0 + 0 = 0,$$

and

$$Cx + Dy = C \cdot 0 + D \cdot 0$$
$$= 0 + 0 = 0,$$

so (0, 0) *must* be a solution of the system for any choices of A, B, C, and D.

27. $3x + 5y = 7$ (1)
$5x + 4y = -10$ (2)

To eliminate x, multiply each side of equation (1) by 5 and each side of equation (2) by -3.

$$15x + 25y = 35$$
$$-15x - 12y = 30$$
$$13y = 65$$
$$y = 5$$

To find the x-value of the solution, substitute 5 for y in equation (1).

$$3x + 5y = 7$$
$$3x + 5(5) = 7$$
$$3x + 25 = 7$$
$$3x = -18$$
$$x = -6$$

The solution is (-6, 5).

31. $24x + 12y = -7$ (1)
$16x - 17 = 18y$ (2)

Rearrange the terms in equation (2) so that the terms can be aligned in columns.

$$24x + 12y = -7 \quad (1)$$
$$16x - 18y = 17 \quad (3)$$

To eliminate y, multiply equation (1) by 3 and equation (3) by 2; then add the results.

$$72x + 36y = -21$$
$$32x - 36y = 34$$
$$104x = 13$$
$$x = \frac{13}{104} = \frac{1}{8}$$

To find the y-value of the solution, substitute 1/8 for x in equation (2).

$$16x - 17 = 18y$$
$$16\left(\frac{1}{8}\right) - 17 = 18y$$
$$2 - 17 = 18y$$
$$-15 = 18y$$
$$y = \frac{-15}{18} = -\frac{5}{6}$$

An alternate method for finding the y-value involves going back to the system formed by equations (1) and (3) and eliminating x. Multiply equation (1) by 2 and equation (3) by -3; then add the results.

$$48x + 24y = -14$$
$$-48x + 54y = -51$$
$$78y = -65$$
$$y = \frac{-65}{78} = -\frac{5 \cdot 13}{6 \cdot 13}$$
$$= -\frac{5}{6}$$

The solution is (1/8, -5/6).

35. $x + y = 7$ (1)
$x + y = -3$ (2)

Multiply equation (2) by -1 and add the result to equation (1).

$$x + y = 7$$
$$-x - y = 3$$
$$0 = 10 \quad \textit{False}$$

The false statement 0 = 10 shows that the given system is self-contradictory, so it has no solution.

39. $5x - 2y = 3$ (1)
$10x - 4y = 5$ (2)

Multiply equation (1) by -2 and add the result to equation (2).

$$-10x + 4y = 6$$
$$\underline{10x - 4y = 5}$$
$$0 = 11 \quad \text{False}$$

The false result indicates that the system has no solution.

43. $2x - 8y = 0$ (1)
$4x + 5y = 0$ (2)

Multiply equation (1) by -2 and add the result to equation (2).

$$-4x + 16y = 0$$
$$\underline{4x + 5y = 0}$$
$$21y = 0$$
$$y = 0$$

Substitute 0 for y in equation (1).

$$2x - 8y = 0$$
$$2x - 8(0) = 0$$
$$2x = 0$$
$$x = 0$$

The solution is (0, 0).

47. $p + 4(6 - 2p) = 24$
$p + 24 - 8p = 24$
$-7p + 24 = 24$
$-7p = 0$
$p = 0$

A check will verify that 0 is the solution.

Section 7.3 Solving Systems of Linear Equations by Substitution

7.3 Margin Exercises

1. **(a)** $3x + 5y = 69$
$y = 4x$

Substitute $4x$ for y in the first equation, and solve for x.

$$3x + 5y = 69$$
$$3x + 5(4x) = 69 \quad \text{Let } y = 4x$$
$$3x + 20x = 69$$
$$23x = 69 \quad \text{Combine terms}$$
$$x = 3 \quad \text{Divide by 23}$$

To find y, use $y = 4x$ and $x = 3$.

$$y = 4x = 4(3) = 12$$

The solution is (3, 12).

(b) $-x + 4y = 26$
$y = -3x$

Substitute $-3x$ for y in the first equation, and solve for x.

$$-x + 4y = 26$$
$$-x + 4(-3x) = 26 \quad \text{Let } y = -3x$$
$$-x + (-12x) = 26$$
$$-13x = 26 \quad \text{Combine terms}$$
$$x = -2 \quad \text{Divide by } -13$$

To find y, use $y = -3x$ and $x = -2$.

$$y = -3x = -3(-2) = 6$$

The solution is $(-2, 6)$.

2. **(a)** $3x - 4y = -11$
$x = y - 2$

Substitute $y - 2$ for x in the first equation, and solve for y.

$$3x - 4y = -11$$
$$3(y - 2) - 4y = -11 \quad \text{Let } x = y - 2$$
$$3y - 6 - 4y = -11$$
$$-y - 6 = -11$$
$$-y = -5$$
$$y = 5$$

Then,
$$x = y - 2 = 5 - 2 = 3.$$

The solution is $(3, 5)$. Check this solution in both of the original equations.

(b) $8x - y = 4$
 $y = 8x + 4$

Substitute $8x + 4$ for y in the first equation, and solve for x.

$$8x - y = 4$$
$$8x - (8x + 4) = 4 \quad \text{Let } y = 8x + 4$$
$$8x - 8x - 4 = 4$$
$$-4 = 4 \quad \textit{False}$$

This false result indicates that the system has no solution.

3. **(a)** $x + 4y = -1$
 $2x - 5y = 11$

Solve the first equation of the system for x.

$$x + 4y = -1$$
$$x = -4y - 1 \quad \text{Subtract } 4y$$

Substitute $-4y - 1$ for x in the second equation of the system, and solve for y.

$$2x - 5y = 11$$
$$2(-4y - 1) - 5y = 11 \quad \text{Let } x = -4y - 1$$
$$-8y - 2 - 5y = 11$$
$$-13y - 2 = 11$$
$$-13y = 13$$
$$y = -1$$

Then,
$$x = -4y - 1 = -4(-1) - 1 = 4 - 1 = 3.$$

The solution is $(3, -1)$. Check this solution in both of the original equations.

(b) $2x + 5y = 4$
 $x + y = -1$

Solve the second equation of the system for x.

$$x + y = -1$$
$$x = -1 - y \quad \text{Subtract } y$$

Substitute $-1 - y$ for x in the first equation of the system, and solve for y.

$$2x + 5y = 4$$
$$2(-1 - y) + 5y = 4 \quad \text{Let } x = -1 - y$$
$$-2 - 2y + 5y = 4$$
$$-2 + 3y = 4$$
$$3y = 6$$
$$y = 2$$

Then,
$$x = -1 - y = -1 - 2 = -3.$$

The solution is $(-3, 2)$. Check this solution in both of the original equations.

4. **(a)** $x = 5 - 3y$ (1)

$2x + 3 = 5x - 4y + 14$ (2)

Simplify equation (2) by subtracting 2x and 14 to get

$3x - 4y = -11.$ (3)

Substitute $5 - 3y$ for x in equation (3), and solve for y.

$3x - 4y = -11$ (3)

$3(5 - 3y) - 4y = -11$ Let $x = 5 - 3y$

$15 - 9y - 4y = -11$

$15 - 13y = -11$

$-13y = -26$

$y = 2$

Then,

$x = 5 - 3y = 5 - 3(2) = 5 - 6 = -1.$

The solution is $(-1, 2)$.

(b) $5x - y = -14 + 2x + y$ (1)

$7x + 9y + 4 = 3x + 8y$ (2)

Simplify equation (1) to get

$3x - 2y = -14.$ (3)

Simplify equation (2) to get

$4x + y = -4,$

and then solve for y.

$y = -4x - 4$ (4)

Substitute $-4x - 4$ for y in equation (3), and solve for x.

$3x - 2y = -14$

$3x - 2(-4x - 4) = -14$ Let $y = -4x - 4$

$3x + 8x + 8 = -14$

$11x + 8 = -14$

$11x = -22$

$x = -2$

Then,

$y = -4x - 4 = -4(-2) - 4 = 8 - 4 = 4.$

The solution is $(-2, 4)$.

5. $12x + y = 8$ (1)

$2x + 3y = -10$ (2)

Multiply equation (1) by -3 and add the result to equation (2).

$-36x - 3y = -24$

$\underline{2x + 3y = -10}$

$-34x = -34$

$x = 1$

To find the y-value, substitute 1 for x in equation (1).

$12x + y = 8$

$12(1) + y = 8$

$12 + y = 8$

$y = -4$

The solution, $(1, -4)$, is the same as that found by the substitution method in Example 6.

6. $\frac{2}{3}x + \frac{1}{2}y = 6$ (1)

$\frac{1}{2}x - \frac{3}{4}y = 0$ (2)

Multiply equation (1) by its LCD, 6, and equation (2) by its LCD, 4, to eliminate the fractions. Then eliminate the y-terms by adding the results.

Chapter 7 Linear Systems

$$4x + 3y = 36 \quad (3)$$
$$\underline{2x - 3y = 0} \quad (4)$$
$$6x = 36 \quad \text{Add}$$
$$x = 6 \quad \text{Divide by 6}$$

Substitute 6 for x in equation (4).

$$2x - 3y = 0$$
$$2(6) - 3y = 0$$
$$12 - 3y = 0$$
$$-3y = -12$$
$$y = 4$$

The solution is (6, 4).

7.3 Section Exercises

3. $x + y = 12 \quad (1)$
    $y = 3x \quad (2)$

Substitute 3x for y in equation (1).

$$x + y = 12$$
$$x + 3x = 12 \quad \text{Let } y = 3x$$
$$4x = 12$$
$$x = 3$$

Because x = 3, we find y by substituting 3 for x in equation (2).

$$y = 3x$$
$$y = 3(3) \quad \text{Let } x = 3$$
$$y = 9$$

The solution is (3, 9).

7. $3x + 5y = 14 \quad (1)$
    $x - 2y = -10 \quad (2)$

Solve equation (2) for x.

$$x - 2y = -10$$
$$x = 2y - 10 \quad (3)$$

Now substitute 2y - 10 for x in equation (1) and solve for y.

$$3x + 5y = 14$$
$$3(2y - 10) + 5y = 14$$
$$6y - 30 + 5y = 14$$
$$11y - 30 = 14$$
$$11y = 44$$
$$y = 4$$

Find x by letting y = 4 in equation (3).

$$x = 2y - 10$$
$$x = 2(4) - 10$$
$$= 8 - 10$$
$$= -2$$

The solution is (-2, 4).

11. $7x + 4y = 13 \quad (1)$
    $x + y = 1 \quad (2)$

Solve equation (2) for y.

$$x + y = 1$$
$$y = 1 - x \quad (3)$$

Substitute 1 - x for y in equation (1).

$$7x + 4y = 13$$
$$7x + 4(1 - x) = 13$$
$$7x + 4 - 4x = 13$$
$$3x + 4 = 13$$
$$3x = 9$$
$$x = 3$$

To find y, let x = 3 in equation (3).

$$y = 1 - x$$
$$y = 1 - 3 = -2$$

The solution is (3, -2).

15. $6x - 8y = 6$ (1)
     $-3x + 2y = -2$ (2)

Solve equation (2) for y.

$$-3x + 2y = -2$$
$$2y = 3x - 2$$
$$y = \frac{3x - 2}{2} \quad (3)$$

Substitute $\frac{3x - 2}{2}$ for y in equation (1) and solve for x.

$$6x - 8y = 6$$
$$6x - 8\left(\frac{3x - 2}{2}\right) = 6$$
$$6x - 4(3x - 2) = 6$$
$$6x - 12x + 8 = 6$$
$$-6x + 8 = 6$$
$$-6x = -2$$
$$x = \frac{-2}{-6} = \frac{1}{3}$$

To find y, let x = 1/3 in equation (3).

$$y = \frac{3x - 2}{2}$$
$$y = \frac{3\left(\frac{1}{3}\right) - 2}{2}$$
$$= \frac{1 - 2}{2} = -\frac{1}{2}$$

The solution is (1/3, -1/2).

19. $12x - 16y = 8$ (1)
     $3x = 4y + 2$ (2)

Solve equation (2) for x.

$$3x = 4y + 2$$
$$x = \frac{4y + 2}{3} \quad (3)$$

Substitute $\frac{4y + 2}{3}$ for x in equation (1).

$$12x - 16y = 8$$
$$12\left(\frac{4y + 2}{3}\right) - 16y = 8$$
$$4(4y + 2) - 16y = 8$$
$$16y + 8 - 16y = 8$$
$$8 = 8 \quad \text{True}$$

This result means that every solution of one equation is also a solution of the other equation, so the system has an infinite number of solutions.

23. $4 + 4x - 3y = 34 + x$ (1)
     $4x = -y - 2 + 3x$ (2)

Simplify equation (1).

$$4 + 4x - 3y = 34 + x$$
$$4x - x - 3y = 34 - 4$$
$$3x - 3y = 30$$
$$x - y = 10 \quad \text{Divide by 3}$$

Simplify equation (2).

$$4x = -y - 2 + 3x$$
$$4x - 3y + y = -2$$
$$x + y = -2$$

The two simplified equations form the system

$$x - y = 10 \quad (3)$$
$$x + y = -2. \quad (4)$$

Solve equation (3) for x.

$$x = y + 10$$

Substitute $y + 10$ for x in equation (4).

$$(y + 10) + y = -2$$
$$2y + 10 = -2$$
$$2y = -12$$
$$y = -6$$

The solution is $(4, -6)$.

27. $\quad -2x + 3y = 12 + 2y \quad (1)$
    $\quad 2x - 5y + 4 = -8 - 4y \quad (2)$

Simplify equation (1).

$$-2x + y = 12 \quad (3)$$

Simplify equation (2).

$$2x - y = -12 \quad (4)$$

To solve the simplified system by the addition method, add equations (3) and (4).

$$-2x + y = 12$$
$$\underline{2x - y = -12}$$
$$0 = 0 \quad \textit{True}$$

This result indicates that the given system has an infinite number of solutions.

31. Yes, they will both get the same answer.

35. $\quad \dfrac{x}{6} + \dfrac{y}{6} = 2 \quad (1)$

$\quad -\dfrac{1}{2}x - \dfrac{1}{3}y = -8 \quad (2)$

Multiply each side of equation (1) by 6 to clear fractions.

$$6\left(\dfrac{x}{6} + \dfrac{y}{6}\right) = 6(2)$$
$$6\left(\dfrac{x}{6}\right) + 6\left(\dfrac{y}{6}\right) = 6(2)$$
$$x + y = 12$$

Multiply each side of equation (2) by the LCD, 6, to clear fractions.

$$6\left(-\dfrac{1}{2}x - \dfrac{1}{3}y\right) = 6(-8)$$
$$6\left(-\dfrac{1}{2}x\right) + 6\left(-\dfrac{1}{3}y\right) = 6(-8)$$
$$-3x - 2y = -48$$

The given system of equations has been simplified as follows.

$$x + y = 12 \quad (3)$$
$$-3x - 2y = -48 \quad (4)$$

Multiply equation (3) by 3 and add the result to equation (4).

$$3x + 3y = 36$$
$$-3x - 2y = -48$$
$$y = -12$$

To find x, let y = -12 in equation (3).

$$x + y = 12$$
$$x + (-12) = 12$$
$$x - 12 = 12$$
$$x = 24$$

The solution is (24, -12).

39. Let x = the number of goals Lemieux had.

Then x + 43 = the number of assists Lemieux had.

Lemieux had a total of 131 points, so

$$x + (x + 43) = 131.$$

Solve this equation.

$$2x + 43 = 131$$
$$2x = 88$$
$$x = 44$$
$$x + 43 = 87$$

Lemieux had 44 goals and 87 assists.

Section 7.4 Applications of Linear Systems

7.4 Margin Exercises

1.
$$x + y = 63$$
$$\underline{x - y = 19}$$
$$2x = 82 \quad \text{Add}$$
$$x = 41$$

Substitute 41 for x in the first equation to find y.

$$x + y = 63$$
$$41 + y = 63 \quad \text{Let } x = 41$$
$$y = 22$$

The solution is x = 41, y = 22.

2. **(a)** Let x = one number;
y = the other number.

Set up a system of equations from the information in the problem.

$$x + y = 97 \quad \textit{The sum is 97}$$
$$x - y = 41 \quad \textit{The difference is 41}$$

(b) Let x = one number;
y = the other number.

Set up a system of equations.

$$x + y = 38 \quad \textit{The sum is 38}$$
$$2x + 3y = 99 \quad \textit{Twice the first added to three times the second is 99}$$

3. **(a)**

Kind of ticket	Number sold	Cost of each	Total value
Regular	r	5	5r
Student	s	3.50	3.50s
Totals	400	—	1850

(b) $\quad r + s = 400 \quad$ *From second column*

$\quad 5r + 3.50s = 1850 \quad$ *From fourth column*

(c) To eliminate the r-terms, multiply the first equation by -5. Then add the result to the second equation.

$$-5r - 5s = -2000$$
$$\underline{5r + 3.50s = 1850}$$
$$-1.50s = -150 \quad Add$$
$$s = 100$$

Substitute 100 for s in the first equation of the original system to find r.

$$r + s = 400$$
$$r + 100 = 400 \quad Let\ s = 100$$
$$r = 300$$

There were 300 regular tickets sold and 100 student tickets sold.

4. $\quad 30x + 80y = 5000$
$\quadx + y = 100$

To eliminate the x-terms, multiply the second equation by -30. Then add the result to the first equation.

$$30x + 80y = 5000$$
$$\underline{-30x - 30y = -3000}$$
$$50y = 2000 \quad Add$$
$$y = 40$$

To find x, substitute 40 for y in the second equation of the original system.

$$x + y = 100$$
$$x + 40 = 100 \quad Let\ y = 40$$
$$x = 60$$

The solution is $x = 60$, $y = 40$.

5. **(a)**

Percent	Liters	Liters of pure alcohol
25	x	.25x
12	y	.12y
15	13	.15(13)

(b) $\quad x + y = 13 \quad$ *From second column*

$\quad .25x + .12y = .15(13) \quad$ *From third column*

To eliminate the x-terms, multiply the first equation by $-.25$. Then add the result to the second equation.

$$-.25x - .25y = -3.25$$
$$\underline{.25x + .12y = 1.95}$$
$$-.13y = -1.3 \quad Add$$
$$y = 10$$

To find x, substitute 10 for y in the first equation of the original system.

$$x + y = 13$$
$$x + 10 = 13 \quad Let\ y = 10$$
$$x = 3$$

3 liters of 25% solution must be mixed with 10 liters of 12% solution.

6. Let x = the amount of 10% solution;
   y = the amount of 25% solution.
   Make a table.

Percent	Cubic centimeters	Pure salt
10	x	.10x
25	y	.25y
20	100	.20(100)

   Set up a system of equations.

   $$x + y = 100$$
   $$.10x + .25y = .20(100)$$

   To eliminate the x-terms, multiply the first equation by $-.10$. Then add the result to the second equation.

   $$-.10x - .10y = -10$$
   $$\underline{.10x + .25y = 20}$$
   $$.15y = 10 \quad Add$$
   $$y = \frac{10}{.15}$$
   $$y = 66\frac{2}{3}$$

   To find x, substitute 66 2/3 for y in the first equation of the original system.

   $$x + y = 100$$
   $$x + 66\frac{2}{3} = 100 \quad Let\ y = 66\ 2/3$$
   $$x = 33\frac{1}{3}$$

   33 1/3 cc of 10% solution and 66 2/3 cc of 25% solution must be mixed.

7. $4x + 4y = 400$
   $x = 20 + y$

   Substitute $20 + y$ for x in the first equation, and solve for y.

   $$4x + 4y = 400$$
   $$4(20 + y) + 4y = 400 \quad Let\ x = 20 + y$$
   $$80 + 4y + 4y = 400$$
   $$80 + 8y = 400$$
   $$8y = 320$$
   $$y = 40$$

   To find x, use $x = 20 + y$ and $y = 40$.

   $$x = 20 + y = 20 + 40 = 60$$

   The solution is $x = 60$, $y = 40$.

8. (a) Let x = the speed of the faster car;
       y = the speed of the slower car.

	r	t	d
Faster car	x	5	5x
Slower car	y	5	5y

   Write a system of equations.

   $5x + 5y = 450$ *Total distance*
   $x = 2y$ *Faster car is twice as fast*

   Substitute 2y for x in the first equation, and solve for y.

   $$5x + 5y = 450$$
   $$5(2y) + 5y = 450 \quad Let\ x = 2y$$
   $$10y + 5y = 450$$
   $$15y = 450$$
   $$y = 30$$

To find x, use x = 2y and y = 30.

$$x = 2y = 2(30) = 60$$

The faster car's speed is 60 miles per hour; the slower car's speed is 30 miles per hour.

(b) Let x = the speed of the current;

y = Ann's speed in still water.

When Ann rows against the current, the current works against her, so the rate of the current is subtracted from her speed. When she rows with the current, the rate of the current is added to her speed. Make a table.

	r	t	d
Against the current	y − x	1	2
With the current	y + x	1	10

Use d = rt to get each equation of the system.

$$(y - x)(1) = 2 \rightarrow -x + y = 2$$
$$(x + y)(1) = 10 \rightarrow \underline{x + y = 10}$$
$$2y = 12 \quad Add$$
$$y = 6$$

Substitute y for 6 in the second equation of the system.

$$x + y = 10$$
$$x + 6 = 10 \quad Let\ y = 6$$
$$x = 4$$

The speed of the current is 4 miles per hour. Ann's speed is 6 miles per hour.

7.4 Section Exercises

3. Let x = the larger number;
   y = the smaller number.

The sum of the numbers is 113, so

$$x + y = 113. \quad (1)$$

The difference between the number is 71, so

$$x - y = 71. \quad (2)$$

Solve the system formed by equations (1) and (2) by the addition method.

$$x + y = 113$$
$$\underline{x - y = 71}$$
$$2x = 184$$
$$x = 92$$

To find y, let x = 92 in equation (1).

$$x + y = 113$$
$$92 + y = 113$$
$$y = 21$$

The numbers are 92 and 21.

7. Let x = the width of the rectangle;
   y = the length.

The length is 8 centimeters longer than the width, so

$$y = x + 8. \quad (1)$$

The perimeter is 36 centimeters, so

$$2x + 2y = 36$$
$$or \quad x + y = 18. \quad (2)$$

Solve the system formed by equations (1) and (2) by the substitution method. Substitute $x + 8$ for y in equation (2).

$$x + y = 18$$
$$x + (x + 8) = 18$$
$$2x + 8 = 18$$
$$2x = 10$$
$$x = 5$$

To find y, let $x = 5$ in equation (1).

$$y = x + 8$$
$$y = 5 + 8 = 13$$

The length of the rectangle is 13 centimeters and the width is 5 centimeters.

11. Let s = the number of student tickets sold;

n = the number of nonstudent tickets sold.

The information given in the problem is summarized in the following table.

Kind of ticket	Number sold	Cost of each (in dollars)	Total value (in dollars)
Student	s	.50	.50s
Nonstudent	n	2.00	2.00n
Total	386	—	523

The total number of tickets sold was 386, so

$$s + n = 386. \quad (1)$$

Because the total value was $523, the right-hand column leads to the equation

$$.50s + 2.00n = 523. \quad (2)$$

These two equations give the following system.

$$s + n = 386 \quad (1)$$
$$.50s + 2.00n = 523 \quad (2)$$

To solve this system by the addition method, multiply equation (1) by $-.50$ and add the result to equation (2).

$$-.50s - .50n = -193$$
$$\underline{.50s + 2.00n = 523}$$
$$1.50n = 330$$
$$n = 220$$

To find s, let $n = 220$ in equation (1).

$$s + n = 386$$
$$s + 220 = 386$$
$$s = 166$$

At the soccer game, 166 student tickets and 220 nonstudent tickets were sold.

15. Let p = the number of paperbacks sold;

h = the number of hardbacks.

The collector bought a total of 20 books, so

$$p + h = 20. \quad (1)$$

The total value was $21.25, so

$$.25p + 1.50h = 21.25. \quad (2)$$

Equation (2) may be simplified by clearing decimals. Multiply each side by 100.

$$100(.25p + 1.50h) = 100(21.25)$$
$$25p + 150h = 2125 \quad (3)$$

We now have the system

$$p + h = 20 \quad (1)$$
$$25p + 150h = 2125. \quad (3)$$

To solve this system by the addition method, multiply equation (1) by -25 and add the result to equation (3).

$$-25p - 25h = -500$$
$$\underline{25p + 150h = 2125}$$
$$125h = 1625$$
$$h = 13$$

To find p, let $h = 13$ in equation (1).

$$p + h = 20$$
$$p + 13 = 20$$
$$p = 7$$

The collector bought 7 paperbacks and 13 hardbacks.

19. Let x = the number of pounds of coffee worth \$6 per pound;

y = the number of pounds of coffee worth \$4 per pound.

Complete the table given in the textbook.

Pounds	Dollars per pound	Cost
x	6	6x
y	3	3y
90	4	360

The mixture contains 90 pounds, so

$$x + y = 90. \quad (1)$$

The cost of the mixture is \$360, so

$$6x + 3y = 360. \quad (2)$$

Equation (3) may be simplified by dividing each side by 3.

$$2x + y = 120 \quad (3)$$

We now have the system

$$x + y = 90 \quad (1)$$
$$2x + y = 120. \quad (3)$$

To solve this system by the addition method, multiply equation (1) by -1 and add the result to equation (2).

$$-x - y = -90$$
$$\underline{2x + y = 120}$$
$$x = 30$$

To find y, let $x = 30$ in equation (1).

$$x + y = 90$$
$$30 + y = 90$$
$$y = 60$$

The merchant will need to mix 30 pounds of coffee at \$6 per pound with 60 pounds at \$3 per pound.

23. Let x = the speed of the boat in still water;

y = the speed of the current.

Use the table and $d = rt$ to write the system of equations.

$36 = (x + y)(3) \quad (1)$ *Distance downstream*

$24 = (x - y)(3) \quad (2)$ *Distance upstream*

Rewrite the equations in the form $Ax + By = C$.

$$36 = (x + y)(3) \quad (1)$$
$$36 = 3x + 3y$$
$$3x + 3y = 36$$
$$\text{or} \quad x + y = 12 \quad (3)$$

$$24 = (x - y)(3) \quad (2)$$
$$24 = 3x - 3y$$
$$3x - 3y = 24$$
$$\text{or} \quad x - y = 8 \quad (4)$$

Add equations (3) and (4).

$$\begin{array}{rl} x + y = 12 & (3) \\ \underline{x - y = 8} & (4) \\ 2x = 20 & \text{Add} \\ x = 10 & \end{array}$$

To find y, substitute 10 for x in equation (3).

$$x + y = 12$$
$$10 + y = 12$$
$$y = 2$$

The speed of the current is 2 miles per hour; the speed of the boat in still water is 10 miles per hour.

27. Let x = the rate of the faster train.

Let y = the rate of the slower train.

Write a system of equations; use $d = rt$ to write the second equation.

$x - y = 20$ (1) *Faster train travels 20 mph faster*

$5x + 5y = 1000$ (2) *Total distance = 1000 miles*

Equation (2) can be simplified by dividing by 5.

$$x + y = 200 \quad (3)$$

Solve the system formed by equations (1) and (3) by the addition method. To eliminate the y-terms, add equations (1) and (3).

$$\begin{array}{rl} x - y = & 20 \\ \underline{x + y = } & \underline{200} \\ 2x = & 220 \\ x = & 110 \end{array}$$

To find y, substitute 110 for x in equation (1).

$$x - y = 20$$
$$110 - y = 20 \quad \text{Let } x = 110$$
$$-y = -90$$
$$y = 90$$

The faster train travels at 110 miles per hour; the slower train travels at 90 miles per hour.

For Exercises 31 and 35, see the graphs in the answer section of the textbook.

31. $x + y \leq 4$

Graph the boundary of the region, the line with equation $x + y = 4$, through its intercepts (4, 0) and (0, 4). Make the line solid because of the $\leq$ sign.

Using (0, 0) as a test point will result in the inequality $0 \leq 4$, which is true, so shade the region containing the origin. The solid line indicates that the boundary is part of the graph.

35. $2x + 4y > 8$

Graph the boundary of the region, the line with equation $2x + 4y = 8$ (which may be simplified to $x + 2y = 4$) through its intercepts $(4, 0)$ and $(0, 2)$. Make this line dashed because of the $>$ sign.

Using $(0, 0)$ as a test point will result in the inequality $0 > 8$, which is false. Shade the region not containing the origin. The dashed line indicates that the boundary is not part of the graph.

Section 7.5 Solving Systems of Linear Inequalities

7.5 Margin Exercises

For Exercises 1–3, refer to the graphs included with the answers to the margin exercises in the textbook.

1. **(a)** $3x + 2y \geq 6$

The solid line is the graph of $3x + 2y = 6$. Use $(0, 0)$ as a test point.

$$3x + 2y \geq 6$$
$$3(0) + 2(0) \geq 6 \quad \text{Let } x = 0, y = 0$$
$$0 \geq 6 \quad \text{False}$$

Since $0 \geq 6$ is false, shade the side of the line that does not contain $(0, 0)$.

(b) $2x - 5y < 10$

x	y
5	0
0	-2

Graph $2x - 5y = 10$ as a dashed line because of the $<$ sign. Use $(0, 0)$ as a test point.

$$2x - 5y < 10$$
$$2(0) - 5(0) < 10 \quad \text{Let } x = 0,$$
$$\qquad\qquad\qquad\qquad y = 0$$
$$0 < 10 \quad \text{True}$$

Since $0 < 10$, shade the side of the line that contains $(0, 0)$.

2. $x - 2y \leq 8$
    $3x + y \geq 6$

The graph of $x - 2y = 8$ has intercepts $(8, 0)$ and $(0, -4)$; the graph of $3x + y = 6$ has intercepts $(2, 0)$ and $(0, 6)$. Both are graphed as solid lines because of the $\leq$ and $\geq$ signs. Use $(0, 0)$ as a test point in each case.

$$x - 2y \leq 8$$
$$0 - 2(0) \leq 8 \quad \text{Let } x = 0, y = 0$$
$$0 \leq 8 \quad \text{True}$$

Shade the side of the graph for $x - 2y = 8$ that contains $(0, 0)$.

$$3x + y \geq 6$$
$$3(0) + 0 \geq 6 \quad \text{Let } x = 0, y = 0$$
$$0 \geq 6 \quad \text{False}$$

Shade the side of the graph for $3x + y = 6$ that does not contain $(0, 0)$. The graph of the system is the overlap of the two shaded regions.

3. $x + 2y < 0$
    $3x - 4y < 12$

    First graph $x + 2y = 0$ as a dashed line through the points listed in the following table of ordered pairs.

x	y
0	0
2	-1

    Use $(0, -1)$ as a test point.

    $x + 2y < 0$
    $0 + 2(-1) < 0$ *Let $x = 0$, $y = -1$*
    $-2 < 0$ *True*

    The solution is the region that includes $(0, -1)$. Then graph $3x - 4y = 12$ as a dashed line through the points found in the following table of ordered pairs.

x	y
0	-3
4	0

    Use $(0, 0)$ as a test point.

    $3x - 4y < 12$
    $3(0) - 4(0) < 12$ *Let $x = 0$, $y = 0$*
    $0 < 12$ *True*

    The solution is the region that includes $(0, 0)$. The graph of the system is the overlap of the two shaded regions.

7.5 Section Exercises

For Exercises 3–15, see the graphs in the answer section of the textbook. (Only the solution region has been shaded in answers.)

3. $x + y \leq 6$
    $x - y \geq 1$

    First graph $x + y = 6$ as a solid line, using the intercepts $(6, 0)$ and $(0, 6)$. Using $(0, 0)$ as a test point will result in the inequality $0 \leq 6$, which is true, so shade the region containing the origin.
    Next, graph $x - y \geq 1$ as a solid line, using the intercepts $(1, 0)$ and $(0, -1)$. Using $(0, 0)$ as a test point will result in the inequality $0 \geq 1$, which is false, so shade the region that does not include the origin.
    The solution of this system is the intersection (overlap) of the two shaded regions, and includes the portions of the two lines that bound the region.

7. $2x + 3y < 6$
    $x - y < 5$

    Graph $2x + 3y = 6$ as a dashed line through $(3, 0)$ and $(0, 2)$. Using $(0, 0)$ as a test point will result in the inequality $0 < 6$, which is true, so shade the region containing the origin.

Now graph $x - y = 5$ as a dashed line through $(5, 0)$ and $(0, -5)$. Using $(0, 0)$ as a test point will result in the inequality $0 < 5$, which is true, so shade the region containing the origin.

The solution of the system is the intersection of the two shaded regions. Because the inequality signs are both $<$, the solution does not include portions of either boundary line.

11. $4x + 3y < 6$
    $x - 2y > 4$

    Graph $4x + 3y = 6$ as a solid line through $(3/2, 0)$ and $(0, 2)$. Using $(0, 0)$ as a test point will result in the true statement $0 < 6$, so shade the region containing the origin.
    Graph $x - 2y = 4$ as a dashed line through $(4, 0)$ and $(0, -2)$. Using $(0, 0)$ as a test point will result in the false statement $0 > 4$, so shade the region not containing the origin.
    The solution of the system is the intersection of the two shaded regions. It does not include portions of the boundary line.

15. $-3x + y \geq 4$
    $6x - 2y \geq -10$

    Graph $-3x + y = 4$ as a solid line through $(-4/3, 0)$ and $(0, 4)$. Using $(0, 0)$ as a test point will result in the false statement $0 \geq 4$, so shade the region not containing the origin. This is the region above the line.
    Graph $6x - 2y = -10$ as a solid line through $(-5/3, 0)$ and $(0, 5)$. Using $(0, 0)$ as a test point will result in the true statement $0 \geq -10$, so shade the region containing the origin. This is the region below the line.
    The solution of the system is the intersection of the two shaded regions. This is the region between the two parallel lines. These boundary lines are included in the solution.

19. $8^2 = 8 \cdot 8 = 64$

23. $a^2 + b^2 = 8^2 + 15^2$ Let $a = 8$, $b = 15$
    $= 64 + 225$
    $= 289$

Chapter 7 Review Exercises

1. $(3, 4)$
   $4x - 2y = 4$
   $5x + y = 19$

   To decide whether $(3, 4)$ is a solution of the system, substitute 3 for x and 4 for y in each equation.

   $4x - 2y = 4$
   $4(3) - 2(4) = 4$?
   $12 - 8 = 4$?
   $4 = 4$ True

$$5x + y = 19$$
$$5(3) + 4 = 19 \quad ?$$
$$15 + 4 = 19 \quad ?$$
$$19 = 19 \quad True$$

Because (3, 4) satisfies both equations, it is a solution of the system.

2. (1, −3)
$$5x + 3y = -4$$
$$2x - 3y = 11$$

Substitute 1 for x and −3 for y in each equation.

$$5x + 3y = -4$$
$$5(1) + 3(-3) = -4 \quad ?$$
$$5 - 9 = -4 \quad ?$$
$$-4 = -4 \quad True$$

$$2x - 3y = 11$$
$$2(1) - 3(-3) = 11 \quad ?$$
$$2 + 9 = 11 \quad ?$$
$$11 = 11 \quad True$$

The ordered pair (1, −3) satisfies both equations, so it is a solution of the system.

3. (−5, 2)
$$x - 4y = -13$$
$$2x + 3y = 4$$

Substitute −5 for x and 2 for y in each equation.

$$x - 5y = -13$$
$$-5 - 5(2) = -13 \quad ?$$
$$-5 - 10 = -13 \quad ?$$
$$-15 = -13 \quad False$$

Because (−5, 2) is not a solution of the first equation, it cannot be a solution of the system.

4. (0, 1)
$$3x + 8y = 8$$
$$4x - 3y = 3$$

Substitute 0 for x and 1 for y in each equation.

$$3x + 8y = 8$$
$$3(0) + 8(1) = 8 \quad ?$$
$$0 + 8 = 8 \quad ?$$
$$8 = 8 \quad True$$

$$4x - 3y = 3$$
$$4(0) - 3(1) = 3 \quad ?$$
$$0 - 3 = 3 \quad ?$$
$$-3 = 3 \quad False$$

The ordered pair (0, 1) does not satisfy *both* equations, so it is not a solution of the system.

5. $x + y = 4$
$2x - y = 5$

To graph the equations, find the intercepts.

$x + y = 4$: Let $y = 0$; then $x = 4$.
Let $x = 0$; then $y = 4$.

Plot the intercepts, (4, 0) and (0, 4), and draw the line through them.

$2x - y = 5$: Let $y = 0$; then $x = 5/2$.
Let $x = 0$; then $y = -5$.

Plot the intercepts, (5/2, 0) and (0, −5), and draw the line through them.

290   Chapter 7   Linear Systems

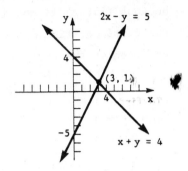

The lines intersect at the point (3, 1). This ordered pair is the solution of the system.

6. $x - 2y = 4$
    $2x + y = -2$

Graph the line $x - 2y = 4$ through its intercepts, (4, 0) and (0, -2). Graph the line through its intercepts, (-1, 0) and (0, -2).

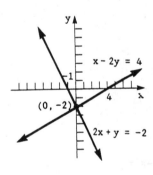

The lines intersect at their common y-intercept, (0, -2). This ordered pair is the solution of the system.

7. $x - 2 = 2y$
    $2x - 4y = 4$

The first equation may be rewritten as $x - 2y = 2$. Find the intercepts for this line.

$x - 2y = 2$: Let $y = 0$; then $x = 2$.
              Let $x = 0$; then $y = -1$.

Draw the line through (2, 0) and (0, -1).
Now find the intercepts for the second line.

$2x - 4y = 4$: Let $y = 0$; then $x = 2$.
               Let $x = 0$; then $y = -1$.

Again, the intercepts are (2, 0) and (0, -1), so the graph is the same line.

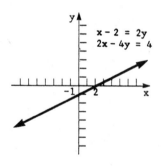

Because the graphs of both equations are the same line, the system has an infinite number of solutions.

8. $2x + 4 = 2y$
    $y - x = -3$

Graph the first line through its intercepts, (-2, 0) and (0, 2).

Graph the second line through its intercepts (3, 0) and (0, -3).

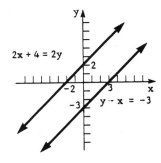

The two lines are parallel. Because there is no point common to both graphs, the system has no solution.

9. Because the x- and y-coordinates of the intersection point are fractions, it would be difficult to read the exact coordinates from the graph.

11. $2x - y = 13$ (1)
    $\underline{x + y = 8}$ (2)
    $3x = 21$ Add
    $x = 7$

To find y, let x = 7 in equation (2).

$$x + y = 8$$
$$7 + y = 8$$
$$y = 1$$

The solution is (7, 1).

12. $3x - y = -13$ (1)
    $x - 2y = -1$ (2)

Multiply equation (2) by -3 and add the result to equation (1).

$3x - y = -13$
$\underline{-3x + 6y = 3}$
$5y = -10$
$y = -2$

To find x, let y = -2 in equation (1).

$$x - 2y = -1$$
$$x - 2(-2) = -1$$
$$x + 4 = -1$$
$$x = -5$$

The solution is (-5, -2).

13. $5x + 4y = -7$ (1)
    $\underline{3x - 4y = -17}$ (2)
    $8x = -24$
    $x = -3$

To find y, let x = -3 in equation (1).

$$5x + 4y = -7$$
$$5(-3) + 4y = -7$$
$$-15 + 4y = -7$$
$$4y = 8$$
$$y = 2$$

The solution is (-3, 2).

14. $-4x + 3y = 25$ (1)
    $6x - 5y = -39$ (2)

Multiply equation (1) by 3 and equation (2) by 2; then add the results.

$-12x + 9y = 75$
$\underline{12x - 10y = -78}$
$-y = -3$
$y = 3$

To find x, let y = 3 in equation (1).

$$-4x + 3y = 25$$
$$-4x + 3(3) = 25$$
$$-4x + 9 = 25$$
$$-4x = 16$$
$$x = -4$$

The solution is (−4, 3).

15. $3x - 4y = 9$ (1)
    $6x - 8y = 18$ (2)

Multiply equation (1) by −2 and add the result to equation (2).

$$-6x + 8y = -18$$
$$\underline{6x - 8y = 18}$$
$$0 = 0 \text{ True}$$

This result indicates that all solutions of equation (1) are also solutions of equation (2). The given system has an infinite number of solutions.

16. $2x + y = 3$ (1)
    $-4x - 2y = 6$ (2)

Multiply equation (1) by 2 and add the result to equation (2).

$$4x + 2y = 6$$
$$\underline{-4x - 2y = 6}$$
$$0 = 12 \text{ False}$$

This result indicates that the given system has no solution.

17. The result "0 = 0" indicates that the system has an infinite number of solutions. In this case, the graphs of the two equations of the system are the same line. Giving the solution as (0, 0) means that there is exactly one solution, x = 0 and y = 0. Thus, the student's answer would be incorrect.

19. $3x + y = 7$ (1)
    $x = 2y$ (2)

Substitute 2y for x in equation (1) and solve the resulting equation for y.

$$3x + y = 7$$
$$3(2y) + y = 7$$
$$6y + y = 7$$
$$7y = 7$$
$$y = 1$$

To find x, let y = 1 in equation (2).

$$x = 2y$$
$$x = 2(1) = 2$$

The solution is (2, 1).

20. $2x - 5y = -19$ (1)
    $y = x + 2$ (2)

Substitute x + 2 for y in equation (1).

$$2x - 5y = -19$$
$$2x - 5(x + 2) = -19$$
$$2x - 5x - 10 = -19$$
$$-3x - 10 = -19$$
$$-3x = -9$$
$$x = 3$$

To find y, let x = 3 in equation (2).

$$y = x + 2$$
$$y = 3 + 2 = 5$$

The solution is (3, 5).

21. $4x + 5y = 44$ (1)
    $x + 2 = 2y$ (2)

Solve equation (2) for x.

$$x + 2 = 2y$$
$$x = 2y - 2 \quad (3)$$

Substitute $2y - 2$ for x in equation (1).

$$4x + 5y = 44$$
$$4(2y - 2) + 5y = 44$$
$$8y - 8 + 5y = 44$$
$$13y - 8 = 44$$
$$13y = 52$$
$$y = 4$$

To find x, let y = 4 in equation (3).

$$x = 2y - 2$$
$$x = 2(4) - 2$$
$$= 8 - 2 = 6$$

The solution is (6, 4).

22. $5x + 15y = 3$ (1)
    $x + 3y = 2$ (2)

Solve equation (2) for x.

$$x + 3y = 2$$
$$x = 2 - 3y \quad (3)$$

Substitute $2 - 3y$ for x in equation (1).

$$5x + 15y = 3$$
$$5(2 - 3y) + 15y = 3$$
$$10 - 15y + 15y = 3$$
$$10 = 3 \quad \textit{False}$$

This result indicates that the system has no solution.

23. $2x + 3y = -5$ (1)
    $3x + 4y = -8$ (2)

Multiply equation (1) by -3 and equation (2) by 2; then add the results.

$$-6x - 9y = 15$$
$$\underline{6x + 8y = -16}$$
$$-y = -1$$
$$y = 1$$

To find x, let y = 1 in equation (1).

$$2x + 3y = -5$$
$$2x + 3(1) = -5$$
$$2x + 3 = -5$$
$$2x = -8$$
$$x = -4$$

The solution is (-4, 1).

24. $6x - 9y = 0$ (1)
    $2x - 3y = 0$ (2)

Multiply equation (2) by -3 and add the result to equation (1).

$$6x - 9y = 0$$
$$\underline{-6x + 9y = 0}$$
$$0 = 0 \quad \textit{True}$$

This result indicates that the system has an infinite number of solutions.

25. $2x + y - x = 3y + 5$ (1)
    $y + 2 = x - 5$ (2)

Simplify equation (1).

$2x + y - x = 3y + 5$
$x - 2y = 5$ (3)

Simplify equation (2).

$y + 2 = x - 5$
$-x + y = -7$ (4)

Add equations (3) and (4) to eliminate x.

$$\begin{aligned} x - 2y &= 5 \\ \underline{-x + y} &= \underline{-7} \\ -y &= -2 \\ y &= 2 \end{aligned}$$

Let $y = 2$ in equation (3) to find x.

$x - 2(2) = 5$
$x - 4 = 5$
$x = 9$

The solution is $(9, 2)$.

26. $5x - 3 + y = 4y + 8$ (1)
    $2y + 1 = x - 3$ (2)

Simplify equation (1).

$5x - 3 + y = 4y + 8$
$5x - 3y - 3 = 8$
$5x - 3y = 11$ (3)

Simplify equation (2).

$2y + 1 = x - 3$
$-x + 2y + 1 = -3$
$-x + 2y = -4$
$-x = -4 - 2y$
$x = 2y + 4$ (4)

Replace x by $2y + 4$ in equation (3) and then solve for y.

$5(2y + 4) - 3y = 11$
$10y + 20 - 3y = 11$
$7y + 20 = 11$
$7y = -9$
$y = -\dfrac{9}{7}$

To find x, let $y = -9/7$ in equation (4).

$x = 2\left(-\dfrac{9}{7}\right) + 4$
$x = -\dfrac{18}{7} + 4$
$x = -\dfrac{18}{7} + \dfrac{28}{7}$
$x = \dfrac{10}{7}$

The solution is $(10/7, -9/7)$.

27. $\dfrac{x}{2} + \dfrac{y}{3} = 7$ (1)

    $\dfrac{x}{4} + \dfrac{2y}{3} = 8$ (2)

Multiply equation (1) by 6 to clear fractions.

$6\left(\dfrac{x}{2} + \dfrac{y}{3}\right) = 6(7)$
$3x + 2y = 42$ (3)

Multiply equation (2) by 12 to clear fractions.

$$12\left(\frac{x}{4} + \frac{2y}{3}\right) = 12$$
$$3x + 8y = 96 \qquad (4)$$

We now have the simplified system

$$3x + 2y = 42 \qquad (3)$$
$$3x + 8y = 96. \qquad (4)$$

To solve this system by the addition method, multiply equation (3) by −1 and add the result to equation (4).

$$-3x - 2y = -42$$
$$\underline{3x + 8y = 96}$$
$$6y = 54$$
$$y = 9$$

To find x, let y = 9 in equation (3).

$$3x + 2y = 42$$
$$3x + 2(9) = 42$$
$$3x + 18 = 42$$
$$3x = 24$$
$$x = 8$$

The solution is (8, 9).

28. $\frac{3x}{4} - \frac{y}{3} = \frac{7}{6}$ (1)

    $\frac{x}{2} + \frac{2y}{3} = \frac{5}{3}$ (2)

Multiply equation (1) by 12 to clear fractions.

$$12\left(\frac{3x}{4}\right) - 12\left(\frac{y}{3}\right) = 12\left(\frac{7}{6}\right)$$
$$9x - 4y = 14 \qquad (3)$$

Multiply equation (2) by 6 to clear fractions.

$$6\left(\frac{x}{2}\right) + 6\left(\frac{2y}{3}\right) = 6\left(\frac{5}{3}\right)$$
$$3x + 4y = 10 \qquad (4)$$

Add equations (3) and (4) to eliminate y.

$$9x - 4y = 14$$
$$\underline{3x + 4y = 10}$$
$$12x = 24$$
$$x = 2$$

To find y, let x = 2 in equation (3).

$$9(2) - 4y = 14$$
$$18 - 4y = 14$$
$$-4y = -4$$
$$y = 1$$

The solution is (2, 1).

29. Let x = the larger number;
    y = the smaller number.

$$x + y = 42 \qquad (1)$$
$$\underline{x - y = 6} \qquad (2)$$
$$2x = 48$$
$$x = 24$$

To find y, let x = 24 in equation (1).

$$x + y = 42$$
$$24 + y = 42$$
$$y = 18$$

The numbers are 24 and 18.

30. Let x = the larger number;
    y = the smaller number.

    The larger number is 2 more than twice as large as the smaller, so

    x = 2y + 2.   (1)

    The sum of the numbers is 11, so

    x + y = 11.   (2)

    To solve this system by substitution, replace x by 2y + 2 in equation (2).

    (2y + 2) + y = 11
    3y + 2 = 11
    3y = 9
    y = 3

    To find x, let y = 3 in equation (1).

    x = 2(3) + 2
    x = 6 + 2
    x = 8

    The two numbers are 3 and 8.

31. Let x = the length of the rectangle;
    y = the width.

    The perimeter is 90 meters, so

    2x + 2y = 90.   (1)

    The length is 1 1/2 (or 3/2) times the width, so

    x = $\frac{3}{2}$y.   (2)

    We have the system

    2x + 2y = 90   (1)
    x = $\frac{3}{2}$y.   (2)

    To solve this system by the substitution method, substitute (3/2)y for x in equation (1); then solve the resulting equation for y.

    2x + 2y = 90
    2($\frac{3}{2}$y) + 2y = 90
    3y + 2y = 90
    5y = 90
    y = 18

    To find x, let y = 18 in equation (2).

    x = $\frac{3}{2}$y

    x = $\frac{3}{2}$(18) = 27

    The length is 27 meters and the width is 18 meters.

32. Let x = the number of $20 bills;
    y = the number of $10 bills.

    The total number of bills is 20, so

    x + y = 20.   (1)

    The total value of the money is $330, so

    20x + 10y = 330.   (2)

    We may simplify equation (2) by dividing each side by 10.

    2x + y = 33   (3)

    We now have the system

    x + y = 20   (1)
    2x + y = 33.   (2)

To solve this equation by the addition method, multiply equation (1) by -1 and add the result to equation (2).

$$-x - y = -20$$
$$\underline{2x + y = 33}$$
$$x = 13$$

To find y, let x = 13 in equation (1).

$$x + y = 20$$
$$13 + y = 20$$
$$y = 7$$

The cashier has 13 twenties and 7 tens.

33. Let x = the number of pounds of $1.30 candy;

    y = the number of pounds of $.90 candy.

Number of pounds	Cost per pound (in dollars)	Total value (in dollars)
x	1.30	1.30x
y	.90	.90y
100	1.00	100(1) = 100

$$x + y = 100 \quad (1)$$
$$1.30x + .90y = 100 \quad (2)$$

Multiply equation (2) by 10 to clear decimals.

$$13x + 9y = 1000 \quad (3)$$

We now have the system

$$x + y = 100 \quad (1)$$
$$13x + 9y = 1000. \quad (3)$$

To solve this system by the substitution method, solve equation (1) for y.

$$y = 100 - x \quad (4)$$

Substitute 100 - x for y in equation (3).

$$13x + 9(100 - x) = 1000$$
$$13x + 900 - 9x = 1000$$
$$4x = 100$$
$$x = 25$$

To find y, let x = 25 in equation (4).

$$y = 100 - 25$$
$$y = 75$$

25 pounds of $1.30 per pound candy and 75 pounds of $.90 per pound candy should be used.

34. Let x = number of liters of 40% solution;

    y = number of liters of 70% solution.

Percent (as a decimal)	Liters of solution	Liters of pure antifreeze
.40	x	.40x
.70	y	.70y
.50	90	.50(90) = 45

From the second and third columns of the table, we obtain the equations

$$x + y = 90 \quad (1)$$
$$.40x + .70y = 45. \quad (2)$$

To clear decimals, multiply both sides of equation (2) by 10.

$$4x + 7y = 450 \quad (3)$$

We now have the system

$$x + y = 90 \quad (1)$$
$$4x + 7y = 450. \quad (3)$$

To solve this system by the addition method, multiply equation (1) by -4 and add the result to equation (3).

$$\begin{aligned} -4x - 4y &= -360 \\ \underline{4x + 7y} &= \underline{450} \\ 3y &= 90 \\ y &= 30 \end{aligned}$$

To find x, let $y = 30$ in equation (1).

$$x + y = 90$$
$$x + 30 = 90$$
$$x = 60$$

To get 90 liters of a 50% antifreeze solution, 60 liters of a 40% solution and 30 liters of a 70% solution will be needed.

35. Let a = the cost of one apple;
b = the cost of one banana.

The cost of 6 apples and 5 bananas is $5.55, so

$$6a + 5b = 5.55. \quad (1)$$

The cost of 12 apples and 2 bananas is $7.50, so

$$12a + 2b = 7.50. \quad (2)$$

To clear decimals, multiply both sides of each equation by 100. This will result in the simplified system

$$600a + 500b = 555 \quad (3)$$
$$1200a + 200b = 750. \quad (4)$$

To solve this system by the addition method, multiply equation (1) by -2 and add the result to equation (4).

$$\begin{aligned} -1200a - 1000b &= -1110 \\ \underline{1200a + 200b} &= \underline{750} \\ -800b &= -360 \\ b &= \frac{-360}{-800} = .45 \end{aligned}$$

To find a, let $b = .45$ in equation (3).

$$600a + 500b = 555$$
$$600a + 500(.45) = 555$$
$$600a + 225 = 555$$
$$600a = 330$$
$$a = \frac{330}{600} = .55$$

The cost of one apple is $.55, and the cost of one banana is $.45.

36. Let x = the speed of the plane in still air;
y = the speed of the wind.

	d	r	t
With wind	540	$x + y$	2
Against wind	690	$x - y$	3

Use the formula $d = rt$.

$$540 = (x + y)(2)$$
$$270 = x + y \quad (1) \text{ Divide by 2}$$
$$690 = (x - y)(3)$$
$$230 = x - y \quad (2) \text{ Divide by 3}$$

Solve the system of equations (1) and (2) by the addition method.

$$270 = x + y \quad (1)$$
$$\underline{230 = x - y \quad (2)}$$
$$500 = 2x \quad \text{Add}$$
$$250 = x$$

$270 = 250 + y$ Let $x = 250$ in (1)
$20 = y$

The speed of the plane in still air is 250 miles per hour and the speed of the wind is 20 miles per hour.

For Exercises 37–40, see the graphs in the answer section of the textbook. (Only the solution region has been shaded in the answers.)

37. $x + y \geq 2$
$x - y \leq 4$

Graph $x + y = 2$ as a solid line through its intercepts, (2, 0) and (0, 2). Using (0, 0) as a test point will result in the false statement $0 \geq 2$, so shade the region not containing the origin.
Graph $x - y \leq 4$ as a solid line through its intercepts, (4, 0) and (0, -4). Using (0, 0) as a test point will result in the true statement $0 \leq 4$, so shade the region containing the origin.
The solution of this system is the intersection of the two shaded regions, and includes the portions of the two lines that bound this region.

38. $y \geq 2x$
$2x + 3y \leq 6$

Graph $y = 2x$ as a solid line through (0, 0) and (1, 2). This line goes through the origin, so a different test point must be used. Choosing (-4, 0) as a test point will result in the true statement $0 \geq -8$, so shade the region containing (-4, 0).
Graph $2x + 3y = 6$ as a solid line through its intercepts, (3, 0) and (0, 2). Choosing (0, 0) as a test point will result in the true statement $0 \leq 6$, so shade the region containing the origin.
The solution of this system is the intersection of the two shaded regions, and includes the portion of the two lines that bound this region.

39. $x + y < 3$
$2x > y$

Graph $x + y = 3$ as a dashed line through (3, 0) and (0, 3). Using (0, 0) as a test point will result in the true statement $0 < 3$, so shade the region containing the origin.
Graph $2x = y$ as a dashed line through (0, 0) and (1, 2). Choosing (0, -3) as a test point will result in the true statement $0 \geq -6$, so shade the region containing (0, -3).

The solution of the system is the intersection of the two shaded regions. It does not contain portions of either boundary line.

40. $y + 2 \leq 2x$
    $x \geq 4$

    Graph $y + 2 = 2x$ as a solid line through its intercepts $(1, 0)$ and $(0, -2)$. Using $(0, 0)$ as the test point will result in the false statement $2 \leq 0$, so shade the side of the line not containing the origin. Graph $x = 4$ as a solid vertical line through $(4, 0)$. Shade the region to the right of this line.
    The solution of the system is the intersection of the two shaded regions, and includes the portions of the two lines that bound this region.

41. $\dfrac{2x}{3} + \dfrac{y}{4} = \dfrac{14}{3}$ (1)
    $\dfrac{x}{2} + \dfrac{y}{12} = \dfrac{8}{3}$ (2)

    To clear fractions, multiply both sides of each equation by 12.

    $12\left(\dfrac{2x}{3} + \dfrac{y}{4}\right) = 12\left(\dfrac{14}{3}\right)$
    $8x + 3y = 56$ (3)

    $12\left(\dfrac{x}{2} + \dfrac{y}{12}\right) = 12\left(\dfrac{8}{3}\right)$
    $6x + y = 32$ (4)

    We now have the simplified system
    $8x + 3y = 56$ (3)
    $6x + y = 32$.   (4)

    To solve this system by the addition method, multiply both sides of equation (4) by -3 and add the result to equation (3).

    $8x + 3y = 56$
    $-18x - 3y = -96$
    $\overline{-10x = -40}$
    $x = 4$

    To find y, let $x = 4$ in equation (4).
    $6x + y = 32$
    $6(4) + y = 32$
    $24 + y = 32$
    $y = 8$

    The solution is $(4, 8)$.

42. $x + y = 2y + 6$ (1)
    $y + 8 = x + 2$ (2)

    Rearrange terms so that like terms are lined up. Equation (1) becomes
    $x - y = 6$,
    and equation (2) becomes
    $-x + y = -6$.

    We now have the simplified system
    $x - y = 6$ (3)
    $-x + y = -6$.   (4)

    To solve this system by the addition method, add equations (3) and (4).

    $x - y = 6$
    $-x + y = -6$
    $\overline{0 = 0}$ *True*

    This result indicates that the system has an infinite number of solutions.

43. $3x + 4y = 6$ (1)
$4x - 5y = 8$ (2)

To solve this system by the addition method, multiply equation (1) by -4 and equation (2) by 3; then add the results.

$$-12x - 16y = -24$$
$$12x - 15y = 24$$
$$-31y = 0$$
$$y = 0$$

To find x, substitute 0 for y in equation (1).

$$3x + 4(0) = 6$$
$$3x = 6$$
$$x = 2$$

The solution is (2, 0).

44. $\dfrac{3x}{2} + \dfrac{y}{5} = -3$ (1)

$4x + \dfrac{y}{3} = -11$ (2)

To clear fractions, multiply each side of equation (1) by 10 and each side of equation (2) by 3.

$$10\left(\dfrac{3x}{2} + \dfrac{y}{5}\right) = 10(-3)$$
$$15x + 2y = -30 \quad (3)$$

$$3\left(4x + \dfrac{y}{3}\right) = 3(-11)$$
$$12x + y = -33 \quad (4)$$

We now have the simplified system

$$15x + 2y = -30 \quad (3)$$
$$12x + y = -33. \quad (4)$$

To solve this system by the substitution method, solve equation (4) for y.

$$y = -33 - 12x \quad (5)$$

Substitute $-33 - 12x$ for y in equation (3) and solve for x.

$$15x + 2y = -30$$
$$15x + 2(-33 - 12x) = -30$$
$$15x - 66 - 24x = -30$$
$$-9x - 66 = -30$$
$$-9x = 36$$
$$x = -4$$

To find y, let $x = -4$ in equation (5).

$$y = -33 - 12(-4)$$
$$y = -33 + 48 = 15$$

The solution is (-4, 15).

For Exercises 45 and 46, see the graphs in the answer section of the textbook. (Only the solution region has been shaded in the answer.)

45. $x + y < 5$
$x - y \geq 2$

Graph $x + y = 5$ as a dashed line through its intercepts, (5, 0) and (0, 5). Using (0, 0) as a test point results in the true statement $0 < 5$, so shade the region containing the origin.
Graph $x - y \geq 2$ as a solid line through its intercepts, (2, 0) and (0, -2). Using (0, 0) as a test point results in the false statement $0 \geq 2$, so shade the region not containing the origin.

The solution of the system is the intersection of the two shaded regions, and includes the portions of the two lines that bound the region.

46. $y \leq 2x$
$x + 2y > 4$

Graph $y = 2x$ as a solid line through (0, 0) and (1, 2). Shade the region below the line.
Graph $x + 2y = 4$ as a dashed line through its intercepts, (4, 0) and (0, 2). Shade the region above this line.
The solution of the system is the intersection of the shaded region. It includes the portion of the line $y = 2x$ that bounds this region, but not the portion of the line $x + 2y = 4$ that bounds this region.

47. $x + 6y = 3$ (1)
$2x + 12y = 2$ (2)

To solve this system by the addition method, multiply equation (1) by -2 and add the result to equation (2).

$$-2x - 12y = -6$$
$$\underline{2x + 12y = 2}$$
$$0 = -4 \quad \textit{False}$$

The result indicates that the system has no solution.

48. Let x = the amount invested at 3%;
y = the amount invested at 4%.

Rate	Amount invested	Annual interest
.03	x	.03x
.04	y	.04y
Totals	18,000	650

From the chart, we obtain the equations

$$x + y = 18,000 \quad (1)$$
$$.03x + .04y = 650. \quad (2)$$

To clear decimals, multiply each side of equation (2) by 100.

$$100(.03x + .04y) = 100(650)$$
$$3x + 4y = 65,000 \quad (3)$$

We now have the system

$$x + y = 18,000 \quad (1)$$
$$3x + 4y = 65,000. \quad (3)$$

To solve this system by the addition method, multiply equation (1) by -3 and add the result to equation (3).

$$-3x - 3y = -54,000$$
$$\underline{3x + 4y = 65,000}$$
$$y = 11,000$$

To find x, let $y = 11,000$ in equation (1).

$$x + y = 18,000$$
$$x + 11,000 = 18,000$$
$$x = 7000$$

He should invest $7000 at 3% and $11,000 at 4%.

49. Let x = the length of each of the two equal sides;

y = the length of the third side.

The perimeter is 29 inches, so

$$x + x + y = 29$$
or $$2x + y = 29. \quad (1)$$

The third side is 5 inches longer than each of the two equal sides, so

$$y = x + 5. \quad (2)$$

We have the system

$$2x + y = 29 \quad (1)$$
$$y = x + 5. \quad (2)$$

To solve this system by the substitution method, substitute x + 5 for y in equation (1).

$$2x + y = 29$$
$$2x + (x + 5) = 29$$
$$3x + 5 = 29$$
$$3x = 24$$
$$x = 8$$

To find y, let x = 8 in equation (2).

$$y = x + 5$$
$$y = 8 + 5 = 13$$

The lengths of the sides of the triangle are 8 inches, 8 inches, and 13 inches.

50. Let x = the larger number;

y = the smaller number.

$$x + y = -12 \quad (1)$$
$$\underline{x - y = 24} \quad (2)$$
$$2x = 12$$
$$x = 6$$

To find y, let x = 6 in equation (1).

$$x + y = -12$$
$$6 + y = -12$$
$$y = -18$$

The two numbers are 6 and −18.

51. Yes, the problem in Exercise 49 can be worked using a single variable. Choose a variable to represent the length of each of the equal sides, and express the length of the third side in terms of this variable. The solution would be as follows:

Let x = the length of each of the two equal sides.

Then x + 5 = the length of the third side.

The perimeter of the triangle is 29 inches, so

$$x + x + (x + 5) = 29.$$

Solve this equation.

$$3x + 5 = 29$$
$$3x = 24$$
$$x = 8$$
$$x + 5 = 13$$

As in Exercises 49, the lengths of the sides of the triangle are 8 inches, 8 inches, and 13 inches.

52. It is impossible for the sum of any two numbers to be both greater than 4 and less than 3. Therefore, system (b) has no solution.

Chapter 7 Test

1. $2x + y = 1$
   $3x - y = 9$

   The intercepts for the line $2x + y = 1$ are $(1/2, 0)$ and $(0, 1)$. Because one of the coordinates is a fraction and the intercepts are close together, it may be difficult to graph the line accurately using only these two points. A third point, such as $(-1, 3)$, may be helpful.

   Graph the line $3x - y = 9$ through its intercepts, $(3, 0)$ and $(0, -9)$.

   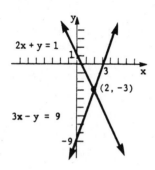

   The two lines intersect at the point $(2, -3)$. This ordered pair is the solution of the system.

2. $x + 2y = 6$
   $-2x + y = -7$

   Graph the line $x + 2y = 6$ though its intercepts, $(6, 0)$ and $(0, 3)$.
   Because the x-intercept of the line $-2x + y = -7$, which is $(7/2, 0)$, has a fractional coordinate, we can graph the line more accurately by using its slope and y-intercept. Rewrite the equation as

   $$y = 2x - 7.$$

   Start by plotting the y-intercept, $(0, -7)$. From this point, go 2 units up and 1 unit to the right to reach the point $(1, -5)$. Draw the line through $(0, -7)$ and $(1, -5)$.

   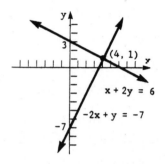

   The two lines intersect at the point $(4, 1)$. This ordered pair is the solution of the system.

3. Answers will vary. The ordered pair $(-3, 4)$ must satisfy the system. One example of such a system is

   $x + y = 1$
   $x - y = -7$.

4. $2x - y = 4$ (1)
   $3x + y = 21$ (2)
   $5x \phantom{{}+y} = 25$ Add
   $x = 5$

To find y, let x = 5 in equation (2).

$$3x + y = 21$$
$$3(5) + y = 21$$
$$15 + y = 21$$
$$y = 6$$

The solution is (5, 6).

5. $4x + 2y = 2$ (1)
   $5x + 4y = 7$ (2)

Multiply equation (1) by -2 and add the result to equation (2).

$$\begin{array}{r} -8x - 4y = -4 \\ \underline{5x + 4y = 7} \\ -3x = 3 \\ x = -1 \end{array}$$

To find y, let x = -1 equation (1).

$$4x + 2y = 2$$
$$4(-1) + 2y = 2$$
$$-4 + 2y = 2$$
$$2y = 6$$
$$y = 3$$

The solution is $(-1, 3)$.

6. $6x + 5y = 13$ (1)
   $3x + 2y = 4$ (2)

Multiply equation (2) by -2 and add the result to equation (1).

$$\begin{array}{r} 6x + 5y = 13 \\ \underline{-6x - 4y = -8} \\ y = 5 \end{array}$$

To find x, let y = 5 in equation (1).

$$6x + 5y = 13$$
$$6x + 5(5) = 13$$
$$6x + 25 = 13$$
$$6x = -12$$
$$x = -2$$

The solution is $(-2, 5)$.

7. $4x + 5y = 2$ (1)
   $-8x - 10y = 6$ (2)

Multiply equation (1) by 2 and add the result to equation (2).

$$\begin{array}{r} 8x + 10y = 4 \\ \underline{-8x - 10y = 6} \\ 0 = 10 \quad True \end{array}$$

This result indicates that the system has no solution.

8. $6x - 5y = 0$ (1)
   $-2x + 3y = 0$ (2)

Multiply equation (2) by 3 and add the result to equation (1).

$$\begin{array}{r} 6x - 5y = 0 \\ \underline{-6x + 9y = 0} \\ 4y = 0 \\ y = 0 \end{array}$$

To find x, let y = 0 in equation (1).

$$6x - 5y = 0$$
$$6x - 5(0) = 0$$
$$6x = 0$$
$$x = 0$$

The solution is (0, 0).

9. $\frac{6}{5}x - \frac{1}{3}y = -20$ (1)

$-\frac{2}{3}x + \frac{1}{6}y = 11$ (2)

To clear fractions, multiply by the LCD for each equation.
Multiply equation (1) by 15.

$$15\left(\frac{6}{5}x - \frac{1}{3}y\right) = 15(-20)$$

$$18x - 5y = -360$$

Multiply equation (2) by 6.

$$6\left(-\frac{2}{3}x + \frac{1}{6}y\right) = 6(11)$$

$$-4x + y = 66$$

We now have the simplified system

$$18x - 5y = -360 \quad (3)$$
$$-4x + y = 66. \quad (4)$$

To solve this system by the addition method, multiply equation (4) by 5 and add the result to equation (3).

$$18x - 5y = -300$$
$$\underline{-20x + 5y = 330}$$
$$-2x = 30$$
$$x = -15$$

To find y, let $x = -15$ in equation (4).

$$-4x + y = 66$$
$$-4(-15) + y = 66$$
$$60 + y = 66$$
$$y = 6$$

The solution is $(-15, 6)$.

10. $2x + y = -4$ (1)

$x = y + 7$ (2)

Substitute $y + 7$ for x in equation (1), and solve for y.

$$2x + y = -4$$
$$2(y + 7) + y = -4$$
$$2y + 14 + y = -4$$
$$3y + 14 = -4$$
$$3y = -18$$
$$y = -6$$

To find x, let $y = -6$ in equation (2).

$$x = y + 7$$
$$x = -6 + 7 = 1$$

The solution is $(1, -6)$.

11. $4x + 3y = -35$ (1)

$x + y = 0$ (2)

Solve equation (2) for y.

$$y = -x \quad (3)$$

Substitute $-x$ for y in equation (1) and solve for x.

$$4x + 3y = -35$$
$$4x + 3(-x) = -35$$
$$4x - 3x = -35$$
$$x = -35$$

To find y, let $x = -35$ in equation (3).

$$y = -x$$
$$= -(-35) = 35$$

The solution is $(-35, 35)$.

12. $8 + 3x - 4y = 14 - 3y$ (1)
    $3x + y + 12 = 9x - y$ (2)

    Simplify each equation by writing it in the form $Ax + By = C$.

    $3x - y = 6$ (3)
    $-6x + 2y = -12$ (4)

    Multiply equation (1) by 2 and add the result to equation (4).

    $6x - 2y = 12$
    $\underline{-6x + 2y = -12}$
    $0 = 0$ True

    This result indicates that the system has an infinite number of solutions.

13. $\frac{x}{2} - \frac{y}{4} = 7$ (1)

    $\frac{2x}{3} + \frac{5y}{4} = 3$ (2)

    To clear fractions, multiply equation (1) by 4 and equation (2) by 12.

    $4\left(\frac{x}{2} - \frac{y}{4}\right) = 4(7)$

    $2x - y = 28$

    $12\left(\frac{2x}{3} + \frac{5y}{4}\right) = 12(3)$

    $8x + 15y = 36$

    We now have the simplified system

    $2x - y = 28$ (3)
    $8x + 15y = 36.$ (4)

    Multiply equation (3) by -4 and add the result to equation (4).

    $-8x + 4y = -112$
    $\underline{8x + 15y = 36}$
    $19y = -76$
    $y = -4$

    To find x, let $y = -4$ in equation (3).

    $2x - y = 28$
    $2x - (-4) = 28$
    $2x + 4 = 28$
    $2x = 24$
    $x = 12$

    The solution is $(12, -4)$.

14. Let x = the larger number;
    y = the smaller number.

    $x + y = 18$ (1)
    $2y = x - 72$ (2)

    Solve equation (1) for y.

    $y = 18 - x$ (3)

    Substitute $18 - x$ for y in equation (2).

    $2y = x - 72$
    $2(18 - x) = x - 72$
    $36 - 2x = x - 72$
    $-3x = -108$
    $x = 36$

    To find y, let $x = 36$ in equation (3).

    $y = 18 - x$
    $y = 18 - 36 = -18$

    The two numbers are -18 and 36.

15. Let r = the number of $12 discs;
    s = the number of $16 discs.

    He bought 15 discs, so

    $r + s = 15.$ (1)

The total cost was $228, so

$$12r + 16s = 228,$$

which can be simplified by dividing both sides by 4 to obtain

$$3r + 4s = 57. \quad (2)$$

We now have the system

$$r + s = 15 \quad (1)$$
$$3r + 4s = 57. \quad (2)$$

Multiply equation (1) by -3 and add the result to equation (2).

$$\begin{array}{r} -3r - 3s = -45 \\ 3r + 4s = 57 \\ \hline s = 12 \end{array}$$

To find r, let $s = 12$ in equation (1).

$$r + s = 15$$
$$r + 12 = 15$$
$$r = 3$$

Wally bought 3 compact discs at $12 each and 12 at $16 each.

16. Let x = the number of liters of 40% solution;
    y = the number of liters of 65% solution.

Percent (as a decimal)	Liters of solution	Liters of pure acid
.40	x	.40x
.65	y	.65y
.45	200	.45(200) = 90

From the chart, we obtain the equations

$$x + y = 200 \quad (1)$$
$$.40x + .65y = 90. \quad (2)$$

To clear decimals, multiply equation (2) by 100. We obtain the new system

$$x + y = 200 \quad (1)$$
$$40x + 65y = 9000. \quad (3)$$

Multiply equation (1) by -40 and add the result to equation (3).

$$\begin{array}{r} -40x - 40y = -8000 \\ 40x + 65y = 9000 \\ \hline 25y = 1000 \\ y = 40 \end{array}$$

To find x, let $y = 40$ in equation (1).

$$x + y = 200$$
$$x + 40 = 200$$
$$x = 160$$

To get 200 liters of a 45% solution, 160 liters of a 40% solution and 40 liters of a 65% solution should be used.

17. Let x = the rate of the slower car;
    y = the rate of the faster car.

One car travels 30 miles per hour faster than the other, so

$$y = x + 30. \quad (1)$$

In 2 1/2 (or 5/2) hours, the slower car travels $(5/2)(x)$ miles and the faster car travels $(5/2)(y)$ miles. The cars will be 225 miles apart, so

$$\frac{5}{2}x + \frac{5}{2}y = 265. \quad (2)$$

Simplify equation (2).

$$2\left(\tfrac{5}{2}x + \tfrac{5}{2}y\right) = 2(265) \quad \text{Multiply by 2}$$
$$5x + 5y = 530$$
$$x + y = 106 \quad \text{Divide by 5}$$

We now have the system

$$y = x + 30 \quad (1)$$
$$x + y = 106. \quad (3)$$

Substitute $x + 30$ for y in equation (3).

$$x + y = 106$$
$$x + (x + 30) = 106$$
$$2x + 30 = 106$$
$$2x = 76$$
$$x = 38$$

To find y, let $x = 38$ in equation (1).

$$y = x + 30$$
$$= 38 + 30 = 68$$

The slower car went 38 miles per hour, and the faster car went 68 miles per hour.

For Exercises 18 and 19, see the graphs in the answer section of the textbook.

18. $2x + 7y \leq 14$
    $x - y \geq 1$

Graph $2x + 7y = 14$ as a solid line through its intercepts, $(7, 0)$ and $(0, 2)$. Choosing $(0, 0)$ as a test point leads to the true statement $0 \leq 14$, so shade the side of the line containing the origin.

Graph $x - y = 1$ as a solid line through its intercepts, $(1, 0)$ and $(0, -1)$. Choosing $(0, 0)$ as a test point leads to the false statement $0 \geq 1$, so shade the side of the line not containing the origin.
The solution of the given system is the intersection of the two shaded regions, and includes the portions of the two lines that bound this region.

19. $2x - y > 6 \quad (1)$
    $4y + 12 \geq -3x \quad (2)$

Graph $2x - y = 6$ as a dashed line through its intercepts, $(3, 0)$ and $(0, -6)$. Choosing $(0, 0)$ as a test point leads to the false statement $0 > 6$, so shade the side of the line not containing the origin.
Graph $4y + 12 = -3x$ as a dashed line through its intercepts, $(-4, 0)$ and $(0, -3)$. Choosing $(0, 0)$ as a test point leads to the true statement $12 \geq 0$, so shade the side of the line containing the origin.
The solution of the given system is the intersection of the two shaded regions. It does not include portions of the lines that bound this region.

20. Answers will vary. One example is the system

$$x + y > 1$$
$$x + y < -1.$$

It is impossible for two numbers to have a sum which is both greater than 1 and less than −1, so this system has no solution.

Cumulative Review: Chapters R–7

1. The integer factors of 40 are −1, 1, −2, 2, −4, 4, −5, 5, −8, 8, −10, 10, −20, 20, −40, and 40.

2. $\dfrac{3x^2 + 2y^2}{10y + 3} = \dfrac{3 \cdot 1^2 + 2 \cdot 5^2}{10(5) + 3}$
   Let $x = 1$, $y = 5$
   $= \dfrac{3 \cdot 1 + 2 \cdot 25}{50 + 3}$
   $= \dfrac{3 + 50}{50 + 3}$
   $= \dfrac{53}{53} = 1$

3. $5 + (-4) = (-4) + 5$

   The order of the numbers has been changed, so this is an example of the commutative property of addition.

4. $r(s - k) = rs - rk$

   This is an example of the distributive property.

5. $-\dfrac{2}{3} + \dfrac{2}{3} = 0$

   The numbers −2/3 and 2/3 are additive inverses (or opposites) of each other. This is an example of the inverse property for addition.

6. $-2 + 6[3 - (4 - 9)] = -2 + 6[3 -(-5)]$
   $= -2 + 6(8)$
   $= -2 + 48$
   $= 46$

7. $2 - 3(6x + 2) = 4(x + 1) + 18$
   $2 - 18x - 6 = 4x + 4 + 18$
   $-18x - 4 = 4x + 22$
   $-18x - 4 - 4x = 4x + 22 - 4x$
   $-22x - 4 = 22$
   $-22x - 4 + 4 = 22 + 4$
   $-22x = 26$
   $x = \dfrac{26}{-22} = -\dfrac{13}{11}$

8. $\dfrac{3}{2}(\dfrac{1}{3}x + 4) = 6(\dfrac{1}{4} + x)$

   Multiply each side by 2.

   $2[\dfrac{3}{2}(\dfrac{1}{3}x + 4)] = 2[6(\dfrac{1}{4} + x)]$

   $3(\dfrac{1}{3}x + 4) = 12(\dfrac{1}{4} + x)$

   Use the distributive property to remove parentheses.

   $x + 12 = 3 + 12x$
   $x + 12 - 12x = 3 + 12x - 12x$
   $-11x + 12 = 3$
   $-11x = -9$
   $x = \dfrac{9}{11}$

9. $-\dfrac{5}{6}x < 15$

   Multiply each side by the reciprocal of −5/6, which is −6/5, and reverse the direction of the inequality symbol.

$$-\frac{5}{6}x < 15$$

$$-\frac{6}{5}\left(-\frac{5}{6}x\right) > -\frac{6}{5}(15)$$

$$x > -18$$

10. $-8 < 2x + 3$

$-8 - 3 < 2x + 3 - 3$

$-11 < 2x$

$\frac{-11}{2} < \frac{2x}{2}$

$-\frac{11}{2} < x$

or $x > -\frac{11}{2}$

11. Let x = the number of yards needed for 45 smocks.

Write a proportion.

$$\frac{x \text{ yards}}{45 \text{ smocks}} = \frac{15 \text{ yards}}{18 \text{ smocks}}$$

$$\frac{x}{45} = \frac{15}{18}$$

$$18x = 15 \cdot 45$$

$$18x = 675$$

$$x = \frac{675}{18} = 37\frac{1}{2}$$

For 45 smocks, 37 1/2 yards would be needed.

12. $-3(-5x^2 + 3x - 10) - (x^2 - 4x + 7)$

$= -3(-5x^2 + 3x - 10) - 1(x^2 - 4x + 7)$

$= 15x^2 - 9x + 30 - x^2 + 4x - 7$

$= 14x^2 - 5x + 23$

13. $(3x - 7)(2y + 4)$

$\ \ \ \ \ \ \ \ \ \ \ \text{F} \ \ \ \ \ \ \ \ \ \ \text{O} \ \ \ \ \ \ \ \ \ \ \text{I} \ \ \ \ \ \ \ \ \ \text{L}$

$= (3x)(2y)+(3x)(4)+(-7)(2y)+(-7)(4)$

$= 6xy + 12x - 14y - 28$

14. $\dfrac{3k^3 + 17k^2 - 27k + 7}{k + 7}$

$$\begin{array}{r}
3k^2 - 4k + 1 \\
k + 7 \overline{\smash{)}3k^3 + 17k^2 - 27k + 7} \\
\underline{3k^3 + 21k^2 } \\
-4k^2 - 27k \\
\underline{-4k^2 - 28k } \\
k + 7 \\
\underline{k + 7} \\
0
\end{array}$$

The remainder is 0, so the answer is the quotient, $3k^2 - 4k + 1$.

15. 36,500,000,000

Move the decimal point to the right of the first nonzero digit, which is 3. Count the number of places the decimal point was moved, which is 10.

Because moving the decimal point made the number smaller, the exponent on 10 will be positive. Thus,

$36{,}500{,}000{,}000 = 3.65 \times 10^{10}$.

16. $\left(\dfrac{x^{-4}y^3}{x^2y^4}\right)^{-1} = \dfrac{x^2y^4}{x^{-4}y^3}$

$= x^{2-(-4)} \cdot y^{4-3}$

$= x^6 y$

17. $10m^2 + 7mp - 12p^2$

Factor this trinomial by trial and error.

$10m^2 + 7mp - 12p^2$

$= (5m - 4p)(2m + 3p)$

18. $64t^2 - 48t + 9$

   This is a perfect square trinomial.

   $64t^2 - 48t + 9$
   $= (8t)^2 - 2(8t)(3) + 3^2$
   $= (8t - 3)^2$

19. $6x^2 - 7x - 3 = 0$
   $(3x + 1)(2x - 3) = 0$
   $3x + 1 = 0$ or $2x - 3 = 0$
   $3x = -1$ $2x = 3$
   $x = -\frac{1}{3}$ or $x = \frac{3}{2}$

   The solutions are $-1/3$ and $3/2$.

20. $r^2 - 121 = 0$
   $(r + 11)(r - 11) = 0$
   $r + 11 = 0$ or $r - 11 = 0$
   $r = -11$ or $r = 11$

   The solutions are -11 and 11.

21. $\dfrac{-3x + 6}{2x + 4} - \dfrac{-3x - 8}{2x + 4}$

   $= \dfrac{(-3x + 6) - (-3x - 8)}{2x + 4}$

   $= \dfrac{-3x + 6 + 3x + 8}{2x + 4}$

   $= \dfrac{14}{2x + 4}$

   $= \dfrac{2 \cdot 7}{2(x + 2)}$

   $= \dfrac{7}{x + 2}$

22. $\dfrac{16k^2 - 9}{8k + 6} \div \dfrac{16k^2 - 24k + 9}{6}$

   $= \dfrac{16k^2 - 9}{8k + 6} \cdot \dfrac{6}{16k^2 - 24k + 9}$

   $= \dfrac{(4k + 3)(4k - 3)}{2(4k + 3)} \cdot \dfrac{6}{(4k - 3)(4k - 3)}$

   $= \dfrac{2 \cdot 3}{2(4k - 3)}$

   $= \dfrac{3}{4k - 3}$

23. $\dfrac{4}{x + 1} + \dfrac{3}{x - 2} = 4$

   Multiply each side by the LCD, $(x + 1)(x - 2)$.

   $(x + 1)(x - 2)\left(\dfrac{4}{x + 1} + \dfrac{3}{x - 2}\right)$
   $= (x + 1)(x - 2)(4)$
   $4(x - 2) + 3(x + 1) = 4(x^2 - x - 2)$
   $4x - 8 + 3x + 3 = 4x^2 - 4x - 8$
   $7x - 5 = 4x^2 - 4x - 8$
   $0 = 4x^2 - 11x - 3$
   $(4x + 1)(x - 3) = 0$
   $4x + 1 = 0$ or $x - 3 = 0$
   $4x = -1$
   $x = -\dfrac{1}{4}$ or $x = 3$

   Checking by substituting these values into the original equation will verify that $-1/4$ and 3 are both solutions.

24. $P = \dfrac{kT}{V}$ for T

   $PV = kT$

   $\dfrac{PV}{k} = \dfrac{kT}{k}$

   $\dfrac{PV}{k} = T$ or $T = \dfrac{PV}{k}$

For Exercises 25–27, see the graphs in the answer section of the textbook.

25. $x - y = 4$

    To graph this line, find the intercepts.

    If $y = 0$, $x = 4$, so the x-intercept is $(4, 0)$.
    If $x = 0$, $y = 4$, so the y-intercept is $(0, 4)$.
    Graph the line through $(4, 0)$ and $(0, 4)$. A third point, such as $(5, 1)$, may be used as a check.

26. $3x + y = 6$

    If $y = 0$, $x = 2$, so the x-intercept is $(2, 0)$.
    If $x = 0$, $y = 6$, so the y-intercept is $(0, 6)$.
    Graph the line through these intercepts. A third point, such as $(1, 3)$, may be used as a check.

27. $x - 3y > 6$

    To graph this inequality, begin by graphing the boundary as a dashed line through $(6, 0)$ and $(0, -2)$. Using $(0, 0)$ as a test point results in the false statement $0 > 6$, so shade the region not containing the origin. Because the symbol $>$, the boundary line is not part of the graph.

28. Through $(-5, 6)$ and $(1, -2)$

    Use the slope formula with $x_1 = -5$, $y_1 = 6$, $x_2 = 1$, and $y_2 = -2$.

    $$m = \frac{y_2 - y_1}{x_2 - x_1}$$
    $$= \frac{-2 - 6}{1 - (-5)}$$
    $$= \frac{-8}{6} = -\frac{4}{3}$$

29. Perpendicular to the line $y = 4x - 3$

    The equation $y = 4x - 3$ is written in $y = mx + b$ form, so the slope is 4. If two lines are perpendicular, the product of their slopes is -1. Since

    $$4\left(-\frac{1}{4}\right) = -1,$$

    the slope of the required line is $-1/4$.

30. Through $(2, -5)$ with slope 3

    Use the point-slope form with $x_1 = 2$, $y_1 = -5$, and $m = 3$.

    $y - y_1 = m(x - x_1)$
    $y - (-5) = 3(x - 2)$
    $y + 5 = 3x - 6$
    $-3x + y = -11$
    $3x - y = 11$ *Multiply by -1*

31. Through the points $(0, 4)$ and $(2, 4)$

    Because the y-coordinates of the two points are the same, this is a horizontal line. For every value of x, the value of y will be 4, so the equation of the line is $y = 4$.

32. **(a)** On the vertical line through (9, −2), the x-coordinate of every point is 9. Therefore, an equation of this line is $x = 9$.

(b) On the horizontal line through (4, −1), the y-coordinate of every point is −1. Therefore, an equation of this line is $y = -1$.

33. $3x + 2y = 14$
$4x + y = -26$

To determine whether (−6, 2) is a solution of this system, substitute −6 for x and 2 for y in each equation.

$$3x + 2y = 14$$
$$3(-6) + 2(2) = 14 \quad ?$$
$$-18 + 4 = 14 \quad ?$$
$$-14 = 14 \quad \text{False}$$

Since (−6, 2) does not satisfy the first equation, it cannot be a solution of the system.

34. $2x - y = -8 \quad (1)$
$x + 2y = 11 \quad (2)$

To solve the system by the addition method, multiply equation (1) by 2 and add the result to equation (2) to eliminate y.

$$4x - 2y = -16$$
$$\underline{x + 2y = 11}$$
$$5x \quad\quad = -5$$
$$x = -1$$

To find y, let $x = -1$ in equation (2).

$$-1 + 2y = 11$$
$$2y = 12$$
$$y = 6$$

The solution is (−1, 6).

35. $4x + 5y = -8 \quad (1)$
$3x + 4y = -7 \quad (2)$

Multiply equation (1) by −3 and equation (2) by 4. Add the resulting equations to eliminate x.

$$-12x - 15y = 24$$
$$\underline{12x + 16y = -28}$$
$$y = -4$$

To find x, let $y = -4$ in equation (1).

$$4x + 5y = -8$$
$$4x + 5(-4) = -8$$
$$4x - 20 = -8$$
$$4x = 12$$
$$x = 3$$

The solution is (3, −4).

36. $3x + 5y = 1 \quad (1)$
$x = y + 3 \quad (2)$

To solve the system by the substitution method, let $x = y + 3$ in equation (1).

$$3x + 5y = 1$$
$$3(y + 3) + 5y = 1$$
$$3y + 9 + 5y = 1$$
$$8y = -8$$
$$y = -1$$

To find x, let y = −1 in equation (2).

$$x = y + 3$$
$$x = -1 + 3 = 2$$

The solution is (2, −1).

37. $3x + 4y = 2$ (1)
    $6x + 8y = 1$ (2)

Multiply equation (1) by −2 and add the result to equation (2).

$$-6x - 8y = -4$$
$$\underline{6x + 8y = 1}$$
$$0 = -3 \quad \text{False}$$

Since 0 = −3 is a false statement, there is no solution.

38. Let x = the larger number;
    y = the smaller number.

$$x + y = 24 \quad \text{Sum is 24}$$
$$\underline{x - y = 6} \quad \text{Difference is 6}$$
$$2x = 30$$
$$x = 15$$

To find y, substitute 15 for x in the first equation.

$$15 + y = 24$$
$$y = 9$$

The numbers are 15 and 9.

39. Let L = length of rectangle,
    W = width of rectangle.

Since the perimeter is 40 meters,

$$2L + 2W = 40. \quad (1)$$

Since the length is 2 meters longer than the width,

$$L = W + 2. \quad (2)$$

Substitute W + 2 for L in equation (1).

$$2(W + 2) + 2W = 40$$
$$2W + 4 + 2W = 40$$
$$4W = 36$$
$$W = 9$$

The width is 9 meters.

40. Let t = the number of $10 bills;
    f = the number of $5 bills.

There are 17 bills, so

$$t + f = 17. \quad (1)$$

The total value of the t $10 bills is 10t and the total value of the f $5 bills is 5f. Since the total value of all the money is $125,

$$10t + 5f = 125. \quad (2)$$

To solve the system by addition, multiply equation (1) by −5 and add the result to equation (2) to eliminate f.

$$-5t - 5f = -85$$
$$\underline{10t + 5f = 125}$$
$$5t = 40$$
$$t = 8$$

To find f, let t = 8 in equation (1).

$$8 + f = 17$$
$$f = 9$$

There are 9 fives and 8 tens.

CHAPTER 8 ROOTS AND RADICALS

Section 8.1 Finding Roots

8.1 Margin Exercises

1. **(a)** The square roots of 100 are 10 and −10 because $10 \cdot 10 = 100$ and $(-10)(-10) = 100$.

   (b) The square roots of 25 are 5 and −5 because $5 \cdot 5 = 25$ and $(-5)(-5) = 25$.

   (c) The square roots of 36 are 6 and −6 because $6 \cdot 6 = 36$ and $(-6)(-6) = 36$.

   (d) The square roots of 64 are 8 and −8 because $8 \cdot 8 = 64$ and $(-8)(-8) = 64$.

2. **(a)** The square of $\sqrt{41}$ is
   $$(\sqrt{41})^2 = 41.$$

   (b) The square of $-\sqrt{39}$ is
   $$(-\sqrt{39})^2 = 39.$$

   (c) The square of $\sqrt{2x^2 + 3}$ is
   $$(\sqrt{2x^2 + 3})^2 = 2x^2 + 3.$$

3. **(a)** $\sqrt{16}$ is the positive square root of 16.
   $$\sqrt{16} = 4$$

   (b) $-\sqrt{169}$ is the negative square root of 169.
   $$-\sqrt{169} = -13$$

   (c) $-\sqrt{225}$ is the negative square root of 225.
   $$-\sqrt{225} = -15$$

   (d) $\sqrt{729}$ is the positive square root of 729.
   $$\sqrt{729} = 27$$

   (e) $\sqrt{\dfrac{36}{25}}$ is the positive square root of $\dfrac{36}{25}$.
   $$\sqrt{\dfrac{36}{25}} = \dfrac{6}{5}$$

4. **(a)** 9 is a perfect square, so $\sqrt{9}$, or 3, is rational.

   (b) 7 is not a perfect square, so $\sqrt{7}$ is irrational.

   (c) $\dfrac{4}{9}$ is a perfect square, so $\sqrt{\dfrac{4}{9}}$, or $\dfrac{2}{3}$, is rational.

   (d) 72 is not a perfect square, so $\sqrt{72}$ is irrational.

   (e) There is no real number whose square is −43, so $\sqrt{-43}$ is not a real number.

5. Use the square root key of a calculator to find a decimal approximation for each square root. Answers are given to the nearest thousandth.

   (a) $\sqrt{28} \approx 5.292$

   (b) $\sqrt{63} \approx 7.937$

   (c) $\sqrt{190} \approx 13.784$

   (d) $\sqrt{1000} \approx 31.623$

6. Substitute the given values in the Pythagorean formula $c^2 = a^2 + b^2$. Then solve for the variable that is not given.

(a)
$$c^2 = a^2 + b^2$$
$$c^2 = 7^2 + 24^2 \quad \text{Let } a = 7, b = 24$$
$$c^2 = 49 + 576 \quad \text{Square}$$
$$c^2 = 625 \quad \text{Add}$$
$$c = \sqrt{625} \quad \text{Take square root; } 25^2 = 625$$
$$c = 25$$

(b)
$$c^2 = a^2 + b^2$$
$$15^2 = a^2 + 13^2 \quad \text{Let } c = 15, b = 13$$
$$225 = a^2 + 169$$
$$56 = a^2 \quad \text{Subtract 169}$$
$$\sqrt{56} = a$$
$$7.483 \approx a \quad \text{Use a calculator}$$

(c)
$$c^2 = a^2 + b^2$$
$$11^2 = 8^2 + b^2 \quad \text{Let } c = 11, a = 8$$
$$121 = 64 + b^2$$
$$57 = b^2 \quad \text{Subtract 64}$$
$$\sqrt{57} = b$$
$$7.550 \approx b \quad \text{Use a calculator}$$

7. Let c = the length of the diagonal. Use the Pythagorean formula, since the width, length, and diagonal of a rectangle form a right triangle.

$$c^2 = a^2 + b^2$$
$$c^2 = 5^2 + 12^2 \quad \text{Let } a = 5, b = 12$$
$$c^2 = 25 + 144$$
$$c^2 = 169$$
$$c = \sqrt{169}$$
$$c = 13$$

The diagonal is 13 feet long.

8. (a) $\sqrt[3]{27} = 3$ because $3^3 = 27$.

(b) $\sqrt[3]{64} = 4$ because $4^3 = 64$.

(c) $\sqrt[3]{-125} = -5$ because $(-5)^3 = -125$.

9. (a) $\sqrt[4]{81} = 3$ because 3 is positive and $3^4 = 81$.

(b) $\sqrt[4]{-81}$ is not a real number because a fourth power of a real number cannot be negative.

(c) $-\sqrt[4]{625}$

First find $\sqrt[4]{625}$. Because $5^4 = 625$, $\sqrt[4]{625} = 5$. Then $-\sqrt[4]{625} = -5$.

(d) $\sqrt[5]{243} = 3$ because $3^5 = 243$.

(e) $\sqrt[5]{-32} = -2$ because $(-2)^5 = -32$.

8.1 Section Exercises

3. Any negative number has no real number square roots because the square of a real number can never be negative.

7. The square roots of 16 are 4 and -4 because $4 \cdot 4 = 16$ and $(-4)(-4) = 16$.

11. The square roots of 25/196 are 5/14 and $-5/14$ because

$$\frac{5}{14} \cdot \frac{5}{14} = \frac{25}{196}$$

and

$$\left(-\frac{5}{14}\right)\left(-\frac{5}{14}\right) = \frac{25}{196}.$$

15. The square of $\sqrt{100}$ is
$$(\sqrt{100})^2 = 100.$$

19. The square of $\sqrt{3x^2 + 4}$ is
$$(\sqrt{3x^2 + 4})^2 = 3x^2 + 4.$$

23. If $\sqrt{a}$ is not a real number, then a must be negative, since any negative number has no real number square roots.

27. $-\sqrt{121}$ is the negative square root of 121.
$$-\sqrt{121} = -11$$

31. $\sqrt{-121}$ is not a real number because there is no real number whose square is -121.

35. $\sqrt{29}$

Because 29 is not a perfect square, $\sqrt{29}$ is irrational. Using the square root key of a calculator gives
$$\sqrt{29} \approx 5.385.$$

39. $-\sqrt{300}$

Because 300 is not a perfect square, $-\sqrt{300}$ is irrational. Using a calculator gives
$$\sqrt{300} \approx 17.321,$$
and so
$$-\sqrt{300} \approx -17.321.$$

43. $-\sqrt{121}$ and $\sqrt{-121}$ are different because $-\sqrt{121}$ is the negative square root of a positive number, while $\sqrt{-121}$ is the square root of a negative number (which is not a real number).

47. $a = 6$, $c = 10$

Substitute the given values in the Pythagorean formula and then solve for b^2.
$$c^2 = a^2 + b^2$$
$$10^2 = 6^2 + b^2$$
$$100 = 36 + b^2$$
$$64 = b^2$$

Now find the positive square root of 64 to get b.
$$b = \sqrt{64} = 8$$

51. The given information involves a right triangle with hypotenuse 25 centimeters and a leg of length 7 centimeters. Let a represent the length of the other leg, and use the Pythagorean formula.
$$c^2 = a^2 + b^2$$
$$25^2 = a^2 + 7^2$$
$$625 = a^2 + 49$$
$$576 = a^2$$
$$\sqrt{576} = a$$
$$a = 24$$

The length of the rectangle is 24 centimeters.

55. Form a right triangle. The distance between the two cars after three hours, which is unknown, is the hypotenuse of the right triangle. The northward leg is a = 3(25) = 75, and the westward leg is b = 3(60) = 180.

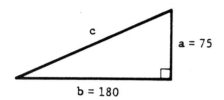

Use the Pythagorean formula to find the hypotenuse c.

$c^2 = a^2 + b^2$
$c^2 = (75)^2 + (180)^2$
$c^2 = 5625 + 32{,}400$
$c^2 = 38{,}025$
$c = \sqrt{38{,}025} = 195$

The cars are 195 miles apart after 3 hours.

59. Answers will vary.
For example, if we choose a = 2 and b = 7,

$\sqrt{a^2 + b^2} = \sqrt{2^2 + 7^2} = \sqrt{53}$,

while

$a + b = 2 + 7 = 9$.

$\sqrt{53} \neq 9$, so we have

$\sqrt{a^2 + b^2} \neq a + b$.

63. $\sqrt[3]{125} = 5$ because $5^3 = 125$.

67. $\sqrt[4]{625} = 5$ because 5 is positive and $5^4 = 625$.

71. $-\sqrt[5]{243}$

First find $\sqrt[5]{243}$. $\sqrt[5]{243} = 3$ because $3^5 = 243$. Therefore, $-\sqrt[5]{243} = -3$.

75. $40 = 2 \cdot 20$
$= 2 \cdot 2 \cdot 10$
$= 2 \cdot 2 \cdot 2 \cdot 5$

In prime factored form, $40 = 2^3 \cdot 5$.

Section 8.2 Multiplication and Division of Radicals

8.2 Margin Exercises

1. (a) $\sqrt{6} \cdot \sqrt{11} = \sqrt{6 \cdot 11} = \sqrt{66}$ *Product rule*

   (b) $\sqrt{2} \cdot \sqrt{5} = \sqrt{10}$ *Product rule*

   (c) $\sqrt{10} \cdot \sqrt{r} = \sqrt{10r}$

   Note that this expression is a real number; since r > 0, 10r is positive.

2. (a) $\sqrt{8} = \sqrt{4 \cdot 2}$ *4 is a perfect square*
   $= \sqrt{4} \cdot \sqrt{2}$ *Product rule*
   $= 2\sqrt{2}$ $\sqrt{4} = 2$

   (b) $\sqrt{27} = \sqrt{9 \cdot 3}$ *9 is a perfect square*
   $= \sqrt{9} \cdot \sqrt{3}$ *Product rule*
   $= 3\sqrt{3}$ $\sqrt{9} = 3$

(c) $\sqrt{50} = \sqrt{25 \cdot 2}$
$= \sqrt{25} \cdot \sqrt{2}$
$= 5\sqrt{2}$

(d) $\sqrt{60} = \sqrt{4 \cdot 15}$
$= \sqrt{4} \cdot \sqrt{15}$
$= 2\sqrt{15}$

3. (a) $\sqrt{3} \cdot \sqrt{15} = \sqrt{3 \cdot 15}$
$= \sqrt{45}$
$= \sqrt{9 \cdot 5}$ 9 is a perfect square
$= \sqrt{9} \cdot \sqrt{5}$ Product rule
$= 3\sqrt{5}$ $\sqrt{9} = 3$

(b) $\sqrt{10} \cdot \sqrt{50} = \sqrt{10 \cdot 50}$
$= \sqrt{10 \cdot 10 \cdot 5}$
$= \sqrt{100 \cdot 5}$
$= \sqrt{100} \cdot \sqrt{5}$
$= 10\sqrt{5}$

(c) $\sqrt{12} \cdot \sqrt{2} = \sqrt{12 \cdot 2}$
$= \sqrt{24}$
$= \sqrt{4 \cdot 6}$
$= \sqrt{4} \cdot \sqrt{6}$
$= 2\sqrt{6}$

(d) $\sqrt{7} \cdot \sqrt{14} = \sqrt{7 \cdot 14}$
$= \sqrt{98}$
$= \sqrt{49 \cdot 2}$
$= \sqrt{49} \cdot \sqrt{2}$
$= 7\sqrt{2}$

4. (a) $\sqrt{\dfrac{81}{16}} = \dfrac{\sqrt{81}}{\sqrt{16}} = \dfrac{9}{4}$ Quotient rule

(b) $\dfrac{\sqrt{192}}{\sqrt{3}} = \sqrt{\dfrac{192}{3}}$
$= \sqrt{64} = 8$

(c) $\sqrt{\dfrac{10}{49}} = \dfrac{\sqrt{10}}{\sqrt{49}} = \dfrac{\sqrt{10}}{7}$

(d) $\dfrac{8\sqrt{50}}{4\sqrt{5}} = \dfrac{8}{4} \cdot \dfrac{\sqrt{50}}{\sqrt{5}} = 2 \cdot \sqrt{\dfrac{50}{5}} = 2\sqrt{10}$

5. (a) $\sqrt{\dfrac{5}{6}} \cdot \sqrt{120} = \dfrac{\sqrt{5}}{\sqrt{6}} \cdot \dfrac{\sqrt{120}}{1}$ Quotient rule
$= \sqrt{\dfrac{600}{6}}$ Product rule
$= \sqrt{100}$
$= 10$

(b) $\sqrt{\dfrac{3}{8}} \cdot \sqrt{\dfrac{7}{2}} = \dfrac{\sqrt{3}}{\sqrt{8}} \cdot \dfrac{\sqrt{7}}{\sqrt{2}}$
$= \dfrac{\sqrt{21}}{\sqrt{16}}$
$= \dfrac{\sqrt{21}}{4}$

6. (a) $\sqrt{36y^6} = \sqrt{36} \cdot \sqrt{y^6} = 6y^3$

(b) $\sqrt{100p^8} = \sqrt{100} \cdot \sqrt{p^8} = 10p^4$

(c) $\sqrt{a^5} = \sqrt{a^4 \cdot a} = \sqrt{a^4} \cdot \sqrt{a} = a^2\sqrt{a}$

7. (a) $\sqrt[3]{108} = \sqrt[3]{27 \cdot 4}$ 27 is a perfect cube
$= \sqrt[3]{27} \cdot \sqrt[3]{4}$
$= 3\sqrt[3]{4}$

(b) $\sqrt[4]{160} = \sqrt[4]{16 \cdot 10}$ 16 is a perfect fourth power
$= \sqrt[4]{16} \cdot \sqrt[4]{10}$
$= 2\sqrt[4]{10}$

(c) $\sqrt[4]{\dfrac{16}{625}} = \dfrac{\sqrt[4]{16}}{\sqrt[4]{625}} = \dfrac{2}{5}$

8.2 Section Exercises

3. $\sqrt{.5} = \sqrt{\frac{1}{2}}$

This statement is true because $.5 = 1/2$. A calculator can be used to show that

$$\sqrt{.5} \approx .707 \text{ and } \sqrt{\frac{1}{2}} \approx .707.$$

7. $\sqrt{3} \cdot \sqrt{27} = \sqrt{3 \cdot 27} = \sqrt{81} = 9$

11. $\sqrt{13} \cdot \sqrt{13} = \sqrt{13 \cdot 13} = 13$

15. (a) $\sqrt{47}$ is in simplified form since 47 has no perfect square factor (other than 1).
The other three choices could be simplified as follows.

$$\sqrt{45} = \sqrt{9 \cdot 5} = 3\sqrt{5}$$
$$\sqrt{48} = \sqrt{16 \cdot 3} = 4\sqrt{3}$$
$$\sqrt{44} = \sqrt{4 \cdot 11} = 2\sqrt{11}$$

19. $\sqrt{90} = \sqrt{9 \cdot 10} = \sqrt{9} \cdot \sqrt{10} = 3\sqrt{10}$

23. $\sqrt{125} = \sqrt{25 \cdot 5} = \sqrt{25} \cdot \sqrt{5} = 5\sqrt{5}$

27. $3\sqrt{27} = 3\sqrt{9 \cdot 3}$
$= 3\sqrt{9} \cdot \sqrt{3}$
$= 3 \cdot 3\sqrt{3}$
$= 9\sqrt{3}$

31. $\sqrt{12} \cdot \sqrt{48} = \sqrt{12 \cdot 48}$
$= \sqrt{12 \cdot 12 \cdot 4}$
$= \sqrt{12 \cdot 12} \cdot \sqrt{4}$
$= 12 \cdot 2$
$= 24$

35. $\sqrt{8} \cdot \sqrt{32}$

First,
$$\sqrt{8} \cdot \sqrt{32} = \sqrt{256}$$
$$= \sqrt{16 \cdot 16}$$
$$= 16.$$

Second,
$$\sqrt{8} \cdot \sqrt{32} = \sqrt{4 \cdot 2} \cdot \sqrt{16 \cdot 2}$$
$$= 2\sqrt{2} \cdot 4\sqrt{2}$$
$$= 2 \cdot 4 \cdot \sqrt{2} \cdot \sqrt{2}$$
$$= 8\sqrt{2 \cdot 2}$$
$$= 8 \cdot 2 = 16.$$

The two methods of simplification give the same answer to this problem. A conjecture that could be made is that, in problems of this type, a product of two square roots, the same answer will be obtained if the two are multiplied into a single square root and then simplified or if the two square roots are simplified individually first and then multiplied.

39. $\sqrt{\frac{7}{16}} = \frac{\sqrt{7}}{\sqrt{16}} = \frac{\sqrt{7}}{4}$

43. $\sqrt{\dfrac{5}{2}} \cdot \sqrt{\dfrac{125}{8}} = \sqrt{\dfrac{5}{2} \cdot \dfrac{125}{8}}$

$= \sqrt{\dfrac{625}{16}}$

$= \dfrac{\sqrt{625}}{\sqrt{16}} = \dfrac{25}{4}$

47. $\sqrt{m^2} = m$

(It is not necessary to write the answer as $|m|$ since the directions told us to assume that the variable represents a nonnegative real number.)

51. $\sqrt{36z^2} = \sqrt{36} \cdot \sqrt{z^2}$
$= 6z$

55. $\sqrt{z^5} = \sqrt{z^4 \cdot z}$
$= \sqrt{z^4} \cdot \sqrt{z}$
$= z^2 \sqrt{z}$

59. $\sqrt[3]{40}$

8 is a perfect cube that is a factor of 40.

$\sqrt[3]{40} = \sqrt[3]{8 \cdot 5}$
$= \sqrt[3]{8} \cdot \sqrt[3]{5} = 2\sqrt[3]{5}$

63. $\sqrt[3]{128}$

8 and 64 are both perfect cubes that are factors of 128. Use 64 since it is larger than 8.

$\sqrt[3]{128} = \sqrt[3]{64 \cdot 2}$
$= \sqrt[3]{64} \cdot \sqrt[3]{2} = 4\sqrt[3]{2}$

67. $\sqrt[3]{\dfrac{8}{27}}$

8 and 27 are both perfect cubes.

$\sqrt[3]{\dfrac{8}{27}} = \dfrac{\sqrt[3]{8}}{\sqrt[3]{27}} = \dfrac{2}{3}$

71. $\sqrt{20}$ and $2\sqrt{5}$

(a) $\sqrt{20} \approx 4.472135955$

(b) $\sqrt{5} \approx 2.236067978$

Multiply both sides of this equation by 2 to obtain

$2\sqrt{5} \approx 4.472135955.$

(c) These approximations suggest that $\sqrt{20}$ is equal to $2\sqrt{5}$, but this cannot be considered a proof of their equality since there is no guarantee that the two decimal approximations have the same digit in all of their corresponding decimal places. We have shown that the first nine digits after the decimal point agree in these two approximations, but what about the tenth, eleventh, and twelfth decimal places, and so on?

75. $2xy + 3x^2y - 9xy + 8x^2y$
$= 2xy - 9xy + 3x^2y + 8x^2y$
$= -7xy + 11x^2y$

Section 8.3 Addition and Subtraction of Radicals

8.3 Margin Exercises

1. (a) $8\sqrt{5} + 2\sqrt{5} = (8 + 2)\sqrt{5} = 10\sqrt{5}$

    (b) $-4\sqrt{3} + 9\sqrt{3} = (-4 + 9)\sqrt{3} = 5\sqrt{3}$

    (c) $12\sqrt{11} - 3\sqrt{11} = (12 - 3)\sqrt{11}$
    $= 9\sqrt{11}$

    (d) $\sqrt{15} + \sqrt{15} = 1\sqrt{15} + 1\sqrt{15}$
    $= (1 + 1)\sqrt{15}$
    $= 2\sqrt{15}$

    (e) $2\sqrt{7} + 2\sqrt{10}$ cannot be simplified further because the $\sqrt{7}$ and $\sqrt{10}$ are unlike radicals.

2. (a) $\sqrt{8} + 4\sqrt{2}$
    $= \sqrt{4} \cdot \sqrt{2} + 4\sqrt{2}$ Product rule
    $= 2\sqrt{2} + 4\sqrt{2}$ $\sqrt{4} = 2$
    $= 6\sqrt{2}$ Add like radicals

    (b) $\sqrt{27} + \sqrt{12} = \sqrt{9} \cdot \sqrt{3} + \sqrt{4} \cdot \sqrt{3}$
    $= 3\sqrt{3} + 2\sqrt{3}$
    $= 5\sqrt{3}$

    (c) $5\sqrt{200} - 6\sqrt{18}$
    $= 5(\sqrt{100} \cdot \sqrt{2}) - 6(\sqrt{9} \cdot \sqrt{2})$
    $= 5(10\sqrt{2}) - 6(3\sqrt{2})$
    $= 50\sqrt{2} - 18\sqrt{2}$
    $= 32\sqrt{2}$

3. (a) $\sqrt{7} \cdot \sqrt{21} + 2\sqrt{27}$
    $= \sqrt{7} \cdot \sqrt{7} \cdot \sqrt{3} + 2(\sqrt{9} \cdot \sqrt{3})$
    $= \sqrt{49} \cdot \sqrt{3} + 2(3\sqrt{3})$
    $= 7\sqrt{3} + 6\sqrt{3}$
    $= 13\sqrt{3}$

    (b) $\sqrt{3r} \cdot \sqrt{6} + \sqrt{8r}$
    $= \sqrt{18r} + \sqrt{8r}$
    $= \sqrt{9 \cdot 2r} + \sqrt{4 \cdot 2r}$
    $= \sqrt{9} \cdot \sqrt{2r} + \sqrt{4} \cdot \sqrt{2r}$
    $= 3\sqrt{2r} + 2\sqrt{2r} = 5\sqrt{2r}$

    (c) $\sqrt[3]{81x^4} + 5\sqrt[3]{24x^4}$
    $= \sqrt[3]{27x^3 \cdot 3x} + 5\sqrt[3]{8x^3 \cdot 3x}$
    $= \sqrt[3]{27x^3} \cdot \sqrt[3]{3x} + 5(\sqrt[3]{8x^3} \cdot \sqrt[3]{3x})$
    $= 3x\sqrt[3]{3x} + 5(2x \cdot \sqrt[3]{3x})$
    $= 3x\sqrt[3]{3x} + 10x\sqrt[3]{3x} = 13x\sqrt[3]{3x}$

8.3 Section Exercises

3. $2\sqrt{3} + 6\sqrt{3} = (2 + 6)\sqrt{3}$ Distributive property
    $= 8\sqrt{3}$

7. $\sqrt{17} + 4\sqrt{17} = 1\sqrt{17} + 4\sqrt{17}$
    $= (1 + 4)\sqrt{17}$
    $= 5\sqrt{17}$

11. $\sqrt{45} + 4\sqrt{20} = \sqrt{9} \cdot \sqrt{5} + 4(\sqrt{4} \cdot \sqrt{5})$
    $= 3\sqrt{5} + 4(2\sqrt{5})$
    $= 3\sqrt{5} + 8\sqrt{5} = 11\sqrt{5}$

15. $-5\sqrt{32} + 2\sqrt{98}$
    $= -5(\sqrt{16} \cdot \sqrt{2}) + 2(\sqrt{49} \cdot \sqrt{2})$
    $= -5(4\sqrt{2}) + 2(7\sqrt{2})$
    $= -20\sqrt{2} + 14\sqrt{2}$
    $= -6\sqrt{2}$

19. $2\sqrt{8} - 5\sqrt{32} - 2\sqrt{48}$
    $= 2(\sqrt{4} \cdot \sqrt{2}) - 5(\sqrt{16} \cdot \sqrt{2})$
    $ - 2(\sqrt{16} \cdot \sqrt{3})$
    $= 2(2\sqrt{2}) - 5(4\sqrt{2}) - 2(4\sqrt{3})$
    $= 4\sqrt{2} - 20\sqrt{2} - 8\sqrt{3}$
    $= -16\sqrt{2} - 8\sqrt{3}$

   (Because $\sqrt{2}$ and $\sqrt{3}$ are unlike radicals, this difference cannot be simplified further.)

23. $\frac{1}{4}\sqrt{288} + \frac{1}{6}\sqrt{72}$
    $= \frac{1}{4}(\sqrt{144} \cdot \sqrt{2}) + \frac{1}{6}(\sqrt{36} \cdot \sqrt{2})$
    $= \frac{1}{4}(12\sqrt{2}) + \frac{1}{6}(6\sqrt{2})$
    $= 3\sqrt{2} + 1\sqrt{2} = 4\sqrt{2}$

27. $\sqrt{6} \cdot \sqrt{2} + 9\sqrt{3} = \sqrt{6 \cdot 2} + 9\sqrt{3}$
    $= \sqrt{12} + 9\sqrt{3}$
    $= \sqrt{4 \cdot 3} + 9\sqrt{3}$
    $= 2\sqrt{3} + 9\sqrt{3} = 11\sqrt{3}$

31. $\sqrt{6x^2} + x\sqrt{24}$
    $= \sqrt{x^2 \cdot 6} + x\sqrt{4 \cdot 6}$
    $= x\sqrt{6} + x \cdot 2\sqrt{6}$
    $= (x + 2x)\sqrt{6} = 3x\sqrt{6}$

35. $-8\sqrt{32k} + 6\sqrt{8k}$
    $= -8(\sqrt{16 \cdot 2k}) + 6(\sqrt{4 \cdot 2k})$
    $= -8(4\sqrt{2k}) + 6(2\sqrt{2k})$
    $= -32\sqrt{2k} + 12\sqrt{2k}$
    $= -20\sqrt{2k}$

39. $4\sqrt[3]{16} - 3\sqrt[3]{54}$

    Recall that 8 and 27 are perfect cubes.

    $4\sqrt[3]{16} - 3\sqrt[3]{54}$
    $= 4(\sqrt[3]{8 \cdot 2}) - 3(\sqrt[3]{27 \cdot 2})$
    $= 4(\sqrt[3]{8} \cdot \sqrt[3]{2}) - 3(\sqrt[3]{27} \cdot \sqrt[3]{2})$
    $= 4(2\sqrt[3]{2}) - 3(3\sqrt[3]{2})$
    $= 8\sqrt[3]{2} - 9\sqrt[3]{2}$
    $= -1\sqrt[3]{2} = -\sqrt[3]{2}$

43. $5\sqrt[4]{m^3} + 8\sqrt[4]{16m^3}$
    $= 5\sqrt[4]{m^3} + 8(\sqrt[4]{16} \cdot \sqrt[4]{m^3})$
    $= 5\sqrt[4]{m^3} + 8(2\sqrt[4]{m^3})$
    $= 5\sqrt[4]{m^3} + 16\sqrt[4]{m^3} = 21\sqrt[4]{m^3}$

47. $(\sqrt{6})^2 = 6$, by the definition of square root.

51. $\sqrt{7500} = \sqrt{2500 \cdot 3}$ 2500 is a perfect square
    $= 50\sqrt{3}$

Section 8.4 Rationalizing the Denominator

8.4 Margin Exercises

1. (a) $\dfrac{3}{\sqrt{5}} = \dfrac{3}{\sqrt{5}} \cdot \dfrac{\sqrt{5}}{\sqrt{5}} = \dfrac{3\sqrt{5}}{5}$

   (b) $\dfrac{-6}{\sqrt{11}} = \dfrac{-6}{\sqrt{11}} \cdot \dfrac{\sqrt{11}}{\sqrt{11}} = \dfrac{-6\sqrt{11}}{11}$

   (c) $-\dfrac{\sqrt{7}}{\sqrt{2}} = -\dfrac{\sqrt{7} \cdot \sqrt{2}}{\sqrt{2} \cdot \sqrt{2}} = -\dfrac{\sqrt{14}}{2}$

   (d) $\dfrac{20}{\sqrt{18}} = \dfrac{20 \cdot \sqrt{2}}{\sqrt{18} \cdot \sqrt{2}}$
   $= \dfrac{20\sqrt{2}}{\sqrt{36}}$
   $= \dfrac{20\sqrt{2}}{6} = \dfrac{10\sqrt{2}}{3}$

$\sqrt{2}$ was chosen because $\sqrt{18} \cdot \sqrt{2} = \sqrt{36} = 6$, a rational number.

2. (a) $\sqrt{\frac{16}{11}} = \frac{\sqrt{16}}{\sqrt{11}} = \frac{\sqrt{16} \cdot \sqrt{11}}{\sqrt{11} \cdot \sqrt{11}} = \frac{4\sqrt{11}}{11}$

 (b) $\sqrt{\frac{5}{18}} = \frac{\sqrt{5}}{\sqrt{18}} = \frac{\sqrt{5} \cdot \sqrt{2}}{\sqrt{18} \cdot \sqrt{2}} = \frac{\sqrt{10}}{\sqrt{36}} = \frac{\sqrt{10}}{6}$

 (c) $\sqrt{\frac{8}{32}}$

When rationalizing the denominator, there are often several ways to approach the problem. Three ways to simplify this radical are shown here.

$\sqrt{\frac{8}{32}} = \frac{\sqrt{8}}{\sqrt{32}} = \frac{\sqrt{8} \cdot \sqrt{2}}{\sqrt{32} \cdot \sqrt{2}}$
$= \frac{\sqrt{16}}{\sqrt{64}} = \frac{4}{8} = \frac{1}{2}$

or $\sqrt{\frac{8}{32}} = \frac{\sqrt{8}}{\sqrt{32}} = \frac{\sqrt{4} \cdot \sqrt{2}}{\sqrt{16} \cdot \sqrt{2}}$
$= \frac{2\sqrt{2}}{4\sqrt{2}} = \frac{2}{4} = \frac{1}{2}$

or $\sqrt{\frac{8}{32}} = \sqrt{\frac{1}{4}} = \frac{1}{2}$

3. (a) $\sqrt{\frac{1}{2}} \cdot \sqrt{\frac{5}{6}} = \sqrt{\frac{1}{2} \cdot \frac{5}{6}} = \sqrt{\frac{5}{12}}$
$= \frac{\sqrt{5}}{\sqrt{12}} = \frac{\sqrt{5}}{\sqrt{12}} \cdot \frac{\sqrt{3}}{\sqrt{3}}$
$= \frac{\sqrt{15}}{\sqrt{36}} = \frac{\sqrt{15}}{6}$

 (b) $\sqrt{\frac{1}{10}} \cdot \sqrt{20} = \sqrt{\frac{1}{10} \cdot \frac{20}{1}} = \sqrt{2}$

 (c) $\sqrt{\frac{5}{8}} \cdot \sqrt{\frac{24}{10}} = \sqrt{\frac{5}{8} \cdot \frac{24}{10}}$
$= \sqrt{\frac{3}{2}} = \frac{\sqrt{3} \cdot \sqrt{2}}{\sqrt{2} \cdot \sqrt{2}}$
$= \frac{\sqrt{6}}{\sqrt{4}} = \frac{\sqrt{6}}{2}$

4. $\frac{\sqrt{5p}}{\sqrt{q}} = \frac{\sqrt{5p} \cdot \sqrt{q}}{\sqrt{q} \cdot \sqrt{q}} = \frac{\sqrt{5pq}}{q}$

5. $\sqrt{\frac{5r^2t^2}{7}} = \frac{\sqrt{5r^2t^2}}{\sqrt{7}}$
$= \frac{rt\sqrt{5}}{\sqrt{7}}$
$= \frac{rt\sqrt{5} \cdot \sqrt{7}}{\sqrt{7} \cdot \sqrt{7}}$
$= \frac{rt\sqrt{35}}{7}$

6. (a) $\sqrt[3]{\frac{5}{7}} = \frac{\sqrt[3]{5}}{\sqrt[3]{7}}$
$= \frac{\sqrt[3]{5} \cdot \sqrt[3]{7^2}}{\sqrt[3]{7} \cdot \sqrt[3]{7^2}}$
$= \frac{\sqrt[3]{5 \cdot 7^2}}{\sqrt[3]{7^3}} = \frac{\sqrt[3]{245}}{7}$

 (b) $\frac{\sqrt[3]{5}}{\sqrt[3]{9}} = \frac{\sqrt[3]{5} \cdot \sqrt[3]{3}}{\sqrt[3]{9} \cdot \sqrt[3]{3}} = \frac{\sqrt[3]{15}}{\sqrt[3]{27}} = \frac{\sqrt[3]{15}}{3}$

8.4 Section Exercises

3. $\frac{7}{\sqrt{5}} = \frac{7 \cdot \sqrt{5}}{\sqrt{5} \cdot \sqrt{5}} = \frac{7\sqrt{5}}{5}$

Observe that $\sqrt{5} \cdot \sqrt{5} = 5$, so we have succeeded in rationalizing the denominator.

7. $\dfrac{-\sqrt{11}}{\sqrt{3}} = \dfrac{-\sqrt{11}\cdot\sqrt{3}}{\sqrt{3}\cdot\sqrt{3}} = \dfrac{-\sqrt{33}}{3}$

11. $\dfrac{24\sqrt{10}}{16\sqrt{3}} = \dfrac{3\sqrt{10}}{2\sqrt{3}}$

$= \dfrac{3\sqrt{10}\cdot\sqrt{3}}{2\sqrt{3}\cdot\sqrt{3}}$

$= \dfrac{3\sqrt{30}}{2\cdot 3} = \dfrac{\sqrt{30}}{2}$

15. $\dfrac{-3}{\sqrt{50}} = \dfrac{-3}{\sqrt{25\cdot 2}}$

$= \dfrac{-3}{5\sqrt{2}}$

$= \dfrac{-3\cdot\sqrt{2}}{5\sqrt{2}\cdot\sqrt{2}}$

$= \dfrac{-3\sqrt{2}}{5\cdot 2} = \dfrac{-3\sqrt{2}}{10}$

19. $\dfrac{\sqrt{24}}{\sqrt{8}}$

When rationalizing the denominator, there are often several ways to approach the problem. Two ways to simplify this expression are shown here.

$\dfrac{\sqrt{24}}{\sqrt{8}} = \dfrac{\sqrt{24}\cdot\sqrt{2}}{\sqrt{8}\cdot\sqrt{2}} = \dfrac{\sqrt{48}}{\sqrt{16}}$

$= \dfrac{\sqrt{16}\cdot\sqrt{3}}{4} = \dfrac{4\sqrt{3}}{4} = \sqrt{3}$

or $\dfrac{\sqrt{24}}{\sqrt{8}} = \sqrt{\dfrac{24}{8}}$ Quotient rule

$= \sqrt{3}$

23. $\sqrt{\dfrac{1}{2}} = \dfrac{\sqrt{1}}{\sqrt{2}}$ Quotient rule

$= \dfrac{1\cdot\sqrt{2}}{\sqrt{2}\cdot\sqrt{2}} = \dfrac{\sqrt{2}}{2}$

27. $\sqrt{\dfrac{7}{13}}\cdot\sqrt{\dfrac{13}{3}} = \sqrt{\dfrac{7}{13}\cdot\dfrac{13}{3}}$

$= \sqrt{\dfrac{7}{3}} = \dfrac{\sqrt{7}}{\sqrt{3}}$

$= \dfrac{\sqrt{7}\cdot\sqrt{3}}{\sqrt{3}\cdot\sqrt{3}} = \dfrac{\sqrt{21}}{3}$

31. $\sqrt{\dfrac{1}{12}}\cdot\sqrt{\dfrac{1}{3}} = \sqrt{\dfrac{1}{12}\cdot\dfrac{1}{3}}$

$= \sqrt{\dfrac{1}{36}} = \dfrac{\sqrt{1}}{\sqrt{36}} = \dfrac{1}{6}$

35. $\sqrt{\dfrac{7}{x}} = \dfrac{\sqrt{7}}{\sqrt{x}}$

$= \dfrac{\sqrt{7}\cdot\sqrt{x}}{\sqrt{x}\cdot\sqrt{x}} = \dfrac{\sqrt{7x}}{x}$

39. $\sqrt{\dfrac{18x^3}{6y}} = \sqrt{\dfrac{3x^3}{y}}$

$= \dfrac{\sqrt{3x^3}}{\sqrt{y}}$

$= \dfrac{\sqrt{3x^3}\cdot\sqrt{y}}{\sqrt{y}\cdot\sqrt{y}}$

$= \dfrac{\sqrt{3x^3 y}}{y}$

$= \dfrac{\sqrt{x^2\cdot 3xy}}{y}$

$= \dfrac{\sqrt{x^2}\cdot\sqrt{3xy}}{y}$

$= \dfrac{x\sqrt{3xy}}{y}$

43. $\dfrac{\sqrt[3]{2}}{\sqrt[3]{5}} = \dfrac{\sqrt[3]{2} \cdot \sqrt[3]{5^2}}{\sqrt[3]{5} \cdot \sqrt[3]{5^2}}$

$= \dfrac{\sqrt[3]{2} \cdot \sqrt[3]{25}}{\sqrt[3]{5^3}} = \dfrac{\sqrt[3]{50}}{5}$

The correct choice for a rationalizing factor in this problem is $\sqrt[3]{5^2} = \sqrt[3]{25}$, which corresponds to choice (b).

47. $\dfrac{\sqrt[3]{4}}{\sqrt[3]{7}} = \dfrac{\sqrt[3]{4} \cdot \sqrt[3]{7^2}}{\sqrt[3]{7} \cdot \sqrt[3]{7^2}}$

$= \dfrac{\sqrt[3]{4} \cdot \sqrt[3]{49}}{\sqrt[3]{7^3}} = \dfrac{\sqrt[3]{196}}{7}$

51. $\dfrac{\sqrt[3]{7m}}{\sqrt[3]{36n}} = \dfrac{\sqrt[3]{7m}}{\sqrt[3]{6^2 n}}$

$= \dfrac{\sqrt[3]{7m} \cdot \sqrt[3]{6n^2}}{\sqrt[3]{6^2 n} \cdot \sqrt[3]{6n^2}}$

$= \dfrac{\sqrt[3]{42mn^2}}{\sqrt[3]{6^3 n^3}} = \dfrac{\sqrt[3]{42mn^2}}{6n}$

55. $(6x - 1)(6x + 1) = (6x)^2 - 1^2$
$= 36x^2 - 1$

Section 8.5 Simplifying Radical Expressions

8.5 Margin Exercises

1. (a) $\sqrt{36} + \sqrt{25} = 6 + 5 = 11$

   (b) $3\sqrt{3} + 2\sqrt{27}$
   $= 3\sqrt{3} + 2(\sqrt{9} \cdot \sqrt{3})$ *9 is a perfect square*
   $= 3\sqrt{3} + 2(3\sqrt{3})$ *$\sqrt{9} = 3$*
   $= 3\sqrt{3} + 6\sqrt{3}$ *Multiply*
   $= 9\sqrt{3}$ *Add like radicals*

   (c) $4\sqrt{8} - 2\sqrt{32}$
   $= 4(\sqrt{4} \cdot \sqrt{2}) - 2(\sqrt{16} \cdot \sqrt{2})$ *Product rule*
   $= 4(2\sqrt{2}) - 2(4\sqrt{2})$ *$\sqrt{4} = 2$, $\sqrt{16} = 4$*
   $= 8\sqrt{2} - 8\sqrt{2}$ *Multiply*
   $= 0$ *Subtract like radicals*

   (d) $2\sqrt{12} - 5\sqrt{48}$
   $= 2(\sqrt{4} \cdot \sqrt{3}) - 5(\sqrt{16} \cdot \sqrt{3})$
   $= 2(2\sqrt{3}) - 5(4\sqrt{3})$
   $= 4\sqrt{3} - 20\sqrt{3}$
   $= -16\sqrt{3}$

2. (a) $\sqrt{7}(\sqrt{2} + \sqrt{5}) = \sqrt{7}(\sqrt{2}) + \sqrt{7}(\sqrt{5})$
   $= \sqrt{14} + \sqrt{35}$

   (b) $\sqrt{2}(\sqrt{8} + \sqrt{20})$
   $= \sqrt{2}(\sqrt{4} \cdot \sqrt{2} + \sqrt{4} \cdot \sqrt{5})$
   $= \sqrt{2}(2\sqrt{2} + 2\sqrt{5})$
   $= \sqrt{2} \cdot 2\sqrt{2} + \sqrt{2} \cdot 2\sqrt{5}$
   $= 2 \cdot 2 + 2\sqrt{10}$
   $= 4 + 2\sqrt{10}$

   (c) $(\sqrt{2} + 5\sqrt{3})(\sqrt{3} - 2\sqrt{2})$
   $= \sqrt{2} \cdot \sqrt{3} - \sqrt{2} \cdot 2\sqrt{2} + 5\sqrt{3} \cdot \sqrt{3}$
   $\quad - 5\sqrt{3} \cdot 2\sqrt{2}$ *FOIL*
   $= \sqrt{6} - 2 \cdot 2 + 5 \cdot 3 - 10\sqrt{6}$
   $= \sqrt{6} - 4 + 15 - 10\sqrt{6}$
   $= 11 - 9\sqrt{6}$

   (d) $(\sqrt{2} - \sqrt{5})(\sqrt{10} + \sqrt{2})$
   $= \sqrt{2} \cdot \sqrt{10} + \sqrt{2} \cdot \sqrt{2} - \sqrt{5} \cdot \sqrt{10}$
   $\quad - \sqrt{5} \cdot \sqrt{2}$ *FOIL*
   $= \sqrt{20} + \sqrt{2} - \sqrt{50} - \sqrt{10}$
   $= \sqrt{4} \cdot \sqrt{5} + 2 - \sqrt{25} \cdot \sqrt{2} - \sqrt{10}$
   $= 2\sqrt{5} + 2 - 5\sqrt{2} - \sqrt{10}$

3. (a) $(3 + \sqrt{5})(3 - \sqrt{5})$
$= 3^2 - (\sqrt{5})^2 \quad (a + b)(a - b)$
$\qquad\qquad\qquad = a^2 - b^2$
$= 9 - 5 \qquad (\sqrt{5})^2 = 5$
$= 4$

(b) $(\sqrt{3} - 2)(\sqrt{3} + 2)$
$= (\sqrt{3})^2 - 2^2$
$= 3 - 4$
$= -1$

(c) $(\sqrt{5} + \sqrt{3})(\sqrt{5} - \sqrt{3})$
$= (\sqrt{5})^2 - (\sqrt{3})^2$
$= 5 - 3$
$= 2$

4. (a) $\dfrac{5}{4 + \sqrt{2}} = \dfrac{5(4 - \sqrt{2})}{(4 + \sqrt{2})(4 - \sqrt{2})}$
$= \dfrac{5(4 - \sqrt{2})}{16 - 2} \quad \begin{array}{l}(a+b)(a-b)\\ = a^2 - b^2\end{array}$
$= \dfrac{5(4 - \sqrt{2})}{14}$

(b) $\dfrac{\sqrt{5} + 3}{2 - \sqrt{5}} = \dfrac{(\sqrt{5} + 3)(2 + \sqrt{5})}{(2 - \sqrt{5})(2 + \sqrt{5})}$
$= \dfrac{2\sqrt{5} + 5 + 6 + 3\sqrt{5}}{4 - 5}$
$= \dfrac{11 + 5\sqrt{5}}{4 - 5} = 11 - 5\sqrt{5}$

(c) $\dfrac{1}{\sqrt{6} + \sqrt{3}} = \dfrac{1(\sqrt{6} - \sqrt{3})}{(\sqrt{6} + \sqrt{3})(\sqrt{6} - \sqrt{3})}$
$= \dfrac{\sqrt{6} - \sqrt{3}}{(\sqrt{6})^2 - (\sqrt{3})^2}$
$= \dfrac{\sqrt{6} - \sqrt{3}}{6 - 3}$
$= \dfrac{\sqrt{6} - \sqrt{3}}{3}$

5. (a) $\dfrac{5\sqrt{3} - 15}{10} = \dfrac{5(\sqrt{3} - 3)}{5 \cdot 2} = \dfrac{\sqrt{3} - 3}{2}$

(b) $\dfrac{8\sqrt{5} + 12}{16} = \dfrac{4(2\sqrt{5} + 3)}{4 \cdot 4} = \dfrac{2\sqrt{5} + 3}{4}$

8.5 Section Exercises

3. $\sqrt{2} \cdot \sqrt{8} = 4$

7. $\sqrt[3]{8} + \sqrt[3]{27} = 5$

11. $8\sqrt{50} - 4\sqrt{72}$
$= 8(\sqrt{25} \cdot \sqrt{2}) - 4(\sqrt{36} \cdot \sqrt{2})$
$= 8(5\sqrt{2}) - 4(6\sqrt{2})$
$= 40\sqrt{2} - 24\sqrt{2}$
$= 16\sqrt{2}$

15. $2\sqrt{5}(\sqrt{2} + 3\sqrt{5})$
$= 2\sqrt{5} \cdot \sqrt{2} + 2\sqrt{5} \cdot 3\sqrt{5}$
$= 2\sqrt{10} + 2 \cdot 3 \cdot \sqrt{5} \cdot \sqrt{5}$
$= 2\sqrt{10} + 6 \cdot 5$
$= 2\sqrt{10} + 30$

19. $(2\sqrt{6} + 3)(3\sqrt{6} + 7)$
$= 2\sqrt{6} \cdot 3\sqrt{6} + 7 \cdot 2\sqrt{6} + 3 \cdot 3\sqrt{6}$
$\quad + 3 \cdot 7 \quad FOIL$
$= 2 \cdot 3 \cdot \sqrt{6} \cdot \sqrt{6} + 14\sqrt{6} + 9\sqrt{6} + 21$
$= 6 \cdot 6 + 23\sqrt{6} + 21$
$= 36 + 23\sqrt{6} + 21$
$= 57 + 23\sqrt{6}$

23. $(2\sqrt{7} + 3)^2$
$= (2\sqrt{7})^2 + 2(2\sqrt{7})(3) + (3)^2$
$\qquad\qquad\qquad\quad$ *Square of a binomial*
$= 4 \cdot 7 + 12\sqrt{7} + 9$
$= 37 + 12\sqrt{7}$

27. $(\sqrt{8} - \sqrt{7})(\sqrt{8} + \sqrt{7})$
$= (\sqrt{8})^2 - (\sqrt{7})^2$ *Product of the sum and difference of two terms*
$= 8 - 7 = 1$

31. $(\sqrt{10} - \sqrt{5})(\sqrt{5} + \sqrt{20})$
$= \sqrt{10} \cdot \sqrt{5} + \sqrt{10} \cdot \sqrt{20} - \sqrt{5} \cdot \sqrt{5}$
$\quad - \sqrt{5} \cdot \sqrt{20}$ *FOIL*
$= \sqrt{50} + \sqrt{200} - 5 - \sqrt{100}$ *Product rule*
$= \sqrt{25 \cdot 2} + \sqrt{100 \cdot 2} - 5 - 10$
$= 5\sqrt{2} + 10\sqrt{2} - 15$
$= 15\sqrt{2} - 15$

35. The expression $-37 - 2\sqrt{15}$ cannot be simplified because the two terms do not involve like radicals. (On the other hand, the expression $-37\sqrt{15} - 2\sqrt{15}$ could be written more simply as $-39\sqrt{15}$ because these two terms are like radical terms and may be combined.)

39. $\dfrac{14}{2 - \sqrt{11}} = \dfrac{14(2 + \sqrt{11})}{(2 - \sqrt{11})(2 + \sqrt{11})}$
Multiply numerator and denominator by conjugate of the denominator
$= \dfrac{14(2 + \sqrt{11})}{(2)^2 - (\sqrt{11})^2}$
$= \dfrac{14(2 + \sqrt{11})}{4 - 11}$
$= \dfrac{14(2 + \sqrt{11})}{-7}$
$= -2(2 + \sqrt{11})$
$= -4 - 2\sqrt{11}$

43. $\dfrac{\sqrt{5}}{\sqrt{2} + \sqrt{3}} = \dfrac{\sqrt{5}(\sqrt{2} - \sqrt{3})}{(\sqrt{2} + \sqrt{3})(\sqrt{2} - \sqrt{3})}$
Multiply by the conjugate
$= \dfrac{\sqrt{5} \cdot \sqrt{2} - \sqrt{5} \cdot \sqrt{3}}{(\sqrt{2})^2 - (\sqrt{3})^2}$
$= \dfrac{\sqrt{10} - \sqrt{15}}{2 - 3}$
$= \dfrac{\sqrt{10} - \sqrt{15}}{-1} = -\sqrt{10} + \sqrt{15}$

47. $\dfrac{\sqrt{5} + 2}{2 - \sqrt{3}} = \dfrac{(\sqrt{5} + 2)(2 + \sqrt{3})}{(2 - \sqrt{3})(2 + \sqrt{3})}$
Multiply by the conjugate
$= \dfrac{2\sqrt{5} + \sqrt{15} + 4 + 2\sqrt{3}}{(2)^2 - (\sqrt{3})^2}$
$= \dfrac{2\sqrt{5} + \sqrt{15} + 4 + 2\sqrt{3}}{4 - 3}$
$= \dfrac{2\sqrt{5} + \sqrt{15} + 4 + 2\sqrt{3}}{1}$
$= 2\sqrt{5} + \sqrt{15} + 4 + 2\sqrt{3}$

51. $\dfrac{2\sqrt{3} + 10}{16} = \dfrac{2(\sqrt{3} + 5)}{2 \cdot 8}$ *Factor numerator and denominator*
$= \dfrac{\sqrt{3} + 5}{8}$ *Lowest terms*

55. $y^2 + 4y + 3 = 0$
$(y + 3)(y + 1) = 0$ *Factor*
$y + 3 = 0$ or $y + 1 = 0$
Set each factor equal to 0
$y = -3$ or $y = -1$
Solve

The solutions are -3 and -1.

Section 8.6 Equations with Radicals
8.6 Margin Exercises

1. (a) $\sqrt{k} = 3$

    $(\sqrt{k})^2 = 3^2$ Square each side

    $k = 9$

    Check

    $\sqrt{k} = 3$

    $\sqrt{9} = 3$?   Let $k = 9$

    $3 = 3$?   True

    The solution is 9.

   (b) $\sqrt{m - 2} = 4$

    $(\sqrt{m - 2})^2 = 4^2$ Square each side

    $m - 2 = 16$

    $m = 18$ Add 2

    Check

    $\sqrt{m - 2} = 4$

    $\sqrt{18 - 2} = 4$?   Let $m = 18$

    $\sqrt{16} = 4$?

    $4 = 4$ True

    The solution is 18.

   (c) $\sqrt{9 - y} = 4$

    $(9 - y)^2 = 4^2$ Square each side

    $9 - y = 16$

    $-y = 7$ Subtract 9

    $y = -7$ Multiply by -1

    Check

    $\sqrt{9 - y} = 4$

    $\sqrt{9 - (-7)} = 4$?   Let $y = -7$

    $\sqrt{16} = 4$?

    $4 = 4$ True

    The solution is -7.

2. (a) $\sqrt{3x + 9} = 2\sqrt{x}$

    $(\sqrt{3x + 9})^2 = (2\sqrt{x})^2$ Square each side

    $3x + 9 = 4x$

    $9 = x$ Subtract $3x$

    Check

    $\sqrt{3x + 9} = 2\sqrt{x}$

    $\sqrt{3(9) + 9} = 2\sqrt{9}$?   Let $x = 9$

    $\sqrt{27 + 9} = 2\sqrt{9}$?

    $\sqrt{36} = 2(3)$?

    $6 = 6$ True

    The solution is 9.

   (b) $5\sqrt{a} = \sqrt{20a + 5}$

    $(5\sqrt{a})^2 = (\sqrt{20a + 5})^2$ Square each side

    $25a = 20a + 5$

    $5a = 5$ Subtract $20a$

    $a = 1$ Divide by 5

    Check

    $5\sqrt{a} = \sqrt{20a + 5}$

    $5\sqrt{1} = \sqrt{20(1) + 5}$?   Let $a = 1$

    $5(1) = \sqrt{25}$?

    $5 = 5$ True

    The solution is 1.

3. (a) $\sqrt{y} + 4 = 0$

    $\sqrt{y} = -4$ Subtract 4

    $(\sqrt{y})^2 = (-4)^2$ Square each side

    $y = 16$

    Check

    $\sqrt{y} + 4 = 0$

    $\sqrt{16} + 4 = 0$?   Let $y = 16$

    $4 + 4 = 0$?

    $8 = 0$ False

Because the statement $8 = 0$ is false, 16 is not a solution. The equation has no solution.

(b) $\quad m = \sqrt{m^2 - 4m - 16}$
$\quad\quad m^2 = (\sqrt{m^2 - 4m - 16})^2$
$\quad\quad\quad\quad\quad\quad$ *Square each side*
$\quad\quad m^2 = m^2 - 4m - 16$
$\quad\quad 0 = -4m - 16 \quad$ *Subtract m^2*
$\quad\quad 16 = -4m \quad\quad\quad$ *Add 16*
$\quad\quad -4 = m \quad\quad\quad\quad$ *Divide by -4*

Check
$\quad m = \sqrt{m^2 - 4m - 16}$?
$\quad -4 = \sqrt{(-4)^2 - 4(-4) - 16}$?
$\quad\quad\quad\quad\quad\quad$ *Let $m = -4$*
$\quad -4 = \sqrt{16 + 16 - 16}$?
$\quad -4 = \sqrt{16}$?
$\quad -4 = 4 \quad\quad\quad\quad$ *False*

The only potential equation does not check, so the equation has no solution.

4. (a) Use the pattern
$\quad (a - b)^2 = a^2 - 2ab + b^2$
with $a = m$ and $b = 5$.
$\quad (m - 5)^2 = m^2 - 2(m)(5) + (5)^2$
$\quad\quad\quad\quad = m^2 - 10m + 25$

(b) $(2k - 5)^2$
$\quad = (2k)^2 - 2(2k)(5) + (5)^2$
$\quad = 4k^2 - 20k + 25$

(c) $(3m - 2p)^2$
$\quad = (3m)^2 - 2(3m)(2p) + (2p)^2$
$\quad = 9m^2 - 12mp + 4p^2$

5. (a) $\quad \sqrt{6w + 6} = w + 1$
$\quad\quad (\sqrt{6w + 6})^2 = (w + 1)^2$
$\quad\quad\quad\quad\quad\quad$ *Square each side*
$\quad\quad 6w + 6 = w^2 + 2w + 1$
$\quad\quad 0 = w^2 - 4w - 5$
$\quad\quad\quad\quad\quad\quad$ *Subtract $6w + 6$*
$\quad\quad 0 = (w - 5)(w + 1)$
$\quad\quad\quad\quad\quad\quad$ *Factor*
$\quad w - 5 = 0 \quad$ or $\quad w + 1 = 0$
$\quad\quad w = 5 \quad$ or $\quad\quad w = -1$

Check both of these potential solutions in the original equation.

$\quad\quad \sqrt{6w + 6} = w + 1$
$\quad \sqrt{6(5) + 6} = 5 + 1 \quad$? *Let $w = 5$*
$\quad\quad \sqrt{30 + 6} = 6 \quad\quad$?
$\quad\quad\quad \sqrt{36} = 6 \quad\quad$?
$\quad\quad\quad\quad 6 = 6 \quad\quad\quad$ *True*

$\quad\quad \sqrt{6w + 6} = w + 1$
$\quad \sqrt{6(-1) + 6} = -1 + 1 \quad$? *Let $w = -1$*
$\quad\quad \sqrt{-6 + 6} = 0 \quad\quad$?
$\quad\quad\quad \sqrt{0} = 0 \quad\quad$?
$\quad\quad\quad\quad 0 = 0 \quad\quad\quad$ *True*

The solutions are 5 and -1.

(b) $\quad 2u - 1 = \sqrt{10u + 9}$
$\quad\quad (2u - 1)^2 = (\sqrt{10u + 9})^2$
$\quad\quad\quad\quad\quad\quad$ *Square each side*
$\quad 4u^2 - 4u + 1 = 10u + 9$
$\quad 4u^2 - 14u - 8 = 0 \quad$ *Subtract $10u + 9$*
$\quad 2(2u^2 - 7u - 4) = 0 \quad$ *Factor*
$\quad 2(2u + 1)(u - 4) = 0 \quad$ *Factor*
$\quad 2u + 1 = 0 \quad$ or $\quad u - 4 = 0$
$\quad\quad 2u = -1$
$\quad\quad u = -\frac{1}{2} \quad$ or $\quad\quad u = 4$

Check

$$2u - 1 = \sqrt{10u + 9}$$
$$2(-\tfrac{1}{2}) - 1 = \sqrt{10(-\tfrac{1}{2}) + 9} \quad ?$$
$$\text{Let } u = -1/2$$
$$-1 - 1 = \sqrt{-5 + 9} \quad ?$$
$$-2 = \sqrt{4} \quad ?$$
$$-2 = 2 \qquad \text{False}$$

$$2u - 1 = \sqrt{10u + 9}$$
$$2(4) - 1 = \sqrt{10(4) + 9}$$
$$\text{Let } u = 4$$
$$8 - 1 = \sqrt{49} \quad ?$$
$$7 = 7 \qquad \text{True}$$

$-1/2$ does not satisfy the original equation. The only solution is 4.

6. (a) $\sqrt{x} - 3 = x - 15$

$$\sqrt{x} = x - 12 \qquad \text{Add 3 to get } \sqrt{x} \text{ alone}$$
$$(\sqrt{x})^2 = (x - 12)^2$$
$$\text{Square each side}$$
$$x = x^2 - 24x + 144$$
$$0 = x^2 - 25x + 144 \qquad \text{Subtract } x$$
$$0 = (x - 16)(x - 9) \qquad \text{Factor}$$
$$x - 16 = 0 \quad \text{or} \quad x - 9 = 0$$
$$x = 16 \quad \text{or} \quad x = 9$$

Check

$$\sqrt{x} - 3 = x - 15$$
$$\sqrt{16} - 3 = 16 - 15 \quad ? \quad \text{Let } x = 16$$
$$4 - 3 = 1$$
$$1 = 1 \qquad \text{True}$$

$$\sqrt{x} - 3 = x - 15$$
$$\sqrt{9} - 3 = 9 - 15 \quad ? \quad \text{Let } x = 9$$
$$3 - 3 = -6$$
$$0 = -6 \qquad \text{False}$$

9 does not satisfy the original equation. The only solution is 16.

(b) $\sqrt{z + 5} + 2 = z + 5$

$$\sqrt{z + 5} = z + 3 \qquad \text{Subtract 2}$$
$$(\sqrt{z + 5})^2 = (z + 3)^2$$
$$\text{Square each side}$$
$$z + 5 = z^2 + 6z + 9$$
$$0 = z^2 + 5z + 4 \qquad \text{Subtract } z + 5$$
$$0 = (z + 4)(z + 1) \qquad \text{Factor}$$
$$z + 4 = 0 \quad \text{or} \quad z + 1 = 0$$
$$z = -4 \quad \text{or} \quad z = -1$$

Check

$$\sqrt{z + 5} + 2 = z + 5$$
$$\sqrt{-4 + 5} + 2 = -4 + 5 \quad \text{Let } z = -4$$
$$\sqrt{1} + 2 = 1$$
$$1 + 2 = 1$$
$$3 = 1 \qquad \text{False}$$

$$\sqrt{z + 5} + 2 = z + 5$$
$$\sqrt{-1 + 5} + 2 = -1 + 5 \quad \text{Let } z = -1$$
$$\sqrt{4} + 2 = 4$$
$$2 + 2 = 4$$
$$4 = 4 \qquad \text{True}$$

-4 does not satisfy the original equation. The only solution is -1.

8.6 Section Exercises

3. $\sqrt{x} = 7$

Use the squaring property of equality to square each side of the equation.

$$(\sqrt{x})^2 = (7)^2$$
$$x = 49$$

Now check this proposed solution in the original equation.

Check

$$\sqrt{x} = 7$$
$$\sqrt{49} = 7 \quad ? \quad \text{Let } x = 49$$
$$7 = 7 \quad \text{True}$$

This number checks in the original equation, so the solution is 49.

7.
$$\sqrt{r - 4} = 9$$
$$(\sqrt{r - 4})^2 = (9)^2 \quad \text{Square each side}$$
$$r - 4 = 81$$
$$r = 85$$

Check

$$\sqrt{r - 4} = 9$$
$$\sqrt{85 - 4} = 9 \quad ? \quad \text{Let } r = 85$$
$$\sqrt{81} = 9 \quad ?$$
$$9 = 9 \quad \text{True}$$

This number checks in the original equation, so the solution is 85.

11.
$$\sqrt{2t + 3} = 0$$
$$(\sqrt{2t + 3})^2 = (0)^2$$
$$2t + 3 = 0$$
$$2t = -3$$
$$t = -\frac{3}{2}$$

Check

$$\sqrt{2t + 3} = 0 \quad ?$$
$$\sqrt{2\left(-\frac{3}{2}\right) + 3} = 0 \quad ? \quad \text{Let } t = -3/2$$
$$\sqrt{-3 + 3} = 0 \quad ?$$
$$\sqrt{0} = 0 \quad ?$$
$$0 = 0 \quad \text{True}$$

This number checks in the original equation, so the solution is -3/2.

15. $\sqrt{m} - 4 = 7$

Add 4 to both sides of the equation before squaring.

$$\sqrt{m} = 11$$
$$(\sqrt{m})^2 = (11)^2$$
$$m = 121$$

Check

$$\sqrt{m} = 11$$
$$\sqrt{121} = 11 \quad ? \quad \text{Let } m = 121$$
$$11 = 11 \quad \text{True}$$

This number checks in the original equation, so the solution is 121.

19.
$$5\sqrt{x} = \sqrt{10x + 15}$$
$$(5\sqrt{x})^2 = (\sqrt{10x + 15})^2$$
$$25x = 10x + 15$$
$$15x = 15$$
$$x = 1$$

Check

$$5\sqrt{x} = \sqrt{10x + 15}$$
$$5\sqrt{1} = \sqrt{10 \cdot 1 + 15} \quad ? \quad \text{Let } x = 1$$
$$5 \cdot 1 = \sqrt{25} \quad ?$$
$$5 = 5 \quad \text{True}$$

This number checks in the original equation, so the solution is 1.

23.
$$k = \sqrt{k^2 - 5k - 15}$$
$$(k)^2 = (\sqrt{k^2 - 5k - 15})^2$$
$$k^2 = k^2 - 5k - 15$$
$$0 = -5k - 15$$
$$5k = -15$$
$$k = -3$$

Check

$$k = \sqrt{k^2 - 5k - 15}$$
$$-3 = \sqrt{(-3)^2 - 5(-3) - 15} \quad ?$$
$$\text{Let } k = -3$$
$$-3 = \sqrt{9 + 15 - 15} \quad ?$$
$$-3 = \sqrt{9} \quad ?$$
$$-3 = 3 \quad \text{False}$$

This number does not check in the original equation, because it makes $k = \sqrt{k^2 - 5k - 15}$ become $-3 = 3$, which is false. Therefore, the original equation has no solution.

27. $\sqrt{2x + 1} = x - 7$

The first step in solving this equation is to square both sides of the equation. The right side is a binomial which must be squared as a quantity, not term by term. The correct square of the right side is

$$(x - 7)^2 = x^2 - 2(x)(7) + (7)^2$$
$$= x^2 - 14x + 49.$$

31. $\sqrt{3k + 10} + 5 = 2k$

$\sqrt{3k + 10} = 2k - 5$ *Subtract 5*

$(\sqrt{3k + 10})^2 = (2k - 5)^2$
                Square each side

$3k + 10 = 4k^2 - 20k + 25$
                Square binomial on the right

$0 = 4k^2 - 23k + 15$
                Standard form

$0 = (k - 5)(4k - 3)$
                Factor

$k - 5 = 0$ or $4k - 3 = 0$
                Set each factor equal to 0

$4k = 3$

$k = 5$ or $k = \dfrac{3}{4}$ *Solve*

Check

$$\sqrt{3k + 10} + 5 = 2k$$
$$\sqrt{3 \cdot 5 + 10} + 5 = 2 \cdot 5 \quad ? \quad \text{Let } k = 5$$
$$\sqrt{25} + 5 = 10 \quad ?$$
$$5 + 5 = 10 \quad ?$$
$$10 = 10 \quad \text{True}$$

$$\sqrt{3k + 10} + 5 = 2k$$
$$\sqrt{3\left(\tfrac{3}{4}\right) + 10} + 5 = 2\left(\tfrac{3}{4}\right) \quad ? \quad \text{Let } k = 3/4$$
$$\sqrt{\tfrac{9}{4} + 10} + 5 = \tfrac{3}{2} \quad ?$$
$$\sqrt{\tfrac{49}{4}} + 5 = \tfrac{3}{2} \quad ?$$
$$\tfrac{7}{2} + 5 = \tfrac{3}{2} \quad ?$$
$$\tfrac{17}{2} = \tfrac{3}{2} \quad \text{False}$$

Of these two numbers, 5 checks in the original equation but 3/4 does not, so the only solution is 5.

35. $\sqrt{6t + 7} + 3 = t + 5$

Begin by subtracting 3 from both sides.

$$\sqrt{6t + 7} = t + 2$$
$$(\sqrt{6t + 7})^2 = (t + 2)^2$$
$$6t + 7 = t^2 + 4t + 4$$
$$0 = t^2 - 2t - 3$$
$$0 = (t - 3)(t + 1)$$
$$t - 3 = 0 \text{ or } t + 1 = 0$$
$$t = 3 \text{ or } t = -1$$

Check

$$\sqrt{6t + 7} + 3 = t + 5$$
$$\sqrt{6 \cdot 3 + 7} + 3 = 3 + 5 \quad ? \quad \text{Let } t = 3$$
$$\sqrt{25} + 3 = 8 \quad ?$$
$$5 + 3 = 8 \quad ?$$
$$8 = 8 \quad \text{True}$$

$$\sqrt{6t+7} + 3 = t + 5$$
$$\sqrt{6(-1)+7} + 3 = -1 + 5 \quad ?$$
$$\text{Let } t = -1$$
$$\sqrt{1} + 3 = 4 \quad ?$$
$$1 + 3 = 4 \quad ?$$
$$4 = 4 \quad \text{True}$$

These numbers both check in the original equation, so the solutions are −1 and 3.

39. $\sqrt{x} + 6 = 2x$
$$\sqrt{x} = 2x - 6$$
$$(\sqrt{x})^2 = (2x - 6)^2$$
$$x = 4x^2 - 24x + 36$$
$$0 = 4x^2 - 25x + 36$$
$$0 = (4x - 9)(x - 4)$$
$$4x - 9 = 0 \quad \text{or} \quad x - 4 = 0$$
$$4x = 9$$
$$x = \frac{9}{4} \quad \text{or} \quad x = 4$$

Check
$$\sqrt{x} + 6 = 2x$$
$$\sqrt{\frac{9}{4}} + 6 = 2 \cdot \frac{9}{4} \quad ? \quad \text{Let } x = 9/4$$
$$\frac{3}{2} + 6 = \frac{9}{2} \quad ?$$
$$\frac{15}{2} = \frac{9}{2} \quad \text{False}$$

$$\sqrt{x} + 6 = 2x$$
$$\sqrt{4} + 6 = 2 \cdot 4 \quad ? \quad \text{Let } x = 4$$
$$2 + 6 = 8 \quad ?$$
$$8 = 8 \quad \text{True}$$

Of these two numbers, 4 checks in the original equation by 9/4 does not, so the only solution is 4.

43. The square roots of 121 are 11 and −11 because $11^2 = 121$ and $(-11)^2 = 121$.

Chapter 8 Review Exercises

1. The square roots of 49 are 7 and −7 because $7^2 = 49$ and $(-7)^2 = 49$.

2. The square roots of 81 are 9 and −9 because $9^2 = 81$ and $(-81)^2 = 81$.

3. The square roots of 196 are 14 and −14 because $14^2 = 196$ and $(-14)^2 = 196$.

4. The square roots of 121 are 11 and −11 because $11^2 = 121$ and $(-11)^2 = 121$.

5. The square roots of 225 are 15 and −15 because $15^2 = 225$ and $(-15)^2 = 225$.

6. The square roots of 729 are 27 and −27 because $27^2 = 729$ and $(-27)^2 = 729$.

7. $\sqrt{16} = 4$ because $4^2 = 16$.

8. $-\sqrt{36} = -6$ because $\sqrt{36} = 6$.

9. $\sqrt[3]{1000} = 10$ because $10^3 = 1000$.

10. $\sqrt[4]{81} = 3$ because 3 is positive and $3^4 = 81$.

11. $\sqrt{-8100}$ is not a real number.

12. $-\sqrt{4225} = -65$ because $\sqrt{4225} = 65$.

13. $\sqrt{\dfrac{49}{36}} = \dfrac{\sqrt{49}}{\sqrt{36}} = \dfrac{7}{6}$

14. $\sqrt{\dfrac{100}{81}} = \dfrac{\sqrt{100}}{\sqrt{81}} = \dfrac{10}{9}$

15. If $\sqrt{a}$ is not a real number, then a must be negative number.

16. Use the Pythagorean formula with $a = 15$, $b = x$, and $c = 17$.

    $$c^2 = a^2 + b^2$$
    $$17^2 = 15^2 + x^2$$
    $$289 = 225 + x^2$$
    $$64 = x^2$$
    $$x = \sqrt{64} = 8$$

17. $\sqrt{23}$

    This number is irrational because 23 is not a perfect square.

    $\sqrt{23} \approx 4.796$

18. $\sqrt{169}$

    This number is rational because 169 is a perfect square.

    $\sqrt{169} = 13$

19. $-\sqrt{25}$

    This number is rational because 25 is a perfect square.

    $-\sqrt{25} = -5$

20. $\sqrt{-4}$

    This is not a real number.

21. $\sqrt{2} \cdot \sqrt{7} = \sqrt{2 \cdot 7} = \sqrt{14}$

22. $\sqrt{12} \cdot \sqrt{3} = \sqrt{12 \cdot 3} = \sqrt{36} = 6$

23. $\sqrt{5} \cdot \sqrt{15} = \sqrt{5 \cdot 15} = \sqrt{75} = \sqrt{25 \cdot 3}$
    $= \sqrt{25} \cdot \sqrt{3} = 5\sqrt{3}$

24. $\sqrt{12} \cdot \sqrt{12} = \sqrt{12 \cdot 12} = \sqrt{144} = 12$

25. $-\sqrt{27} = -\sqrt{9 \cdot 3} = -\sqrt{9} \cdot \sqrt{3} = -3\sqrt{3}$

26. $\sqrt{48} = \sqrt{16 \cdot 3} = \sqrt{16} \cdot \sqrt{3} = 4\sqrt{3}$

27. $\sqrt{160} = \sqrt{16 \cdot 10} = \sqrt{16} \cdot \sqrt{10} = 4\sqrt{10}$

28. $\sqrt[3]{-125} = -5$ because $(-5)^3 = -125$.

29. $\sqrt[3]{1728} = \sqrt[3]{64 \cdot 27} = \sqrt[3]{64} \cdot \sqrt[3]{27}$
    $= 4 \cdot 3 = 12$

30. $\sqrt{12} \cdot \sqrt{27} = \sqrt{12 \cdot 27} = \sqrt{324}$
    $= \sqrt{4 \cdot 81} = \sqrt{4} \cdot \sqrt{81}$
    $= 2 \cdot 9 = 18$

31. $\sqrt{32} \cdot \sqrt{48} = \sqrt{32 \cdot 48} = \sqrt{16 \cdot 2 \cdot 16 \cdot 3}$
    $= \sqrt{16 \cdot 16 \cdot 2 \cdot 3}$
    $= \sqrt{16 \cdot 16} \cdot \sqrt{2 \cdot 3} = 16\sqrt{6}$

32. $\sqrt{50} \cdot \sqrt{125} = \sqrt{50 \cdot 125} = \sqrt{25 \cdot 2 \cdot 25 \cdot 5}$
    $= \sqrt{25 \cdot 25 \cdot 2 \cdot 5}$
    $= \sqrt{25 \cdot 25} \cdot \sqrt{2 \cdot 5}$
    $= 25\sqrt{10}$

33. $\sqrt{\dfrac{9}{4}} = \dfrac{\sqrt{9}}{\sqrt{4}} = \dfrac{3}{2}$

34. $-\sqrt{\dfrac{121}{400}} = -\dfrac{\sqrt{121}}{\sqrt{400}} = -\dfrac{11}{20}$

35. $\sqrt{\dfrac{3}{49}} = \dfrac{\sqrt{3}}{\sqrt{49}} = \dfrac{\sqrt{3}}{7}$

36. $\sqrt{\dfrac{7}{169}} = \dfrac{\sqrt{7}}{\sqrt{169}} = \dfrac{\sqrt{7}}{13}$

37. $\sqrt{\dfrac{1}{6}} \cdot \sqrt{\dfrac{5}{6}} = \sqrt{\dfrac{1}{6} \cdot \dfrac{5}{6}}$
    $= \sqrt{\dfrac{5}{36}}$
    $= \dfrac{\sqrt{5}}{\sqrt{36}} = \dfrac{\sqrt{5}}{6}$

38. $\sqrt{\dfrac{2}{5}} \cdot \sqrt{\dfrac{2}{45}} = \sqrt{\dfrac{2}{5} \cdot \dfrac{2}{45}}$
    $= \sqrt{\dfrac{4}{225}}$
    $= \dfrac{\sqrt{4}}{\sqrt{225}} = \dfrac{2}{15}$

39. $\dfrac{3\sqrt{10}}{\sqrt{5}} = \dfrac{3\sqrt{10} \cdot \sqrt{5}}{\sqrt{5} \cdot \sqrt{5}}$
    $= \dfrac{3\sqrt{50}}{\sqrt{25}}$
    $= \dfrac{3\sqrt{25 \cdot 2}}{5}$
    $= \dfrac{3\sqrt{25} \cdot \sqrt{2}}{5}$
    $= \dfrac{3 \cdot 5 \cdot \sqrt{2}}{5} = 3\sqrt{2}$

40. $\dfrac{24\sqrt{12}}{6\sqrt{3}} = \dfrac{24\sqrt{12} \cdot \sqrt{3}}{6\sqrt{3} \cdot \sqrt{3}}$
    $= \dfrac{24\sqrt{36}}{6 \cdot 3}$
    $= \dfrac{24 \cdot 6}{6 \cdot 3} = 8$

41. $\dfrac{8\sqrt{150}}{4\sqrt{75}} = \dfrac{2\sqrt{150}}{\sqrt{75}}$
    $= \dfrac{2\sqrt{25 \cdot 6}}{\sqrt{25 \cdot 3}}$
    $= \dfrac{2 \cdot 5\sqrt{6}}{5\sqrt{3}}$
    $= \dfrac{2\sqrt{6} \cdot \sqrt{3}}{\sqrt{3} \cdot \sqrt{3}}$
    $= \dfrac{2\sqrt{18}}{3}$
    $= \dfrac{2\sqrt{9 \cdot 2}}{3}$
    $= \dfrac{2 \cdot 3\sqrt{2}}{3} = 2\sqrt{2}$

42. $\sqrt{p} \cdot \sqrt{p} = p$

43. $\sqrt{k} \cdot \sqrt{m} = \sqrt{km}$

44. $\sqrt{r^{18}} = r^9$ because $(r^9)^2 = r^{18}$.

45. $\sqrt{x^{10}y^{16}} = x^5 y^8$ because
    $(x^5 y^8)^2 = x^{10} y^{16}$.

46. $\sqrt{x^9} = \sqrt{x^8 \cdot x} = \sqrt{x^8} \cdot \sqrt{x} = x^4\sqrt{x}$

47. $\sqrt{\dfrac{36}{p^2}}$ $(p \neq 0) = \dfrac{\sqrt{36}}{\sqrt{p^2}} = \dfrac{6}{p}$

48. $\sqrt{a^{15}b^{21}} = \sqrt{a^{14}b^{20} \cdot ab}$
    $= \sqrt{a^{14}b^{20}} \cdot \sqrt{ab}$
    $= a^7 b^{10} \sqrt{ab}$

49. $\sqrt{121 x^6 y^{10}} = 11 x^3 y^5$ because
    $(11 x^3 y^5)^2 = 121 x^6 y^{10}$.

50. Using a calculator,

$$\sqrt{.5} \approx .7071067812$$

$$\frac{\sqrt{2}}{2} \approx \frac{1.414213562}{2}$$

$$= .7071067812$$

It looks like these two expressions represent the same number. In fact, they do represent the same number because

$$\sqrt{.5} = \sqrt{\frac{5}{10}} = \sqrt{\frac{1}{2}} = \frac{\sqrt{1}}{\sqrt{2}}$$

$$= \frac{1 \cdot \sqrt{2}}{\sqrt{2} \cdot \sqrt{2}} = \frac{\sqrt{2}}{2}.$$

51. $\sqrt{11} + \sqrt{11} = 1\sqrt{11} + 1\sqrt{11} = 2\sqrt{11}$

52. $3\sqrt{2} + 6\sqrt{2} = (3 + 6)\sqrt{2} = 9\sqrt{2}$

53. $3\sqrt{75} + 2\sqrt{27}$
    $= 3(\sqrt{25} \cdot \sqrt{3}) + 2(\sqrt{9} \cdot \sqrt{3})$
    $= 3(5\sqrt{3}) + 2(3\sqrt{3})$
    $= 15\sqrt{3} + 6\sqrt{3} = 21\sqrt{3}$

54. $4\sqrt{12} + \sqrt{48}$
    $= 4(\sqrt{4} \cdot \sqrt{3}) + \sqrt{16} \cdot \sqrt{3}$
    $= 4(2\sqrt{3}) + 4\sqrt{3}$
    $= 8\sqrt{3} + 4\sqrt{3} = 12\sqrt{3}$

55. $4\sqrt{24} - 3\sqrt{54} + \sqrt{6}$
    $= 4(\sqrt{4} \cdot \sqrt{6}) - 3(\sqrt{9} \cdot \sqrt{6}) + \sqrt{6}$
    $= 4(2\sqrt{6}) - 3(3\sqrt{6}) + \sqrt{6}$
    $= 8\sqrt{6} - 9\sqrt{6} + 1\sqrt{6}$
    $= 0\sqrt{6} = 0$

56. $2\sqrt{7} - 4\sqrt{28} + 3\sqrt{63}$
    $= 2\sqrt{7} - 4(\sqrt{4} \cdot \sqrt{7}) + 3(\sqrt{9} \cdot \sqrt{7})$
    $= 2\sqrt{7} - 4(2\sqrt{7}) + 3(3\sqrt{7})$
    $= 2\sqrt{7} - 8\sqrt{7} + 9\sqrt{7} = 3\sqrt{7}$

57. $\frac{2}{5}\sqrt{75} + \frac{3}{4}\sqrt{160}$
    $= \frac{2}{5}(\sqrt{25} \cdot \sqrt{3}) + \frac{3}{4}(\sqrt{16} \cdot \sqrt{10})$
    $= \frac{2}{5}(5\sqrt{3}) + \frac{3}{4}(4\sqrt{10})$
    $= 2\sqrt{3} + 3\sqrt{10}$

58. $\frac{1}{3}\sqrt{18} + \frac{1}{4}\sqrt{32}$
    $= \frac{1}{3}(\sqrt{9} \cdot \sqrt{2}) + \frac{1}{4}(\sqrt{16} \cdot \sqrt{2})$
    $= \frac{1}{3}(3\sqrt{2}) + \frac{1}{4}(4\sqrt{2})$
    $= 1\sqrt{2} + 1\sqrt{2} = 2\sqrt{2}$

59. $\sqrt{15} \cdot \sqrt{2} + 5\sqrt{30} = \sqrt{30} + 5\sqrt{30}$
    $= 1\sqrt{30} + 5\sqrt{30}$
    $= 6\sqrt{30}$

60. $\sqrt{4x} + \sqrt{36x} - \sqrt{9x}$
    $= \sqrt{4}\sqrt{x} + \sqrt{36}\sqrt{x} - \sqrt{9}\sqrt{x}$
    $= 2\sqrt{x} + 6\sqrt{x} - 3\sqrt{x} = 5\sqrt{x}$

61. $\sqrt{16p} + 3\sqrt{p} - \sqrt{49p}$
    $= \sqrt{16}\sqrt{p} + 3\sqrt{p} - \sqrt{49}\sqrt{p}$
    $= 4\sqrt{p} + 3\sqrt{p} - 7\sqrt{p}$
    $= 0\sqrt{p} = 0$

62. $\sqrt{20m^2} - m\sqrt{45}$
    $= \sqrt{4m^2 \cdot 5} - m(\sqrt{9} \cdot \sqrt{5})$
    $= \sqrt{4m^2} \sqrt{5} - m(3\sqrt{5})$
    $= 2m\sqrt{5} - 3m\sqrt{5}$
    $= -1m\sqrt{5} = -m\sqrt{5}$

63. $3k\sqrt{8k^2n} + 5k^2\sqrt{2n}$
    $= 3k(\sqrt{4k^2} \cdot \sqrt{2n}) + 5k^2\sqrt{2n}$
    $= 3k(2k\sqrt{2n}) + 5k^2\sqrt{2n}$
    $= 6k^2\sqrt{2n} + 5k^2\sqrt{2n}$
    $= (6k^2 + 5k^2)\sqrt{2n}$
    $= 11k^2\sqrt{2n}$

64. $\dfrac{10}{\sqrt{3}} = \dfrac{10 \cdot \sqrt{3}}{\sqrt{3} \cdot \sqrt{3}} = \dfrac{10\sqrt{3}}{3}$

65. $\dfrac{15}{\sqrt{2}} = \dfrac{15 \cdot \sqrt{2}}{\sqrt{2} \cdot \sqrt{2}} = \dfrac{15\sqrt{2}}{2}$

66. $\dfrac{8\sqrt{2}}{\sqrt{5}} = \dfrac{8\sqrt{2} \cdot \sqrt{5}}{\sqrt{5} \cdot \sqrt{5}} = \dfrac{8\sqrt{10}}{5}$

67. $\dfrac{5}{\sqrt{5}} = \dfrac{5 \cdot \sqrt{5}}{\sqrt{5} \cdot \sqrt{5}} = \dfrac{5\sqrt{5}}{5} = \sqrt{5}$

68. $\dfrac{12}{\sqrt{24}} = \dfrac{12}{\sqrt{4 \cdot 6}} = \dfrac{12}{2\sqrt{6}}$
    $= \dfrac{12 \cdot \sqrt{6}}{2\sqrt{6} \cdot \sqrt{6}} = \dfrac{12\sqrt{6}}{2 \cdot 6}$
    $= \dfrac{12\sqrt{6}}{12} = \sqrt{6}$

69. $\dfrac{\sqrt{2}}{\sqrt{15}} = \dfrac{\sqrt{2} \cdot \sqrt{15}}{\sqrt{15} \cdot \sqrt{15}} = \dfrac{\sqrt{30}}{15}$

70. $\sqrt{\dfrac{2}{5}} = \dfrac{\sqrt{2}}{\sqrt{5}} = \dfrac{\sqrt{2} \cdot \sqrt{5}}{\sqrt{5} \cdot \sqrt{5}} = \dfrac{\sqrt{10}}{5}$

71. $\sqrt{\dfrac{5}{14}} \cdot \sqrt{28} = \sqrt{\dfrac{5}{14} \cdot 28}$
    $= \sqrt{\dfrac{140}{14}} = \sqrt{10}$

72. $\sqrt{\dfrac{2}{7}} \cdot \sqrt{\dfrac{1}{3}} = \sqrt{\dfrac{2}{7} \cdot \dfrac{1}{3}}$
    $= \sqrt{\dfrac{2}{21}} = \dfrac{\sqrt{2}}{\sqrt{21}}$
    $= \dfrac{\sqrt{2} \cdot \sqrt{21}}{\sqrt{21} \cdot \sqrt{21}} = \dfrac{\sqrt{42}}{21}$

73. $\sqrt{\dfrac{r^2}{16x}}$ $(x \neq 0) = \dfrac{\sqrt{r^2}}{\sqrt{16x}}$
    $= \dfrac{r \cdot \sqrt{x}}{\sqrt{16x} \cdot \sqrt{x}}$
    $= \dfrac{r\sqrt{x}}{\sqrt{16x^2}} = \dfrac{r\sqrt{x}}{4x}$

74. $\sqrt[3]{\dfrac{1}{3}} = \dfrac{\sqrt[3]{1}}{\sqrt[3]{3}} = \dfrac{1 \cdot \sqrt[3]{3^2}}{\sqrt[3]{3} \cdot \sqrt[3]{3^2}}$
    $= \dfrac{\sqrt[3]{3^2}}{\sqrt[3]{3^3}} = \dfrac{\sqrt[3]{9}}{3}$

75. $\sqrt[3]{\dfrac{2}{7}} = \dfrac{\sqrt[3]{2}}{\sqrt[3]{7}} = \dfrac{\sqrt[3]{2} \cdot \sqrt[3]{7^2}}{\sqrt[3]{7} \cdot \sqrt[3]{7^2}}$
    $= \dfrac{\sqrt[3]{2 \cdot 7^2}}{\sqrt[3]{7^3}} = \dfrac{\sqrt[3]{98}}{7}$

76. $\dfrac{\sqrt{6}}{4}$ and $\sqrt{\dfrac{48}{128}}$

To show that these expressions represent the same number, work with the second expression in the following way: First simplify the fraction inside the radical, then use the quotient rule to split the expression into two radicals, rationalize the denominator, and simplify the result.

$$\sqrt{\frac{48}{128}} = \sqrt{\frac{3}{8}} = \frac{\sqrt{3}}{\sqrt{8}} = \frac{\sqrt{3} \cdot \sqrt{2}}{\sqrt{8} \cdot \sqrt{2}}$$
$$= \frac{\sqrt{6}}{\sqrt{16}} = \frac{\sqrt{6}}{4}$$

77. $-\sqrt{3}(\sqrt{5} + \sqrt{27})$
$$= -\sqrt{3}(\sqrt{5}) + (-\sqrt{3})(\sqrt{27})$$
$$= -\sqrt{3 \cdot 5} - \sqrt{3 \cdot 27}$$
$$= -\sqrt{15} - \sqrt{81}$$
$$= -\sqrt{15} - 9$$

78. $3\sqrt{2}(\sqrt{3} + 2\sqrt{2})$
$$= 3\sqrt{2}(\sqrt{3}) + 3\sqrt{2}(2\sqrt{2})$$
$$= 3\sqrt{6} + 6 \cdot 2$$
$$= 3\sqrt{6} + 12$$

79. $(2\sqrt{3} - 4)(5\sqrt{3} + 2)$
$$= 2\sqrt{3}(5\sqrt{3}) + (2\sqrt{3})(2) - 4(5\sqrt{3})$$
$$\quad - 4(2) \quad \text{FOIL}$$
$$= 10 \cdot 3 + 4\sqrt{3} - 20\sqrt{3} - 8$$
$$= 30 - 16\sqrt{3} - 8$$
$$= 22 - 16\sqrt{3}$$

80. $(5\sqrt{7} + 2)^2$
$$= (5\sqrt{7})^2 + 2(5\sqrt{7})(2) + 2^2$$
$$\quad \text{Square of a binomial}$$
$$= 25 \cdot 7 + 20\sqrt{7} + 4$$
$$= 175 + 20\sqrt{7} + 4$$
$$= 179 + 20\sqrt{7}$$

81. $(\sqrt{5} - \sqrt{7})(\sqrt{5} + \sqrt{7})$
$$= (\sqrt{5})^2 - (\sqrt{7})^2$$
$$= 5 - 7 = -2$$

82. $(2\sqrt{3} + 5)(2\sqrt{3} - 5)$
$$= (2\sqrt{3})^2 - (5)^2$$
$$= 4 \cdot 3 - 25$$
$$= 12 - 25 = -13$$

83. $(\sqrt{7} + 2\sqrt{6})(\sqrt{12} - \sqrt{2})$
$$= \sqrt{7}(\sqrt{12}) - \sqrt{7}(\sqrt{2}) + 2\sqrt{6}(\sqrt{12})$$
$$\quad - \sqrt{2}(2\sqrt{6})$$
$$= \sqrt{84} - \sqrt{14} + 2\sqrt{72} - 2\sqrt{12}$$
$$= \sqrt{4 \cdot 21} - \sqrt{14} + 2\sqrt{36 \cdot 2} - 2\sqrt{4 \cdot 3}$$
$$= 2\sqrt{21} - \sqrt{14} + 2 \cdot 6\sqrt{2} - 2 \cdot 2\sqrt{3}$$
$$= 2\sqrt{21} - \sqrt{14} + 12\sqrt{2} - 4\sqrt{3}$$

84. $\dfrac{1}{2 + \sqrt{5}}$
$$= \frac{1(2 - \sqrt{5})}{(2 + \sqrt{5})(2 - \sqrt{5})} \quad \text{Multiply by the conjugate}$$
$$= \frac{2 - \sqrt{5}}{(2)^2 - (\sqrt{5})^2}$$
$$= \frac{2 - \sqrt{5}}{4 - 5}$$
$$= \frac{2 - \sqrt{5}}{-1} = -2 + \sqrt{5}$$

85. $\dfrac{2}{\sqrt{2} - 3} = \dfrac{2(\sqrt{2} + 3)}{(\sqrt{2} - 3)(\sqrt{2} + 3)}$
$$= \frac{2\sqrt{2} + 6}{(\sqrt{2})^2 - (3)^2}$$
$$= \frac{2\sqrt{2} + 6}{2 - 9}$$
$$= \frac{2\sqrt{2} + 6}{-7}$$
$$= \frac{(2\sqrt{2} + 6)(-1)}{-7(-1)}$$
$$= \frac{-2\sqrt{2} - 6}{7}$$

86. $\dfrac{\sqrt{8}}{\sqrt{2} + 6}$
$$= \frac{\sqrt{8}(\sqrt{2} - 6)}{(\sqrt{2} + 6)(\sqrt{2} - 6)} \quad \text{Multiply by the conjugate}$$
$$= \frac{\sqrt{16} - 6\sqrt{8}}{2 - 36}$$
$$= \frac{4 - 6 \cdot \sqrt{4 \cdot 2}}{-34}$$

$$= \frac{4 - 6 \cdot 2\sqrt{2}}{-34}$$

$$= \frac{4 - 12\sqrt{2}}{-34}$$

$$= \frac{-2(-2 + 6\sqrt{2})}{-2(17)} \quad \text{Factor numerator and denominator}$$

$$= \frac{-2 + 6\sqrt{2}}{17} \quad \text{Lowest terms}$$

87. $\dfrac{\sqrt{3}}{1 + \sqrt{3}} = \dfrac{\sqrt{3}(1 - \sqrt{3})}{(1 + \sqrt{3})(1 - \sqrt{3})}$

$$= \frac{\sqrt{3} - \sqrt{9}}{1 - 3}$$

$$= \frac{\sqrt{3} - 3}{-2}$$

$$= \frac{(\sqrt{3} - 3)(-1)}{-2(-1)}$$

$$= \frac{-\sqrt{3} + 3}{2}$$

88. $\dfrac{\sqrt{5} - 1}{\sqrt{2} + 3} = \dfrac{(\sqrt{5} - 1)(\sqrt{2} - 3)}{(\sqrt{2} + 3)(\sqrt{2} - 3)}$

$$= \frac{\sqrt{10} - 3\sqrt{5} - \sqrt{2} + 3}{2 - 9}$$

$$= \frac{\sqrt{10} - 3\sqrt{5} - \sqrt{2} + 3}{-7}$$

$$= \frac{-\sqrt{10} + 3\sqrt{5} + \sqrt{2} - 3}{7}$$

89. $\dfrac{2 + \sqrt{6}}{\sqrt{3} - 1} = \dfrac{(2 + \sqrt{6})(\sqrt{3} + 1)}{(\sqrt{3} - 1)(\sqrt{3} + 1)}$

$$= \frac{2\sqrt{3} + 2 + \sqrt{18} + \sqrt{6}}{3 - 1}$$

$$= \frac{2\sqrt{3} + 2 + \sqrt{9 \cdot 2} + \sqrt{6}}{2}$$

$$= \frac{2\sqrt{3} + 2 + 3\sqrt{2} + \sqrt{6}}{2}$$

90. $\dfrac{15 + 10\sqrt{6}}{15} = \dfrac{5(3 + 2\sqrt{6})}{5(3)} \quad \text{Factor}$

$$= \frac{3 + 2\sqrt{6}}{3} \quad \text{Lowest terms}$$

91. $\dfrac{3 + 9\sqrt{7}}{12} = \dfrac{3(1 + 3\sqrt{7})}{3(4)}$

$$= \frac{1 + 3\sqrt{7}}{4}$$

92. $\dfrac{6 + \sqrt{192}}{2} = \dfrac{6 + \sqrt{64 \cdot 3}}{2}$

$$= \frac{6 + 8\sqrt{3}}{2}$$

$$= \frac{2(3 + 4\sqrt{3})}{2}$$

$$= 3 + 4\sqrt{3}$$

93. $\sqrt{m} + 5 = 0$

Subtract 5 from both sides before squaring.

$$\sqrt{m} = -5$$
$$(\sqrt{m})^2 = (-5)^2$$
$$m = 25$$

This number does not check in the original equation since $\sqrt{25} + 5$ is 10, not 0, so the original equation has no solution.

94. $\sqrt{p} + 4 = 0$
$$\sqrt{p} = -4$$
$$(\sqrt{p})^2 = (-4)^2$$
$$p = 16$$

This number does not check in the original equation, so there is no solution.

95. $\sqrt{k+1} = 7$
    $(\sqrt{k+1})^2 = (7)^2$
    $k + 1 = 49$
    $k = 48$

    This number checks in the original equation, so the solution is 48.

96. $\sqrt{5m+4} = 3\sqrt{m}$
    $(\sqrt{5m+4})^2 = (3\sqrt{m})^2$
    $5m + 4 = 9m$
    $4 = 4m$
    $1 = m$

    This number checks in the original equation, so the solution is 1.

97. $\sqrt{2p+3} = \sqrt{5p-3}$
    $(\sqrt{2p+3})^2 = (\sqrt{5p-3})^2$
    $2p + 3 = 5p - 3$
    $6 = 3p$
    $2 = p$

    This number makes the original equation look like $\sqrt{7} = \sqrt{7}$, which is true, so the solution is 2.

98. $\sqrt{4y+1} = y - 1$
    $(\sqrt{4y+1})^2 = (y-1)^2$
    $4y + 1 = y^2 - 2y + 1$
    $0 = y^2 - 6y$
    $0 = y(y - 6)$
    $y = 0$ or $y - 6 = 0$
    $y = 0$ or $y = 6$

    Of these two numbers, 6 checks in the original equation but 0 does not, so the only solution is 6.

99. $\sqrt{-2k-4} = k + 2$
    $(\sqrt{-2k-4})^2 = (k+2)^2$
    $-2k - 4 = k^2 + 4k + 4$
    $0 = k^2 + 6k + 8$
    $0 = (k+2)(k+4)$
    $k + 2 = 0$ or $k + 4 = 0$
    $k = -2$ or $k = -4$

    Of these two numbers, -2 checks in the original equation but -4 does not, so the only solution is -2.

100. $\sqrt{2-x} + 3 = x + 7$
     $\sqrt{2-x} = x + 4$
     $(\sqrt{2-x})^2 = (x+4)^2$
     $2 - x = x^2 + 8x + 16$
     $0 = x^2 + 9x + 14$
     $0 = (x+2)(x+7)$
     $x + 2 = 0$ or $x + 7 = 0$
     $x = -2$ or $x = -7$

     Of these two numbers, -2 checks in the original equation but -7 does not, so the only solution is -2.

101. $\sqrt{x} - x + 2 = 0$
     $\sqrt{x} = x - 2$
     $(\sqrt{x})^2 = (x-2)^2$
     $x = x^2 - 4x + 4$
     $0 = x^2 - 5x + 4$
     $0 = (x-4)(x-1)$
     $x - 4 = 0$ or $x - 1 = 0$
     $x = 4$ or $x = 1$

     In the original equation, 4 checks but 1 does not. The only solution is 4.

102. $\sqrt{2-x} + x = 0$
$\sqrt{2-x} = -x$
$(\sqrt{2-x})^2 = (-x)^2$
$2 - x = x^2$
$0 = x^2 + x - 2$
$0 = (x+2)(x-1)$
$x + 2 = 0$ or $x - 1 = 0$
$x = -2$ or $x = 1$

In the original equation, -2 checks but 1 does not. The only solution is -2.

103. $\sqrt{4y - 2} = \sqrt{3y + 1}$
$(\sqrt{4y-2})^2 = (\sqrt{3y+1})^2$
$4y - 2 = 3y + 1$
$y = 3$

This number checks in the original equation, so the solution is 3.

104. $\sqrt{2x + 3} = x + 2$
$(\sqrt{2x+3})^2 = (x+2)^2$
$2x + 3 = x^2 + 4x + 4$
$0 = x^2 + 2x + 1$
$0 = (x+1)(x+1)$
$x + 1 = 0$
$x = -1$

This number checks in the original equation, so the solution is -1.

105. $\sqrt{3} \cdot \sqrt{27} = \sqrt{81} = 9$

106. $2\sqrt{27} + 3\sqrt{75} - \sqrt{300}$
$= 2\sqrt{9 \cdot 3} + 3\sqrt{25 \cdot 3} - \sqrt{100 \cdot 3}$
$= 2 \cdot 3\sqrt{3} + 3 \cdot 5\sqrt{3} - 10\sqrt{3}$
$= 6\sqrt{3} + 15\sqrt{3} - 10\sqrt{3}$
$= 11\sqrt{3}$

107. $\sqrt{\dfrac{121}{t^2}}$ $(t \neq 0) = \dfrac{\sqrt{121}}{\sqrt{t^2}} = \dfrac{11}{t}$

108. $\dfrac{1}{5 + \sqrt{2}} = \dfrac{1(5 - \sqrt{2})}{(5 + \sqrt{2})(5 - \sqrt{2})}$
$= \dfrac{5 - \sqrt{2}}{(5)^2 - (\sqrt{2})^2}$
$= \dfrac{5 - \sqrt{2}}{25 - 2}$
$= \dfrac{5 - \sqrt{2}}{23}$

109. $\sqrt{\dfrac{1}{3}} \cdot \sqrt{\dfrac{24}{5}} = \sqrt{\dfrac{1}{3} \cdot \dfrac{24}{5}} = \sqrt{\dfrac{8}{5}} = \dfrac{\sqrt{8}}{\sqrt{5}}$
$= \dfrac{\sqrt{8} \cdot \sqrt{5}}{\sqrt{5} \cdot \sqrt{5}} = \dfrac{\sqrt{40}}{5}$
$= \dfrac{\sqrt{4 \cdot 10}}{5} = \dfrac{2\sqrt{10}}{5}$

110. $\sqrt{50y^2} = \sqrt{25y^2 \cdot 2}$
$= \sqrt{25y^2} \cdot \sqrt{2}$
$= 5y\sqrt{2}$

111. $\sqrt[3]{-125} = -5$ because $(-5)^3 = -125$.

112. $-\sqrt{5}(\sqrt{2} + \sqrt{75})$
$= -\sqrt{5}(\sqrt{2}) + (-\sqrt{5})(\sqrt{75})$
$= -\sqrt{10} - \sqrt{375}$
$= -\sqrt{10} - \sqrt{25 \cdot 15}$
$= -\sqrt{10} - 5\sqrt{15}$

113. $\sqrt{\dfrac{16r^3}{3s}}$ $(s \neq 0) = \dfrac{\sqrt{16r^3}}{\sqrt{3s}} = \dfrac{\sqrt{16r^2} \cdot \sqrt{r}}{\sqrt{3s}}$
$= \dfrac{4r\sqrt{r}}{\sqrt{3s}} = \dfrac{4r\sqrt{r} \cdot \sqrt{3s}}{\sqrt{3s} \cdot \sqrt{3s}}$
$= \dfrac{4r\sqrt{3rs}}{3s}$

344 Chapter 8 Roots and Radicals

114. $\dfrac{12 + 6\sqrt{13}}{12} = \dfrac{6(2 + \sqrt{13})}{6(2)}$

$= \dfrac{2 + \sqrt{13}}{2}$

115. $-\sqrt{162} + \sqrt{8} = -\sqrt{81 \cdot 2} + \sqrt{4 \cdot 2}$

$= -9\sqrt{2} + 2\sqrt{2}$

$= -7\sqrt{2}$

116. $(\sqrt{5} - \sqrt{2})^2$

$= (\sqrt{5})^2 - 2\sqrt{5}\sqrt{2} + (\sqrt{2})^2$

   Square of a binomial

$= 5 - 2\sqrt{10} + 2$

$= 7 - 2\sqrt{10}$

117. $(6\sqrt{7} + 2)(4\sqrt{7} - 1)$

$= 6\sqrt{7}(4\sqrt{7}) - 1(6\sqrt{7})$

$\quad + 2(4\sqrt{7}) + 2(-1)$

$= 24 \cdot 7 - 6\sqrt{7} + 8\sqrt{7} - 2$

$= 168 - 2 + 2\sqrt{7}$

$= 166 + 2\sqrt{7}$

118. $-\sqrt{121} = -11$

119. $\sqrt{98} = \sqrt{49 \cdot 2} = \sqrt{49} \cdot \sqrt{2} = 7\sqrt{2}$

120. $\sqrt{x + 2} = x - 4$

$(\sqrt{x + 2})^2 = (x - 4)^2$

$x + 2 = x^2 - 8x + 16$

$0 = x^2 - 9x + 14$

$0 = (x - 2)(x - 7)$

$x - 2 = 0$ or $x - 7 = 0$

$x = 2$ or $\quad x = 7$

In the original equation, 7 checks but 2 does not. The only solution is 7.

121. $\sqrt{k} + 3 = 0$

$\sqrt{k} = -3$

$(\sqrt{k})^2 = (-3)^2$

$k = 9$

This number does not check in the original equation, since $\sqrt{9} + 3 = 0$ is false. The original equation has no solution.

122. $\sqrt{1 + 3t} - t = -3$

$\sqrt{1 + 3t} = t - 3$

$(\sqrt{1 + 3t})^2 = (t - 3)^2$

$1 + 3t = t^2 - 6t + 9$

$0 = t^2 - 9t + 8$

$0 = (t - 1)(t - 8)$

$t - 1 = 0$ or $t - 8 = 0$

$t = 1$ or $\quad t = 8$

In the original equation, 8 checks but 1 does not. The only solution is 8.

Chapter 8 Test

1. The square roots of 196 are 14 and -14 because $14^2 = 196$ and $(-14)^2 = 196$.

2. (a) $\sqrt{142}$ is irrational because 142 is not a perfect square.
   (b) $\sqrt{142} \approx 11.916$

3. $\sqrt[3]{216} = \sqrt[3]{8 \cdot 27}$

$= \sqrt[3]{8} \cdot \sqrt[3]{27}$

$= 2 \cdot 3 = 6$

Chapter 8 Test

4. $-\sqrt{27} = -\sqrt{9 \cdot 3} = -\sqrt{9} \cdot \sqrt{3} = -3\sqrt{3}$

5. $\sqrt{\dfrac{128}{25}} = \dfrac{\sqrt{128}}{\sqrt{25}} = \dfrac{\sqrt{64 \cdot 2}}{5} = \dfrac{8\sqrt{2}}{5}$

6. $\sqrt[3]{32} = \sqrt[3]{8 \cdot 4} = \sqrt[3]{8} \cdot \sqrt[3]{4} = 2\sqrt[3]{4}$

7. $\dfrac{20\sqrt{18}}{5\sqrt{3}} = \dfrac{4\sqrt{9 \cdot 2}}{\sqrt{3}}$
    $= \dfrac{4 \cdot 3\sqrt{2}}{\sqrt{3}}$
    $= \dfrac{12\sqrt{2} \cdot \sqrt{3}}{\sqrt{3} \cdot \sqrt{3}}$
    $= \dfrac{12\sqrt{6}}{3} = 4\sqrt{6}$

8. $3\sqrt{28} + \sqrt{63} = 3(\sqrt{4 \cdot 7}) + \sqrt{9 \cdot 7}$
    $= 3(2\sqrt{7}) + 3\sqrt{7}$
    $= 6\sqrt{7} + 3\sqrt{7} = 9\sqrt{7}$

9. $3\sqrt{27x} - 4\sqrt{48x} + 2\sqrt{3x}$
    $= 3(\sqrt{9 \cdot 3x}) - 4(\sqrt{16 \cdot 3x}) + 2\sqrt{3x}$
    $= 3(3\sqrt{3x}) - 4(4\sqrt{3x}) + 2\sqrt{3x}$
    $= 9\sqrt{3x} - 16\sqrt{3x} + 2\sqrt{3x} = -5\sqrt{3x}$

10. $\sqrt[3]{32x^2y^3} = \sqrt[3]{8y^3 \cdot 4x^2}$
    $= \sqrt[3]{8y^3} \cdot \sqrt[3]{4x^2}$
    $= 2y\sqrt[3]{4x^2}$

11. $(6 - \sqrt{5})(6 + \sqrt{5})$
    $= (6)^2 - (\sqrt{5})^2$
    $= 36 - 5 = 31$

12. $(2 - \sqrt{7})(3\sqrt{2} + 1)$
    $= 2(3\sqrt{2}) + 2(1) - \sqrt{7}(3\sqrt{2}) - \sqrt{7}(1)$
    $= 6\sqrt{2} + 2 - 3\sqrt{14} - \sqrt{7}$

13. $(\sqrt{5} + \sqrt{6})^2$
    $= (\sqrt{5})^2 + 2(\sqrt{5})(\sqrt{6}) + (\sqrt{6})^2$
    $= 5 + 2\sqrt{30} + 6$
    $= 11 + 2\sqrt{30}$

14. Use the Pythagorean formula with $c = 9$ and $b = 3$.
    $$c^2 = a^2 + b^2$$
    $$9^2 = a^2 + 3^2$$
    $$81 = a^2 + 9$$
    $$72 = a^2$$
    $$\sqrt{72} = a$$
    (a) $a = \sqrt{72} = \sqrt{36 \cdot 2} = 6\sqrt{2}$ inches
    (b) $a = \sqrt{72} \approx 8.485$ inches

15. $\dfrac{5\sqrt{2}}{\sqrt{7}} = \dfrac{5\sqrt{2} \cdot \sqrt{7}}{\sqrt{7} \cdot \sqrt{7}} = \dfrac{5\sqrt{14}}{7}$

16. $\sqrt{\dfrac{2}{3x}} \;(x \neq 0) = \dfrac{\sqrt{2}}{\sqrt{3x}} = \dfrac{\sqrt{2} \cdot \sqrt{3x}}{\sqrt{3x} \cdot \sqrt{3x}} = \dfrac{\sqrt{6x}}{3x}$

17. $\dfrac{-2}{\sqrt[3]{4}} = \dfrac{-2 \cdot \sqrt[3]{2}}{\sqrt[3]{4} \cdot \sqrt[3]{2}} = \dfrac{-2\sqrt[3]{2}}{\sqrt[3]{8}}$
    $= \dfrac{-2\sqrt[3]{2}}{2} = -\sqrt[3]{2}$

18. $\dfrac{-3}{4 - \sqrt{3}} = \dfrac{-3(4 + \sqrt{3})}{(4 - \sqrt{3})(4 + \sqrt{3})}$
    $= \dfrac{-12 - 3\sqrt{3}}{(4)^2 - (\sqrt{3})^2}$
    $= \dfrac{-12 - 3\sqrt{3}}{16 - 3}$
    $= \dfrac{-12 - 3\sqrt{3}}{13}$

19. $\sqrt{x + 1} = 5 - x$
    $(\sqrt{x + 1})^2 = (5 - x)^2$
    $x + 1 = 25 - 10x + x^2$
    $0 = x^2 - 11x + 24$
    $0 = (x - 3)(x - 8)$
    $x - 3 = 0$ or $x - 8 = 0$
    $x = 3$ or $x = 8$

    In the original equation, 3 checks but 8 does not. The only solution is 3.

20. $3\sqrt{x} - 1 = 2x$
    $3\sqrt{x} = 2x + 1$
    $(3\sqrt{x})^2 = (2x + 1)^2$
    $9x = 4x^2 + 4x + 1$
    $0 = 4x^2 - 5x + 1$
    $0 = (4x - 1)(x - 1)$
    $4x - 1 = 0$ or $x - 1 = 0$
    $4x = 1$
    $x = \frac{1}{4}$ or $x = 1$

    Both of these numbers check in the original equation. The solutions are 1/4 and 1.

Cumulative Review: Chapters R–8

1. $3(6 + 7) + 6 \cdot 4 - 3^2$
    $= 3(13) + 6 \cdot 4 - 3^2$
    $= 3(13) + 6 \cdot 4 - 9$
    $= 39 + 24 - 9$
    $= 63 - 9 = 54$

2. $\dfrac{3(6 + 7) + 3}{2(4) - 1} = \dfrac{3(13) + 3}{8 - 1}$
    $= \dfrac{39 + 3}{7}$
    $= \dfrac{42}{7} = 6$

3. $|-6| - |-3| = 6 - 3 = 3$

4. $-9 + 14 + 11 + (-3 + 5)$
    $= -9 + 14 + 11 + 2$
    $= 5 + 11 + 2$
    $= 16 + 2 = 18$

5. $13 - [-4 - (-2)]$
    $= 13 - (-4 + 2)$
    $= 13 - (-2)$
    $= 13 + 2 = 15$

6. $-2.523 + 8.674 - 1.928$
    $= 6.151 - 1.928$
    $= 4.223$

7. $5(k - 4) - k = k - 11$
    $5k - 20 - k = k - 11$
    $4k - 20 = k - 11$
    $3k = 9$
    $k = 3$

8. $-\frac{3}{4}y \leq 12$
    $-\frac{4}{3}(-\frac{3}{4}y) \geq -\frac{4}{3}(12)$
    $y \geq -16$

9. $5z + 3 - 4 > 2z + 9 + z$
    $5z - 1 > 3z + 9$
    $2z > 10$
    $z > 5$

10. Let w = the width of the rectangle.
    Then $w + 7$ = the length of the rectangle.

Using $P = 2L + 2W$, the formula for the perimeter of a rectangle, obtain the following equation.

$$56 = 2(w + 7) + 2(w)$$
$$56 = 2w + 14 + 2w$$
$$56 = 4w + 14$$
$$42 = 4w$$
$$\frac{42}{4} = w$$
$$w = \frac{21}{2} = 10\frac{1}{2}$$

The width is 10 1/2 meters and the length is 10 1/2 + 7 = 17 1/2 meters.

11. $(3x^6)(2x^2y)^2$
  $= (3x^6)(2)^2(x^2)^2(y)^2$
  $= (3x^6) \cdot 4x^4y^2$
  $= 12x^{10}y^2$

12. $\left(\frac{3^2 y^{-2}}{2^{-1} y^3}\right)^{-3} = \frac{(3^2 y^{-2})^{-3}}{(2^{-1} y^3)^{-3}}$
  $= \frac{(3^2)^{-3}(y^{-2})^{-3}}{(2^{-1})^{-3}(y^3)^{-3}}$
  $= \frac{3^{-6} y^6}{2^3 y^{-9}}$
  $= \frac{1}{3^6 \cdot 2^3} \cdot y^{6-(-9)}$
  $= \frac{1}{3^6 \cdot 2^3} \cdot y^{15}$
  $= \frac{1}{729 \cdot 8} \cdot \frac{y^{15}}{1} = \frac{y^{15}}{5832}$

13. Subtract $7x^3 - 8x^2 + 4$ from $10x^3 + 3x^2 - 9$.

  $(10x^3 + 3x^2 - 9) - (7x^3 - 8x^2 + 4)$
  $= 10x^3 + 3x^2 - 9 - 7x^3 + 8x^2 - 4$
  $= 3x^3 + 11x^2 - 13$

14. $(8t^3 - 4t^2 - 14t + 15) \div (2t + 3)$

$$
\begin{array}{r}
4t^2 - 8t + 5 \\
2t+3{\overline{\smash{\big)}\,8t^3 - 4t^2 - 14t + 15}} \\
\underline{8t^3 + 12t^2} \\
-16t^2 - 14t \\
\underline{-16t^2 - 24t} \\
10t + 15 \\
\underline{10t + 15} \\
0
\end{array}
$$

The remainder is 0, so the answer is the quotient, $4t^2 - 8t + 5$.

15. $m^2 + 12m + 32 = (m + 8)(m + 4)$

16. $25t^4 - 36 = (5t^2 + 6)(5t^2 - 6)$

17. $12a^2 + 4ab - 5b^2 = (6a + 5b)(2a - b)$

18. $81z^2 + 72z + 16$
  $= (9z)^2 + 2(9z)(4) + 4^2$
  $= (9z + 4)^2$

19. $$x^2 - 7x = -12$$
  $$x^2 - 7x + 12 = 0$$
  $$(x - 3)(x - 4) = 0$$
  $$x - 3 = 0 \text{ or } x - 4 = 0$$
  $$x = 3 \text{ or } x = 4$$

  The solutions are 3 and 4.

20. $(x + 4)(x - 1) = -6$
  $x^2 + 3x - 4 = -6$
  $x^2 + 3x + 2 = 0$
  $(x + 2)(x + 1) = 0$
  $x + 2 = 0 \text{ or } x + 1 = 0$
  $x = -2 \text{ or } x = -1$

  The solutions are -2 and -1.

21. $\dfrac{3}{x^2 + 5x - 14} = \dfrac{3}{(x + 7)(x - 2)}$

   The expression is undefined when x is -7 or 2, because those values make the denominator equal zero.

22. $\dfrac{x^2 - 3x - 4}{x^2 + 3x} \cdot \dfrac{x^2 + 2x - 3}{x^2 - 5x + 4}$

   $= \dfrac{(x - 4)(x + 1)}{x(x + 3)} \cdot \dfrac{(x - 1)(x + 3)}{(x - 4)(x - 1)}$ Factor

   $= \dfrac{x + 1}{x}$ Lowest terms

23. $\dfrac{t^2 + 4t - 5}{t + 5} \div \dfrac{t - 1}{t^2 + 8t + 15}$

   $= \dfrac{t^2 + 4t - 5}{t + 5} \cdot \dfrac{t^2 + 8t + 15}{t - 1}$

                   Multiply by the reciprocal

   $= \dfrac{(t + 5)(t - 1)}{t + 5} \cdot \dfrac{(t + 5)(t + 3)}{t - 1}$ Factor

   $= (t + 5)(t + 3)$ Lowest terms

24. $\dfrac{\frac{2}{3} + \frac{1}{2}}{\frac{1}{9} - \frac{1}{6}} = \dfrac{\frac{4}{6} + \frac{3}{6}}{\frac{2}{18} - \frac{3}{18}}$

   $= \dfrac{\frac{7}{6}}{\frac{-1}{18}}$

   $= \dfrac{7}{6} \div \dfrac{-1}{18}$

   $= \dfrac{7}{6} \cdot \dfrac{18}{-1}$

   $= \dfrac{126}{-6} = -21$

25. $\dfrac{y}{y^2 - 1} + \dfrac{y}{y + 1}$

   $= \dfrac{y}{(y + 1)(y - 1)} + \dfrac{y(y - 1)}{(y + 1)(y - 1)}$

   $= \dfrac{y + y(y - 1)}{(y + 1)(y - 1)}$

   $= \dfrac{y + y^2 - y}{(y + 1)(y - 1)} = \dfrac{y^2}{(y + 1)(y - 1)}$

26. $\dfrac{2}{x + 3} - \dfrac{4}{x - 1}$

   $= \dfrac{2(x - 1)}{(x + 3)(x - 1)} - \dfrac{4(x + 3)}{(x - 1)(x + 3)}$

   $= \dfrac{2(x - 1) - 4(x + 3)}{(x + 3)(x - 1)}$

   $= \dfrac{2x - 2 - 4x - 12}{(x + 3)(x - 1)}$

   $= \dfrac{-2x - 14}{(x + 3)(x - 1)}$

For Exercises 27–29, see the graphs in the answer section of the textbook.

27. $-4x + 5y = -20$

   Find the intercepts.

   If $y = 0$, $x = 5$, so the x-intercept is $(5, 0)$.

   If $x = 0$, $y = -4$, so the y-intercept is $(0, -4)$.

   Draw the line that passes through the points $(5, 0)$ and $(0, -4)$.

28. $x = 2$

   $x = 0y + 2$ is an equivalent equation.

   Make a table of ordered pairs.

x	y
2	5
2	8

Draw the line that passes through (2, 5) and (2, 8), which is a vertical line.

29. $2x - 5y > 10$

    The boundary, $2x - 5y = 10$ is the line that passes through (5, 0) and (0, -2); draw it as a dashed line because of the > symbol. Use (0, 0) as a test point. Because

    $$2(0) - 5(0) > 10$$

    is a false statement, shade the side of the dashed boundary that does not include the origin, (0, 0).

30. (9, -2), (-3, 8)

    The slope of the line through these points is

    $$m = \frac{y_2 - y_1}{x_2 - x_1} = \frac{8 - (-2)}{-3 - 9}$$
    $$= \frac{10}{-12} = -\frac{5}{6}.$$

31. $4x - y = 19$ (1)
    $3x + 2y = -5$ (2)

    We will solve this system by the addition method. Multiply both sides of equation (1) by 2, and then add the result to equation (2).

    $$\begin{aligned} 8x - 2y &= 38 \\ 3x + 2y &= -5 \\ \hline 11x &= 33 \\ x &= 3 \end{aligned}$$

Let $x = 3$ in equation (1).

$$4(3) - y = 19$$
$$12 - y = 19$$
$$-y = 7$$
$$y = -7$$

The solution of the system of equations is the ordered pair (3, -7).

32. $2x - y = 6$ (1)
    $3y = 6x - 18$ (2)

    We will solve this system by the substitution method. Solve equation (2) for y by dividing both sides by (3).

    $$y = 2x - 6$$

    Substitute $2x - 6$ for y in the first equation.

    $$2x - (2x - 6) = 6$$
    $$2x - 2x + 6 = 6$$
    $$6 = 6$$

    Obtaining this true statement indicates that the two original equations are both describing the same line. This system has an infinite number of solutions.

33. Let x = number of ten-dollar bills;
    y = number of twenty-dollar bills.

    There are 20 bills, so

    $$x + y = 20. \quad (1)$$

The total value of the money is $250, so

$$10x + 20y = 250 \quad (2)$$

or

$$x + 2y = 25. \quad (3)$$

We will work with the system formed by equations (1) and (3).

$$x + y = 20 \quad (1)$$
$$x + 2y = 25 \quad (3)$$

To solve this system by the addition method, multiply equation (1) by -1 and add the result to equation (3).

$$-x - y = -20$$
$$\underline{x + 2y = 25}$$
$$y = 5$$

To find the value of x, substitute 5 for x in equation (1).

$$x + 5 = 20$$
$$x = 15$$

The cashier has 15 tens and 5 twenties.

34. $\dfrac{\sqrt{56}}{\sqrt{7}} = \sqrt{\dfrac{56}{7}} = \sqrt{8} = \sqrt{4 \cdot 2} = 2\sqrt{2}$

35. $\sqrt{27} - 2\sqrt{12} + 6\sqrt{75}$
    $= \sqrt{9} \cdot \sqrt{3} - 2\sqrt{4} \cdot \sqrt{3} + 6\sqrt{25} \cdot \sqrt{3}$
    $= 3\sqrt{3} - 2(2\sqrt{3}) + 6(5\sqrt{3})$
    $= 3\sqrt{3} - 4\sqrt{3} + 30\sqrt{3} = 29\sqrt{3}$

36. $\dfrac{2}{\sqrt{3} + \sqrt{5}} = \dfrac{2(\sqrt{3} - \sqrt{5})}{(\sqrt{3} + \sqrt{5})(\sqrt{3} - \sqrt{5})}$
    $= \dfrac{2(\sqrt{3} - \sqrt{5})}{3 - 5}$
    $= \dfrac{2(\sqrt{3} - \sqrt{5})}{-2}$
    $= \dfrac{\sqrt{3} - \sqrt{5}}{-1} = -\sqrt{3} + \sqrt{5}$

37. $\sqrt{200x^2y^5} = \sqrt{100x^2y^4 \cdot 2y}$
    $= \sqrt{100x^2y^4} \cdot \sqrt{2y}$
    $= 10xy^2\sqrt{2y}$

38. $\dfrac{5 + \sqrt{75}}{10} = \dfrac{5 + \sqrt{25 \cdot 3}}{10}$
    $= \dfrac{5 + 5\sqrt{3}}{10}$
    $= \dfrac{5(1 + \sqrt{3})}{5(2)} = \dfrac{1 + \sqrt{3}}{2}$

39. $(3\sqrt{2} + 1)(4\sqrt{2} - 3)$
    $= 3\sqrt{2}(4\sqrt{2}) - 3\sqrt{2}(3) + 1(4\sqrt{2})$
    $\quad + 1(-3)$
    $= 12 \cdot 2 - 9\sqrt{2} + 4\sqrt{2} - 3$
    $= 24 - 3 - 5\sqrt{2}$
    $= 21 - 5\sqrt{2}$

40. $\sqrt{x} + 2 = x - 10$
    $\sqrt{x} = x - 12$
    $(\sqrt{x})^2 = (x - 12)^2$
    $x = x^2 - 24x + 144$
    $0 = x^2 - 25x + 144$
    $0 = (x - 16)(x - 9)$
    $x - 16 = 0 \text{ or } x - 9 = 0$
    $x = 16 \text{ or } \quad x = 9$

In the original equation, 16 checks but 9 does not, so the only solution is 16.

CHAPTER 9 QUADRATIC EQUATIONS

Section 9.1 Solving Quadratic Equations by the Square Root Property

9.1 Margin Exercises

1. **(a)** $k^2 = 49$

    Solve by the square root property.

    $k = \sqrt{49}$ or $k = -\sqrt{49}$
    $k = 7$ or $k = -7$

    The solutions are 7 and -7, which may be written ± 7.

    (b) $b^2 = 11$

    By the square root property,

    $b = \sqrt{11}$ or $b = -\sqrt{11}$.

    The solutions are $\sqrt{11}$ and $-\sqrt{11}$.

    (c) $c^2 = 12$

    By the square root property,

    $c = \sqrt{12}$ or $c = -\sqrt{12}$
    $c = \sqrt{4} \cdot \sqrt{3}$ $c = -\sqrt{4} \cdot \sqrt{3}$
    $c = 2\sqrt{3}$ or $c = -2\sqrt{3}$.

    The solutions are $2\sqrt{3}$ and $-2\sqrt{3}$.

    (d) $x^2 = -9$

    The square of a real number cannot be negative. (The square root property cannot be used because b must be positive.) Thus, there is no real number solution for this equation.

2. **(a)** $(m + 2)^2 = 36$

    By the square root property,

    $m + 2 = 6$ or $m + 2 = -6$
    $m = 4$ or $m = -8$.

    The solutions are 4 and -8.

    (b) $(p - 4)^2 = 3$

    $p - 4 = \sqrt{3}$ or $p - 4 = -\sqrt{3}$
    $p = 4 + \sqrt{3}$ or $p = 4 - \sqrt{3}$

    The solutions are $4 + \sqrt{3}$ and $4 - \sqrt{3}$.

3. **(a)** $(2x - 5)^2 = 18$

    $2x - 5 = \sqrt{18}$ or $2x - 5 = -\sqrt{18}$
    Square root property
    $2x = 5 + \sqrt{18}$ $2x = 5 - \sqrt{18}$
    $x = \dfrac{5 + \sqrt{18}}{2}$ $x = \dfrac{5 - \sqrt{18}}{2}$
    $x = \dfrac{5 + \sqrt{9} \cdot \sqrt{2}}{2}$ $x = \dfrac{5 - \sqrt{9} \cdot \sqrt{2}}{2}$
    $x = \dfrac{5 + 3\sqrt{2}}{2}$ or $x = \dfrac{5 - 3\sqrt{2}}{2}$

    The solutions are $\dfrac{5 + 3\sqrt{2}}{2}$ and $\dfrac{5 - 3\sqrt{2}}{2}$.

    (b) $(7z - 1)^2 = -1$

    Since the square root of -1 is not a real number, there are no real number solutions.

9.1 Section Exercises

3. If $k = 0$, then $x^2 = k$ has no real solutions.

    This statement is false because $x = 0$ is a solution of $x^2 = 0$.

7. It is not correct to say that the solution of $x^2 = 81$ is 9, because -9 also satisfies the equation.

When we solve an equation, we want to find *all* values of the variable that satisfy the equation. The completely correct answer is that the solutions of $x^2 = 81$ are 9 and -9.

11. $k^2 = 14$

   Use the square root property to get

   $k = \sqrt{14}$ or $k = -\sqrt{14}$.

   The solutions are $\sqrt{14}$ and $-\sqrt{14}$.

15. $y^2 = \frac{25}{4}$

   $y = \sqrt{\frac{25}{4}}$ or $y = -\sqrt{\frac{25}{4}}$

   $y = \frac{5}{2}$ or $y = -\frac{5}{2}$

   The solutions are 5/2 and $-5/2$.

19. $r^2 - 3 = 0$
   $r^2 = 3$
   $r = \sqrt{3}$ or $r = -\sqrt{3}$

   The solutions are $\sqrt{3}$ and $-\sqrt{3}$.

23. $(z + 5)^2 = -13$

   The square root of -13 is not a real number, so there is no real solution for this equation.

27. $(3k + 2)^2 = 49$

   $3k + 2 = \sqrt{49}$ or $3k + 2 = -\sqrt{49}$
   $3k + 2 = 7$ or $3k + 2 = -7$
   $3k = 5$ or $3k = -9$
   $k = \frac{5}{3}$ or $k = -3$

   The solutions are 5/3 and -3.

31. $(5 - 2x)^2 = 30$

   $5 - 2x = \sqrt{30}$ or $5 - 2x = -\sqrt{30}$
   $-2x = -5 + \sqrt{30}$ or $-2x = -5 - \sqrt{30}$
   $x = \frac{-5 + \sqrt{30}}{-2}$ or $x = \frac{-5 - \sqrt{30}}{-2}$
   $x = \frac{-5 + \sqrt{30}}{-2} \cdot \frac{-1}{-1}$ or $x = \frac{-5 - \sqrt{30}}{-2} \cdot \frac{-1}{-1}$
   $x = \frac{5 - \sqrt{30}}{2}$ or $x = \frac{5 + \sqrt{30}}{2}$

   The solutions are

   $\frac{5 + \sqrt{30}}{2}$ and $\frac{5 - \sqrt{30}}{2}$.

35. $(\frac{1}{2}x + 5)^2 = 12$

   Begin by using the square root property.

   $\frac{1}{2}x + 5 = \sqrt{12}$ or $\frac{1}{2}x + 5 = -\sqrt{12}$

   Now simplify the radical.

   $\sqrt{12} = \sqrt{4 \cdot 3} = \sqrt{4} \cdot \sqrt{3} = 2\sqrt{3}$

   $\frac{1}{2}x + 5 = 2\sqrt{3}$ or $\frac{1}{2}x + 5 = -2\sqrt{3}$

   $\frac{1}{2}x = -5 + 2\sqrt{3}$ or $\frac{1}{2}x = -5 - 2\sqrt{3}$

   Multiply both equations by 2.

   $x = -10 + 4\sqrt{3}$ or $x = -10 - 4\sqrt{3}$

   The solutions are $-10 + 4\sqrt{3}$ and $-10 - 4\sqrt{3}$.

39. Linda's and Johnny's answers are equivalent. To show this, multiply the numerators and denominators of Linda's answers by -1.

$$\frac{(-5 + \sqrt{30})(-1)}{(-2)(-1)} = \frac{5 - \sqrt{30}}{2}$$

$$\frac{(-5 - \sqrt{30})(-1)}{(-2)(-1)} = \frac{5 + \sqrt{30}}{2}$$

Observe that these are Johnny's answers.

43. $A = \pi r^2$

Replace A by the given area, 36π.

$$36\pi = \pi r^2$$

Divide by π.

$$36 = r^2$$

This is the same as

$$r^2 = 36.$$

Use the square root property.

$$r = 6 \quad \text{or} \quad r = -6$$

A radius cannot be negative, however, so reject -6. The radius is 6 inches.

47. $\dfrac{6 + \sqrt{24}}{8} = \dfrac{6 + \sqrt{4 \cdot 6}}{8}$

$= \dfrac{6 + 2\sqrt{6}}{8}$

$= \dfrac{2(3 + \sqrt{6})}{2(4)}$

$= \dfrac{3 + \sqrt{6}}{4}$

Section 9.2 Solving Quadratic Equations by Completing the Square

9.2 Margin Exercises

1. (a) $x^2 + 6x + 9$
$= (x)^2 + 2(x)(3) + (3)^2$
$= (x + 3)^2$

(b) $q^2 - 20q + 100 = (q - 10)^2$

2. $a^2 + 4a = 1$

Take half of the coefficient of a and square the result.

$$\frac{1}{2}(4) = 2, \text{ and } 2^2 = 4.$$

Add 4 to each side of the equation, and write the left side as a perfect square.

$$a^2 + 4a + 4 = 1 + 4$$
$$(a + 2)^2 = 5$$

Use the square root property.

$a + 2 = \sqrt{5}$ or $a + 2 = -\sqrt{5}$
$a = -2 + \sqrt{5}$ or $a = -2 - \sqrt{5}$

The solutions are $-2 + \sqrt{5}$ and $-2 - \sqrt{5}$.

3. (a) $9m^2 + 18m + 5 = 0$

Divide each side by 9 to get 1 as the coefficient of the squared term.

$$m^2 + 2m + \frac{5}{9} = 0 \quad \textit{Divide by 9}$$

Rewrite the equation to get the variable terms on one side of the equals sign and the constants on the other side.

$$m^2 + 2m = -\frac{5}{9} \quad \textit{Subtract 5/9}$$

Chapter 9 Quadratic Equations

Take half the coefficient of m, or $(\frac{1}{2})(2) = 1$, and square the result: $1^2 = 1$. Then add 1 to each side.

$$m^2 + 2m + 1 = -\frac{5}{9} + 1 \quad \text{Add 1}$$

$$m^2 + 2m + 1 = \frac{4}{9} \quad 1 = 9/9$$

$$(m + 1)^2 = \frac{4}{9} \quad \text{Factor}$$

Apply the square root property, and solve for m.

$$m + 1 = \sqrt{\frac{4}{9}} \quad \text{or} \quad m + 1 = -\sqrt{\frac{4}{9}}$$

$$m + 1 = \frac{2}{3} \qquad\qquad m + 1 = -\frac{2}{3}$$

$$m = -1 + \frac{2}{3} \qquad m = -1 - \frac{2}{3}$$

$$m = -\frac{1}{3} \quad \text{or} \quad m = -\frac{5}{3}$$

The solutions are $-1/3$ and $-5/3$.

(b) $4k^2 - 24k + 11 = 0$

Divide each side of the equation by 4 to get 1 as the coefficient of the squared term.

$$k^2 - 6k + \frac{11}{4} = 0 \quad \text{Divide by 4}$$

$$k^2 - 6k = -\frac{11}{4} \quad \text{Subtract } 11/4$$

Square half the coefficient of k.

$$\left[\frac{1}{2}(-6)\right]^2 = (-3)^2 = 9$$

Then add 9 to both sides.

$$k^2 - 6k + 9 = -\frac{11}{4} + 9$$

$$\qquad\qquad\qquad\qquad \text{Add 9}$$

$$k^2 - 6k + 9 = \frac{25}{4} \quad 9 = 36/4$$

$$(k - 3)^2 = \frac{25}{4} \quad \text{Factor}$$

Use the square root property.

$$k - 3 = \sqrt{\frac{25}{4}} \quad \text{or} \quad k - 3 = -\sqrt{\frac{25}{4}}$$

$$k - 3 = \frac{5}{2} \qquad\qquad k - 3 = -\frac{5}{2}$$

$$k = 3 + \frac{5}{2} \qquad\qquad k = 3 - \frac{5}{2}$$

$$k = \frac{11}{2} \quad \text{or} \quad k = \frac{1}{2}$$

The solutions are 11/2 and 1/2.

4. $3x^2 + 5x - 2 = 0$

$$x^2 + \frac{5x}{3} - \frac{2}{3} = 0 \quad \text{Divide by 3}$$

$$x^2 + \frac{5x}{3} = \frac{2}{3} \quad \text{Add 2/3}$$

$$x^2 + \frac{5x}{3} + \frac{25}{36} = \frac{2}{3} + \frac{25}{36}$$

$$\qquad\qquad \text{Add } (\frac{1}{2} \cdot \frac{5}{3})^2 = \frac{25}{36}$$

$$(x + \frac{5}{6})^2 = \frac{49}{36} \quad 2/3 = 24/36$$

$$x + \frac{5}{6} = \sqrt{\frac{49}{36}} \quad \text{or} \quad x + \frac{5}{6} = -\sqrt{\frac{49}{36}} \quad \text{Square root property}$$

$$x + \frac{5}{6} = \frac{7}{6} \qquad\qquad x + \frac{5}{6} = -\frac{7}{6}$$

$$x = \frac{2}{6} \qquad\qquad\qquad x = -\frac{12}{6}$$

$$x = \frac{1}{3} \quad \text{or} \quad x = -2$$

The solutions are 1/3 and -2.

5. $5v^2 + 3v + 1 = 0$

$$v^2 + \frac{3v}{5} + \frac{1}{5} = 0 \quad \text{Divide by 5}$$

$$v^2 + \frac{3v}{5} = -\frac{1}{5} \quad \text{Subtract 1/5}$$

$$v^2 + \frac{3v}{5} + \frac{9}{100} = -\frac{1}{5} + \frac{9}{100}$$

$$\qquad\qquad \text{Add } (\frac{1}{2} \cdot \frac{3}{5})^2 = \frac{9}{100}$$

$$(v + \frac{3}{10})^2 = -\frac{11}{100}$$

The square root of −11/100 is not a real number, so the square root property does not apply. This equation has no real number solution.

6. **(a)** $r^2 + 1 = 3r$

$r^2 - 3r = -1$ Subtract $3r + 1$

$r^2 - 3r + \frac{9}{4} = -1 + \frac{9}{4}$ Add $\left(-3 \cdot \frac{1}{2}\right)^2 = \frac{9}{4}$

$\left(r - \frac{3}{2}\right)^2 = \frac{5}{4}$

$r - \frac{3}{2} = \sqrt{\frac{5}{4}}$ or $r - \frac{3}{2} = -\sqrt{\frac{5}{4}}$

Square root property

$r - \frac{3}{2} = \frac{\sqrt{5}}{2}$ $r - \frac{3}{2} = -\frac{\sqrt{5}}{2}$

$r = \frac{3 + \sqrt{5}}{2}$ or $r = \frac{3 - \sqrt{5}}{2}$

The solutions are $\frac{3 + \sqrt{5}}{2}$ and $\frac{3 - \sqrt{5}}{2}$.

(b) $(x + 2)(x + 1) = 5$

$x^2 + 3x + 2 = 5$

$x^2 + 3x = 3$

$\frac{1}{2}(3) = \frac{3}{2}$, and $\left(\frac{3}{2}\right)^2 = \frac{9}{4}$.

Add 9/4 to each side.

$x^2 + 3x + \frac{9}{4} = 3 + \frac{9}{4}$

$\left(x + \frac{3}{2}\right)^2 = \frac{21}{4}$

$x + \frac{3}{2} = \sqrt{\frac{21}{4}}$ or $x + \frac{3}{2} = -\sqrt{\frac{21}{4}}$

$x + \frac{3}{2} = \frac{\sqrt{21}}{2}$ or $x + \frac{3}{2} = -\frac{\sqrt{21}}{2}$

$x = -\frac{3}{2} + \frac{\sqrt{21}}{2}$ or $x = -\frac{3}{2} - \frac{\sqrt{21}}{2}$

The solutions are $\frac{-3 + \sqrt{21}}{2}$ and $\frac{-3 - \sqrt{21}}{2}$.

7. $s = -16t^2 + 128t$

Let $s = 48$ and solve for t.

$48 = -16t^2 + 128t$

$16t^2 - 128t + 48 = 0$

Divide by 16.

$t^2 - 8t + 3 = 0$

$t^2 - 8t = -3$

$t^2 - 8t + 16 = -3 + 16$

$(t - 4)^2 = 13$

$t - 4 = \sqrt{13}$ or $t - 4 = -\sqrt{13}$

$t = 4 + \sqrt{13}$ or $t = 4 - \sqrt{13}$

$t \approx 7.6$ or $t \approx .4$

The ball will be 48 feet above the ground after about .4 seconds and again after about 7.6 seconds.

9.2 Section Exercises

3. $x^2 - 4x = -3$

Take half of the coefficient of x and square it. Half of −4 is −2, and $(-2)^2 = 4$. Add 4 to each side of the equation, and write the left side as a perfect square.

$x^2 - 4x + 4 = -3 + 4$

$(x - 2)^2 = 1$

Use the square root property.

$x - 2 = \sqrt{1}$ or $x - 2 = -\sqrt{1}$

$x - 2 = 1$ or $x - 2 = -1$

$x = 3$ or $x = 1$

These answers check in the original equation. The solutions are 1 and 3.

7. $z^2 + 6z + 9 = 0$

   Subtract 9 from each side.

   $$z^2 + 6z = -9$$

   The coefficient of z is 6. Take half of 6, square the result, and add this square to each side. The left-hand side can then be written as a perfect square.

   $$z^2 + 6z + 9 = -9 + 9$$
   $$(z + 3)^2 = 0$$

   Use the square root property.

   $$z + 3 = \sqrt{0} \quad \text{or} \quad z + 3 = -\sqrt{0}$$
   $$z + 3 = 0 \quad \text{or} \quad z + 3 = 0$$
   $$z = -3 \quad \text{or} \quad z = -3$$

   Both cases give the same answer, and it checks in the original equation. The only solution is −3.

11. $k^2 - 5k$

    Take half of the coefficient of k and square it.

    $$\tfrac{1}{2}(-5) = -\tfrac{5}{2}, \text{ and } \left(-\tfrac{5}{2}\right)^2 = \tfrac{25}{4}.$$

    Add 25/4 to the expression $k^2 - 5k$ to make it a perfect square.

15. $2x^2 - 4x = 9$

    Before completing the square, the coefficient of x^2 must be 1. Dividing each side of the equation by 2 is the correct way to begin solving the equation, and this corresponds to choice (d).

19. $4y^2 + 4y = 3$

    Divide each side by 4.

    $$y^2 + y = \tfrac{3}{4}$$

    The coefficient of y is 1. Take half of 1, square the result, and add this square to each side. The left-hand side can then be written as a perfect square.

    $$y^2 + y + \tfrac{1}{4} = \tfrac{3}{4} + \tfrac{1}{4}$$
    $$\left(y + \tfrac{1}{2}\right)^2 = 1$$
    $$y + \tfrac{1}{2} = 1 \quad \text{or} \quad y + \tfrac{1}{2} = -1$$
    $$y = -\tfrac{1}{2} + 1 \quad \text{or} \quad y = -\tfrac{1}{2} - 1$$
    $$y = \tfrac{1}{2} \quad \text{or} \quad y = -\tfrac{3}{2}$$

    The solutions are 1/2 and −3/2.

23. $3k^2 + 7k = 4$

    Divide each side by 3.

    $$k^2 + \tfrac{7}{3}k = \tfrac{4}{3}$$

    Take half of the coefficient of k and square it.

    $$\tfrac{1}{2}\left(\tfrac{7}{3}\right) = \tfrac{7}{6}, \text{ and } \left(\tfrac{7}{6}\right)^2 = \tfrac{49}{36}.$$

    Add 49/36 to each side of the equation.

    $$k^2 + \tfrac{7}{3}k + \tfrac{49}{36} = \tfrac{4}{3} + \tfrac{49}{36}$$
    $$\left(k + \tfrac{7}{6}\right)^2 = \tfrac{97}{36}$$

Use the square root property.

$k + \frac{7}{6} = \sqrt{\frac{97}{36}}$ or $k + \frac{7}{6} = -\sqrt{\frac{97}{36}}$

$k + \frac{7}{6} = \frac{\sqrt{97}}{6}$ or $k + \frac{7}{6} = -\frac{\sqrt{97}}{6}$

$k = -\frac{7}{6} + \frac{\sqrt{97}}{6}$ or $k = -\frac{7}{6} - \frac{\sqrt{97}}{6}$

$k = \frac{-7 + \sqrt{97}}{6}$ or $k = \frac{-7 - \sqrt{97}}{6}$

The solutions are

$\frac{-7 + \sqrt{97}}{6}$ and $\frac{-7 - \sqrt{97}}{6}$.

27. $-x^2 + 2x = -5$

Divide each side by -1.

$x^2 - 2x = 5$

Take half of the coefficient of x and square it. Half of -2 is -1, and $(-1)^2 = 1$. Add 1 to each side of the equation.

$x^2 - 2x + 1 = 5 + 1$

$(x - 1)^2 = 6$

Use the square root property.

$x - 1 = \sqrt{6}$ or $x - 1 = -\sqrt{6}$
$x = 1 + \sqrt{6}$ or $x = 1 - \sqrt{6}$

The solutions are $1 + \sqrt{6}$ and $1 - \sqrt{6}$.

31. $\frac{8 - 6\sqrt{3}}{6} = \frac{2(4 - 3\sqrt{3})}{2(3)}$ *Factor*

$= \frac{4 - 3\sqrt{3}}{3}$ *Lowest terms*

Section 9.3 Solving Quadratic Equations by the Quadratic Formula

9.3 Margin Exercises

1. (a) $5x^2 + 2x - 1 = 0$ has the form of the standard quadratic equation,

$ax^2 + bx + c = 0$.

Thus, $a = 5$, $b = 2$, and $c = -1$.

(b) $3m^2 = m - 2$

$3m^2 - m + 2 = 0$
           Rewrite in $ax^2 + bx + c = 0$ form

Then $a = 3$, $b = -1$, and $c = 2$.

(c) $p(p + 5) = 4$

$p^2 + 5p = 4$ *Distributive property*

$p^2 + 5p - 4 = 0$ $ax^2 + bx + c = 0$ *form*

Then $a = 1$, $b = 5$, and $c = -4$.

2. (a) $-\frac{c}{a} + \frac{b^2}{4a^2} = \frac{b^2}{4a^2} - \underline{\quad?\quad}$

By the commutative property of addition,

$-\frac{c}{a} + \frac{b^2}{4a^2} = \frac{b^2}{4a^2} + \left(-\frac{c}{a}\right)$

$= \frac{b^2}{4a^2} - \frac{c}{a}$.

? is $\frac{c}{a}$.

(b) $\frac{b^2}{4a^2} - \frac{c}{a} = \frac{b^2}{4a^2} - \frac{?}{4a^2}$

Since the least common denominator is $4a^2$, multiply the second fraction, c/a, by $4a/4a$.

358 Chapter 9 Quadratic Equations

$$\frac{b^2}{4a^2} - \frac{c}{a} = \frac{b^2}{4a^2} - \frac{c}{a} \cdot \frac{4a}{4a}$$

$$= \frac{b^2}{4a^2} - \frac{4ac}{4a^2}$$

? is $4ac$.

(c) $\dfrac{b^2}{4a^2} - \dfrac{4ac}{4a^2} = \dfrac{?}{4a^2}$

$$\frac{b^2}{4a^2} - \frac{4ac}{4a^2} = \frac{b^2 - 4ac}{4a^2} \quad \text{Subtract numerators}$$

? is $b^2 - 4ac$.

3. (a) $2x^2 + 3x - 5 = 0$

The quadratic equation is in standard form, so $a = 2$, $b = 3$, and $c = -5$. Substitute these values into the quadratic formula.

$$x = \frac{-b \pm \sqrt{b^2 - 4ac}}{2a}$$

$$x = \frac{-3 \pm \sqrt{3^2 - 4(2)(-5)}}{2(2)}$$

$$= \frac{-3 \pm \sqrt{9 + 40}}{4}$$

$$= \frac{-3 \pm \sqrt{49}}{4}$$

$$= \frac{-3 \pm 7}{4}$$

$x = \dfrac{-3 + 7}{4}$ or $x = \dfrac{-3 - 7}{4}$

$= \dfrac{4}{4}$ $= \dfrac{-10}{4}$

$x = 1$ or $x = -\dfrac{5}{2}$

The solutions are 1 and $-5/2$.

(b) $6p^2 + p = 1$

Rewrite the equation in standard form.

$$6p^2 + p - 1 = 0 \quad \text{Subtract 1}$$

Then $a = 6$, $b = 1$, and $c = -1$. Substitute these values into the quadratic formula.

$$p = \frac{-b \pm \sqrt{b^2 - 4ac}}{2a}$$

$$p = \frac{-1 \pm \sqrt{1^2 - 4(6)(-1)}}{2(6)}$$

$$= \frac{-1 \pm \sqrt{1 + 24}}{12}$$

$$= \frac{-1 \pm 5}{12}$$

$p = \dfrac{-1 + 5}{12}$ or $p = \dfrac{-1 - 5}{12}$

$= \dfrac{4}{12}$ $= \dfrac{-6}{12}$

$p = \dfrac{1}{3}$ or $p = -\dfrac{1}{2}$

The solutions are $1/3$ and $-1/2$.

4. $-y^2 = 8y + 1$

Rewrite the equation in standard form.

$$0 = y^2 + 8y + 1 \quad \text{Add } y^2$$

Then $a = 1$, $b = 8$, and $c = 1$. Substitute these values into the quadratic formula.

$$y = \frac{-b \pm \sqrt{b^2 - 4ac}}{2a}$$

$$= \frac{-8 \pm \sqrt{8^2 - 4(1)(1)}}{2(1)}$$

$$= \frac{-8 \pm \sqrt{64 - 4}}{2}$$

$$= \frac{-8 \pm \sqrt{60}}{2}$$

$$= \frac{-8 \pm \sqrt{4} \cdot \sqrt{15}}{2}$$

$$= \frac{-8 \pm 2\sqrt{15}}{2}$$

$$= \frac{2(-4 \pm \sqrt{15})}{2}$$

$$= -4 \pm \sqrt{15}$$

The solutions are $-4 + \sqrt{15}$ and $-4 - \sqrt{15}$.

5. $9y^2 - 12y + 4 = 0$

    Here, $a = 9$, $b = -12$, and $c = 4$. Substitute these values into the quadratic formula.

    $$y = \frac{-b \pm \sqrt{b^2 - 4ac}}{2a}$$

    $$= \frac{-(-12) \pm \sqrt{(-12)^2 - 4(9)(4)}}{2(9)}$$

    $$= \frac{12 \pm \sqrt{144 - 144}}{18}$$

    $$= \frac{12}{18} = \frac{2}{3}$$

    Since there is just one solution, 2/3, the trinomial $9y^2 - 12y + 4$ is a perfect square.

6. $x^2 - \frac{4}{3}x + \frac{2}{3} = 0$

    $3x^2 - 4x + 2 = 0$ *Multiply by 3*

    Then $a = 3$, $b = -4$, and $c = 2$. Substitute these values into the quadratic formula.

    $$x = \frac{-b \pm \sqrt{b^2 - 4ac}}{2a}$$

    $$= \frac{-(-4) \pm \sqrt{(-4)^2 - 4(3)(2)}}{2(3)}$$

    $$= \frac{4 \pm \sqrt{16 - 24}}{6}$$

    $$= \frac{4 \pm \sqrt{-8}}{6}$$

    There is no real number solution because $\sqrt{-8}$ is not a real number.

9.3 Section Exercises

3. $3x^2 = 4x + 2$

    First write the equation in the form $ax^2 + bx + c = 0$.

    $$3x^2 - 4x - 2 = 0$$

    The coefficients are $a = 3$, $b = -4$, and $c = -2$.

7. $3x^2 + 5x + 1 = 0$

    Here, $a = 3$, $b = 5$, and $c = 1$. Substitute these numbers into the quadratic formula, and simplify the result.

360   Chapter 9   Quadratic Equations

$$x = \frac{-b \pm \sqrt{b^2 - 4ac}}{2a}$$

$$x = \frac{-5 \pm \sqrt{5^2 - 4(3)(1)}}{2(3)}$$

$$x = \frac{-5 \pm \sqrt{25 - 12}}{6}$$

$$x = \frac{-5 \pm \sqrt{13}}{6}$$

The solutions are

$$\frac{-5 + \sqrt{13}}{6} \text{ and } \frac{-5 - \sqrt{13}}{6}.$$

11. $p^2 - 4p + 4 = 0$

Here, $a = 1$, $b = -4$, and $c = 4$.
By the quadratic formula,

$$p = \frac{-b \pm \sqrt{b^2 - 4ac}}{2a}$$

$$= \frac{-(-4) \pm \sqrt{(-4)^2 - 4(1)(4)}}{2(1)}$$

$$= \frac{4 \pm \sqrt{16 - 16}}{2}$$

$$= \frac{4 \pm 0}{2}$$

$$= \frac{4}{2} = 2.$$

The only solution is 2.

15. $2y^2 = 5 + 3y$

Subtract 5 and 3y from each side of the equation.

$$2y^2 - 3y - 5 = 0$$

Here, $a = 2$, $b = -3$, and $c = -5$.
By the quadratic formula,

$$y = \frac{-(-3) \pm \sqrt{(-3)^2 - 4(2)(-5)}}{2(2)}$$

$$= \frac{3 \pm \sqrt{9 + 40}}{4}$$

$$= \frac{3 \pm \sqrt{49}}{4} = \frac{3 \pm 7}{4}$$

$$y = \frac{3 + 7}{4} \text{ or } y = \frac{3 - 7}{4}$$

$$y = \frac{10}{4} = \frac{5}{2} \text{ or } y = \frac{-4}{4} = -1.$$

The solutions are 5/2 and -1.

19. $7x^2 = 12x$

Subtract 12x from both sides.

$$7x^2 - 12x = 0$$

Here, $a = 7$, $b = -12$, and $c = 0$.
By the quadratic formula,

$$x = \frac{-(-12) \pm \sqrt{(-12)^2 - 4(7)(0)}}{2(7)}$$

$$= \frac{12 \pm \sqrt{144 - 0}}{14}$$

$$= \frac{12 \pm \sqrt{144}}{14}$$

$$= \frac{12 \pm 12}{14}$$

$$x = \frac{12 + 12}{14} \text{ or } x = \frac{12 - 12}{14}$$

$$x = \frac{24}{14} = \frac{12}{7} \text{ or } x = \frac{0}{14} = 0.$$

The solutions are 12/7 and 0.

23. $25x^2 - 4 = 0$

Here $a = 25$, $b = 0$, and $c = -4$.
By the quadratic formula,

$$x = \frac{-0 \pm \sqrt{0^2 - 4(25)(-4)}}{2(25)}$$

$$= \frac{0 \pm \sqrt{0 + 400}}{50}$$

$$= \frac{\pm\sqrt{400}}{50}$$

$$x = \frac{\pm 20}{50} = \pm\frac{2}{5}.$$

The solutions are 2/5 and -2/5.

27. $-2x^2 = -3x + 2$

Add 3x and subtract 2 on each side.

$$-2x^2 + 3x - 2 = 0$$

Here, $a = -2$, $b = 3$, and $c = -2$.
By the quadratic formula,

$$x = \frac{-3 \pm \sqrt{3^2 - 4(-2)(-2)}}{2(-2)}$$

$$= \frac{-3 \pm \sqrt{9 - 16}}{-4}$$

$$x = \frac{-3 \pm \sqrt{-7}}{-4}.$$

The radical $\sqrt{-7}$ is not a real number, so the equation has no real number solution.

31. If $b^2 - 4ac$, the radicand in the quadratic formula, is negative, then the original equation has no real number solutions. (The square root of a negative number is not a real number.)

35. $\frac{1}{2}x^2 + \frac{1}{6}x = 1$

Eliminate the denominators by multiplying each side of the equation by the common denominator, 6.

$$6\left(\frac{1}{2}x^2 + \frac{1}{6}x\right) = 6(1)$$

$$3x^2 + x = 6$$

Subtract 6 on each side.

$$3x^2 + x - 6 = 0$$

From this equation, identify $a = 3$, $b = 1$, and $c = -6$.

Use the quadratic formula to complete the solution.

$$x = \frac{-1 \pm \sqrt{1^2 - 4(3)(-6)}}{2(3)}$$

$$= \frac{-1 \pm \sqrt{1 + 72}}{6}$$

$$= \frac{-1 \pm \sqrt{73}}{6}$$

The solutions are

$$\frac{-1 + \sqrt{73}}{6} \text{ and } \frac{-1 - \sqrt{73}}{6}.$$

39. $\frac{3}{8}x^2 - x + \frac{17}{24} = 0$

Multiply each side by the common denominator, 24.

$$24\left(\frac{3}{8}x^2 - x + \frac{17}{24}\right) = 24(0)$$

$$9x^2 - 24x + 17 = 0$$

Here, $a = 9$, $b = -24$, and $c = 17$.
Use the quadratic formula to complete the solution.

$$x = \frac{-(-24) \pm \sqrt{(-24)^2 - 4(9)(17)}}{2(9)}$$

$$= \frac{24 \pm \sqrt{576 - 612}}{18}$$

$$= \frac{24 \pm \sqrt{-36}}{18}$$

The radical $\sqrt{-36}$ is not a real number, so the equation has no real number solution.

43. Use the Pythagorean formula with legs x and x + 1 and hypotenuse x + 4, and simplify.

$$a^2 + b^2 = c^2$$
$$x^2 + (x + 1)^2 = (x + 4)^2$$
$$x^2 + x^2 + 2x + 1 = x^2 + 8x + 16$$
$$2x^2 + 2x + 1 = x^2 + 8x + 16$$

Move all terms on the right-hand side to the left-hand side by subtracting.

$$x^2 - 6x - 15 = 0$$

Here, a = 1, b = -6, and c = -15. Use the quadratic formula.

$$x = \frac{-(-6) \pm \sqrt{(-6)^2 - 4(1)(-15)}}{2(1)}$$
$$= \frac{6 \pm \sqrt{36 + 60}}{2}$$
$$x = \frac{6 \pm \sqrt{96}}{2}$$

$\sqrt{96} \approx 9.798$, so

$$x \approx \frac{6 + 9.798}{2} \quad \text{or} \quad x \approx \frac{6 - 9.798}{2}$$
$$x \approx 7.899 \quad \text{or} \quad x \approx -1.899.$$

Reject the negative solution, because a side of a triangle cannot have a negative length. Use the positive solution for x to find the values of x, x + 1, and x + 4. The lengths of the sides of the right triangle are approximately 7.899, 8.899, and 11.899.

47. 3x + 5y = 15

Find the intercepts.

If y = 0, then x = 5, so the x-intercept is (5, 0).

If x = 0, then y = 3, so the y-intercept is (0, 3).

Draw the line that passes through the points (5, 0) and (0, 3). See the graph in the answer section of the textbook.

Summary Exercises on Quadratic Equations

3. $$y^2 - \frac{100}{81} = 0$$
$$y^2 = \frac{100}{81}$$

Use the square root property.

$$y = \sqrt{\frac{100}{81}} \quad \text{or} \quad y = -\sqrt{\frac{100}{81}}$$
$$y = \frac{10}{9} \quad \text{or} \quad y = -\frac{10}{9}$$

The solutions are 10/9 and -10/9.

7. $$z(z - 9) = -20$$
$$z^2 - 9z = -20$$
$$z^2 - 9z + 20 = 0$$

Solve this equation by factoring.

$$(z - 4)(z - 5) = 0$$
$$z - 4 = 0 \quad \text{or} \quad z - 5 = 0$$
$$z = 4 \quad \text{or} \quad z = 5$$

The solutions are 4 and 5.

11. $(x + 6)^2 = 121$

Use the square root property.

$$x + 6 = \sqrt{121} \quad \text{or} \quad x + 6 = -\sqrt{121}$$
$$x + 6 = 11 \quad \text{or} \quad x + 6 = -11$$
$$x = 5 \quad \text{or} \quad x = -17$$

The solutions are 5 and -17.

15. $(5x - 8)^2 = -6$

The square root of -6 is not a real number, so the square root property does not apply. This equation has no real number solution.

19. $\qquad 8z^2 = 15 + 2z$
$8z^2 - 2z - 15 = 0$

Solve this equation by factoring.

$$(4x + 5)(2x - 3) = 0$$
$$4x + 5 = 0 \quad \text{or} \quad 2x - 3 = 0$$
$$4x = -5 \quad \text{or} \qquad 2x = 3$$
$$x = -\frac{5}{4} \quad \text{or} \qquad x = \frac{3}{2}$$

The solutions are $-5/4$ and $3/2$.

23. $\qquad 5y^2 - 22y = -8$
$5y^2 - 22y + 8 = 0$

Solve by factoring.

$$(5y - 2)(y - 4) = 0$$
$$5y - 2 = 0 \quad \text{or} \quad y - 4 = 0$$
$$5y = 2$$
$$y = \frac{2}{5} \quad \text{or} \qquad y = 4$$

The solutions are $2/5$ and 4.

27. $\qquad 4x^2 = -1 + 5x$
$4x^2 - 5x + 1 = 0$

Solve by factoring.

$$(x - 1)(4x - 1) = 0$$
$$x - 1 = 0 \quad \text{or} \quad 4x - 1 = 0$$
$$x = 1 \quad \text{or} \qquad 4x = 1$$
$$x = 1 \quad \text{or} \qquad x = \frac{1}{4}$$

The solutions are 1 and $1/4$.

31. $\dfrac{r^2}{2} + \dfrac{7r}{4} + \dfrac{11}{8} = 0$

Multiply both sides by the least common denominator, 8.

$$8\left(\frac{r^2}{2} + \frac{7r}{4} + \frac{11}{8}\right) = 8(0)$$
$$4r^2 + 14r + 11 = 0$$

Use the quadratic formula with $a = 4$, $b = 14$, and $c = 11$.

$$x = \frac{-14 \pm \sqrt{14^2 - 4(4)(11)}}{2(4)}$$
$$= \frac{-14 \pm \sqrt{196 - 176}}{8}$$
$$= \frac{-14 \pm \sqrt{20}}{8}$$

Note that $\sqrt{20} = \sqrt{4 \cdot 5} = 2\sqrt{5}$.

$$x = \frac{-14 \pm 2\sqrt{5}}{8}$$
$$= \frac{2(-7 \pm \sqrt{5})}{2(4)}$$
$$x = \frac{-7 \pm \sqrt{5}}{4}$$

The solutions are $\dfrac{-7 + \sqrt{5}}{4}$ and $\dfrac{-7 - \sqrt{5}}{4}$.

35. $y^2 - y + 3 = 0$

Use the quadratic formula with $a = 1$, $b = -1$, and $c = 3$.

$$x = \frac{-(-1) \pm \sqrt{(-1)^2 - 4(1)(3)}}{2(1)}$$
$$= \frac{1 \pm \sqrt{1 - 12}}{2}$$
$$= \frac{1 \pm \sqrt{-11}}{2}$$

The radical $\sqrt{-11}$ is not a real number, so the equation has no real number solution.

39.
$$5k^2 + 19k = 2k + 12$$
$$5k^2 + 17k - 12 = 0$$

Solve this equation by factoring.

$$(5k - 3)(k + 4) = 0$$
$$5k - 3 = 0 \quad \text{or} \quad k + 4 = 9$$
$$5k = 3$$
$$k = \frac{3}{5} \quad \text{or} \quad k = -4$$

The solutions are 3/5 and −4.

Section 9.4 Graphing Quadratic Equations in Two Variables

9.4 Margin Exercises

1. $y = x^2$

   To find the y-values, substitute each x-value into $y = x^2$.

x	y
3	9
2	4
1	1
0	0
-1	1
-2	4
-3	9

2. $y = -x^2 + 3$

   To find the y-values, substitute each x-value into $y = -x^2 + 3$.

   If $x = -2$, $y = -(-2)^2 + 3 = -1$ gives (−2, −1).
   If $x = -1$, $y = -(-1)^2 + 3 = 2$ gives (−1, 2).
   If $x = 1$, $y = -(1)^2 + 3 = 2$ gives (1, 2).
   If $x = 2$, $y = -(2)^2 + 3 = -1$ gives (2, −1).

3. $y = -x^2 - 1$

   Here, $a = -1$, $b = 0$, and $c = -1$. The x-value of the vertex is

   $$x = -\frac{b}{2a} = -\frac{0}{2(-1)} = 0.$$

   The y-value of the vertex is

   $$y = -0^2 - 1 = -1,$$

   so the vertex is (0, −1).
   Make a table of ordered pairs.

x	y
-2	-5
-1	-2
0	-1
1	-2
2	-5

   Plot these five points and connect them with a smooth curve.
   Refer to the graph included with the answer to this margin exercise in the textbook.

4. $y = x^2 + 2x - 8$

   Find any x-intercepts by substituting 0 for y in the equation.

$y = x^2 + 2x - 8$
$0 = x^2 + 2x - 8$ Let $y = 0$
$0 = (x + 4)(x - 2)$ Factor
$x + 4 = 0$ or $x - 2 = 0$
$x = -4$ or $x = 2$

The x-intercepts are (-4, 0) and (2, 0). Now find any y-intercepts by substituting 0 for x.

$y = x^2 + 2x - 8$
$y = 0^2 + 2(0) - 8$ Let $x = 0$
$y = -8$

The y-intercept is (0, -8). The x-value of the vertex is halfway between the x-intercepts, (-4, 0) and (2, 0).

$x = \frac{1}{2}(-4 + 2) = \frac{1}{2}(-2) = -1$

Find the y-value of the vertex by substituting -1 for x in the given equation.

$y = x^2 + 2x - 8$
$y = (-1)^2 + 2(-1) - 8$ Let $x = -1$
$y = 1 - 2 - 8$
$y = -9$

The vertex is (-1, -9).

x	y	
-4	0	
-3	-5	
-2	-8	
-1	-9	← Vertex
0	-8	
1	-5	
2	0	

Plot the intercepts, vertex, and additional points shown in the table of values and connect them with a smooth curve. Refer to the graph included with the answer to this margin exercise in the textbook.

5. $y = x^2 - 4x + 1$

If $x = 5$,
$y = 5^2 - 4(5) + 1 = 6$,
giving the ordered pair (5, 6).

If $x = 1$,
$y = 1^2 - 4(1) + 1 = -2$,
giving the ordered pair (1, -2).

If $x = 4$,
$y = 4^2 - 4(4) + 1 = 1$,
giving the ordered pair (4, 1).

If $x = 3$,
$y = 3^2 - 4(3) + 1 = -2$,
giving the ordered pair (3, -2).

If $x = -1$,
$y = (-1)^2 - 4(-1) + 1 = 6$,
giving the ordered pair (-1, 6).

6. $y = -x^2 + 2x + 4$

Here, $a = -1$, $b = 2$, and $c = 4$. The x-value of the vertex is

$x = -\frac{b}{2a} = -\frac{2}{2(-1)} = 1.$

The y-value of the vertex is

$y = -1^2 + 2(1) + 4 = 5,$

so the vertex is (1, 5).
Make a table of ordered pairs.

x	y
-1	1
0	4
1	5
2	4
3	1

Plot these five points and connect them with a smooth curve. Refer to the graph included with the answer to this margin exercise in the textbook.

9.4 Section Exercises

For Exercises 3–15, see the graphs in the answer section of the textbook.

3. $y = 2x^2$

   Let $x = 0$ to get $y = 2 \cdot 0^2 = 0$; the y-intercept is $(0, 0)$.
   Let $y = 0$ to get $0 = 2x^2$, which implies $x = 0$; the only x-intercept is $(0, 0)$.
   In $y = 2x^2$, $a = 2$, $b = 0$, and $c = 0$.
   The x-value of the vertex is

   $$x = -\frac{b}{2a} = -\frac{0}{2(2)} = 0.$$

   The y-value of the vertex is

   $$y = 2(0)^2 = 0,$$

   so the vertex is $(0, 0)$.
   Make a table of ordered pairs.

x	y
-2	8
-1	2
0	0
1	2
2	8

   Plot these five ordered pairs and connect them with a smooth curve.

7. $y = x^2 - 4$

   Let $x = 0$ to get $y = 0^2 - 4 = -4$; the y-intercept is $(0, -4)$. Let $y = 0$ and solve for x.

   $$0 = x^2 - 4$$
   $$0 = (x + 2)(x - 2)$$
   $$x + 2 = 0 \quad \text{or} \quad x - 2 = 0$$
   $$x = -2 \quad \text{or} \quad x = 2$$

   The x-intercepts are $(-2, 0)$ and $(2, 0)$.
   In $y = x^2 - 4$, $a = 1$, $b = 0$, and $c = -4$.
   The x-value of the vertex is

   $$x = -\frac{b}{2a} = -\frac{0}{2(1)} = 0.$$

   The y-value of the vertex is

   $$y = 0^2 - 4 = -4,$$

   so the vertex is $(0, -4)$.
   Make a table of ordered pairs.

x	y
-3	5
-2	0
-1	-3
0	-4
1	-3
2	0
3	5

   Plot these seven ordered pairs and connect them with a smooth curve.

11. $y = (x + 3)^2$

    Let $x = 0$ to get $y = (0 + 3)^2 = 9$; the y-intercept is $(0, 9)$.
    Let $y = 0$ and solve for x.

    $$0 = (x + 3)^2$$
    $$x + 3 = 0$$
    $$x = -3$$

    The one x-intercept is $(-3, 0)$.

In $y = (x + 3)^2 = x^2 + 6x + 9$,
$a = 1$, $b = 6$, and $c = 9$.
The x-value of the vertex is

$$x = -\frac{b}{2a} = -\frac{6}{2(1)} = -\frac{6}{2} = -3.$$

The y-value of the vertex is

$$y = (-3 + 3)^2 = 0^2 = 0,$$

so the vertex is $(-3, 0)$.
Make a table of ordered pairs whose x-values are on either side of the vertex's x-value of $x = -3$.

x	y
-5	4
-4	1
-3	0
-2	1
-1	4
0	9

Plot these six ordered pairs and connect them with a smooth curve.

15. $y = -x^2 + 4x - 4$

Let $x = 0$ to get

$$y = -(0)^2 + 4(0) - 4 = -4.$$

The y-intercept is $(0, -4)$.
Let $y = 0$ and solve for x.

$$0 = -x^2 + 4x - 4$$
$$x^2 - 4x + 4 = 0$$
$$(x - 2)^2 = 0$$
$$x - 2 = 0$$
$$x = 2$$

The one x-intercept is $(2, 0)$.
In $y = -x^2 + 4x - 4$, $a = -1$, $b = 4$, and $c = -4$.
The x-value of the vertex is

$$x = -\frac{b}{2a} = -\frac{4}{2(-1)} = \frac{4}{-2} = 2.$$

The y-value of the vertex is

$$y = -(2)^2 + 4(2) - 4 = 0,$$

so the vertex is $(2, 0)$.
Make a table of ordered pairs.

x	y
0	-4
1	-1
2	0
3	-1
4	-4

Plot these five ordered pairs and connect them with a smooth curve; observe that this parabola opens downward.

Chapter 9 Review Exercises

1. $y^2 = 144$
   $y = \sqrt{144}$ or $y = -\sqrt{144}$
   $y = 12$ or $y = -12$

   The solutions are 12 and -12.

2. $x^2 = 37$
   $x = \sqrt{37}$ or $x = -\sqrt{37}$
   The solutions are $\sqrt{37}$ and $-\sqrt{37}$.

3. $m^2 = 128$
   $m = \sqrt{128}$ or $m = -\sqrt{128}$

   Observe that

   $$\sqrt{128} = \sqrt{64 \cdot 2} = 8\sqrt{2},$$

   so

   $$m = 8\sqrt{2} \text{ or } m = -8\sqrt{2}.$$

   The solutions are $8\sqrt{2}$ and $-8\sqrt{2}$.

4. $(k + 2)^2 = 25$

   $k + 2 = \sqrt{25}$ or $k + 2 = -\sqrt{25}$

   $k + 2 = 5$ or $k + 2 = -5$

   $k = 3$ or $k = -7$

   The solutions are 3 and −7.

5. $(r - 3)^2 = 10$

   $r - 3 = \sqrt{10}$ or $r - 3 = -\sqrt{10}$

   $r = 3 + \sqrt{10}$ or $r = 3 - \sqrt{10}$

   The solutions are $3 + \sqrt{10}$ and $3 - \sqrt{10}$.

6. $(2p + 1)^2 = 14$

   $2p + 1 = \sqrt{14}$ or $2p + 1 = -\sqrt{14}$

   $2p = -1 + \sqrt{14}$ or $2p = -1 - \sqrt{14}$

   $p = \dfrac{-1 + \sqrt{14}}{2}$ or $p = \dfrac{-1 - \sqrt{14}}{2}$

   The solutions are

   $\dfrac{-1 + \sqrt{14}}{2}$ and $p = \dfrac{-1 - \sqrt{14}}{2}$.

7. $(3k + 2)^2 = -3$

   $3k + 2 = \sqrt{-3}$ or $3k + 2 = -\sqrt{-3}$

   The radical $\sqrt{-3}$ is not a real number, so the equation has no real number solution.

8. The square root property can be applied only to equations of the form $(ax + b)^2 = $ a number.

9. $m^2 + 6m + 5 = 0$

   Rewrite the equation with the variable terms on one side and the constant on the other side.

   $m^2 + 6m = -5$

   Take half the coefficient of m and square it.

   $\dfrac{1}{2}(6) = 3$, and $(3)^2 = 9$.

   Add 9 to each side of the equation.

   $m^2 + 6m + 9 = -5 + 9$

   $m^2 + 6m + 9 = 4$

   $(m + 3)^2 = 4$ *Factor*

   $m + 3 = \sqrt{4}$ or $m + 3 = -\sqrt{4}$

   $m + 3 = 2$ or $m + 3 = -2$

   $m = -1$ or $m = -5$

   The solutions are −1 and −5.

10. $p^2 + 4p = 7$

    Take half the coefficient of p and square it.

    $\dfrac{1}{2}(4) = 2$, and $(2)^2 = 4$.

    Add 4 to each side of the equation.

    $p^2 + 4p + 4 = 7 + 4$

    $(p + 2)^2 = 11$

    $p + 2 = \sqrt{11}$ or $p + 2 = -\sqrt{11}$

    $p = -2 + \sqrt{11}$ or $p = -2 - \sqrt{11}$

    The solutions are $-2 + \sqrt{11}$ and $-2 - \sqrt{11}$.

11. $-x^2 + 5 = 2x$

    Divide each side of the equation by −1 to make the coefficient of the squared term equal to 1.

    $-1(-x^2 + 5) = -1(2x)$

    $x^2 - 5 = -2x$

Rewrite the equation with the variable terms on one side and the constant on the other side.

$$x^2 + 2x = 5$$

Take half the coefficient of x and square it.

$$\tfrac{1}{2}(2) = 1, \text{ and } 1^2 = 1.$$

Add 1 to both sides of the equation.

$$x^2 + 2x + 1 = 5 + 1$$
$$(x + 1)^2 = 6$$
$$x + 1 = \sqrt{6} \quad \text{or} \quad x + 1 = -\sqrt{6}$$
$$x = -1 + \sqrt{6} \quad \text{or} \quad x = -1 - \sqrt{6}$$

The solutions are $-1 + \sqrt{6}$ and $-1 - \sqrt{6}$.

12. $2y^2 - 3 = -8y$
$2y^2 + 8y = 3$

Divide both sides by 2 to get the y^2 coefficient equal to 1.

$$\tfrac{1}{2}(2y^2 + 8y) = \tfrac{1}{2}(3)$$
$$y^2 + 4y = \tfrac{3}{2}$$

Take half the coefficient of y and square it.

$$\tfrac{1}{2}(4) = 2, \text{ and } 2^2 = 4.$$

Add 4 to each side of the equation.

$$y^2 + 4y + 4 = \tfrac{3}{2} + 4$$
$$(y + 2)^2 = \tfrac{11}{2}$$

$$y + 2 = \sqrt{\tfrac{11}{2}} \quad \text{or} \quad y + 2 = -\sqrt{\tfrac{11}{2}}$$
$$y + 2 = \tfrac{\sqrt{11}}{\sqrt{2}} \cdot \tfrac{\sqrt{2}}{\sqrt{2}} \quad \text{or} \quad y + 2 = -\tfrac{\sqrt{11}}{\sqrt{2}} \cdot \tfrac{\sqrt{2}}{\sqrt{2}}$$
$$y + 2 = \tfrac{\sqrt{22}}{2} \quad \text{or} \quad y + 2 = -\tfrac{\sqrt{22}}{2}$$
$$y = -2 + \tfrac{\sqrt{22}}{2} \quad \text{or} \quad y = -2 - \tfrac{\sqrt{22}}{2}$$
$$y = \tfrac{-4}{2} + \tfrac{\sqrt{22}}{2} \quad \text{or} \quad y = \tfrac{-4}{2} - \tfrac{\sqrt{22}}{2}$$
$$y = \tfrac{-4 + \sqrt{22}}{2} \quad \text{or} \quad y = \tfrac{-4 - \sqrt{22}}{2}$$

The solutions are

$$\tfrac{-4 + \sqrt{22}}{2} \text{ and } \tfrac{-4 - \sqrt{22}}{2}.$$

13. $5k^2 - 3k - 2 = 0$

Divide both sides by 5 to get the k^2 coefficient equal to 1.

$$k^2 - \tfrac{3}{5}k - \tfrac{2}{5} = 0$$

Rewrite the equation with the variable terms on one side and the constant on the other side.

$$k^2 - \tfrac{3}{5}k = \tfrac{2}{5}$$

Take half the coefficient of k and square it.

$$\tfrac{1}{2}\left(-\tfrac{3}{5}\right) = -\tfrac{3}{10}, \text{ and } \left(-\tfrac{3}{10}\right)^2 = \tfrac{9}{100}$$

$$k^2 - \tfrac{3}{5}k + \tfrac{9}{100} = \tfrac{2}{5} + \tfrac{9}{100}$$
$$\left(k - \tfrac{3}{10}\right)^2 = \tfrac{40}{100} + \tfrac{9}{100}$$
$$\left(k - \tfrac{3}{10}\right)^2 = \tfrac{49}{100}$$

$k - \frac{3}{10} = \sqrt{\frac{49}{100}}$ or $k - \frac{3}{10} = -\sqrt{\frac{49}{100}}$

$k - \frac{3}{10} = \frac{7}{10}$ or $k - \frac{3}{10} = -\frac{7}{10}$

$k = \frac{10}{10}$ or $k = -\frac{4}{10}$

$k = 1$ or $k = -\frac{2}{5}$

The solutions are 1 and −2/5.

14. $(4a + 1)(a - 1) = -7$

Divide on the left side and then simplify. Get all variable terms on one side and the constant on the othe side.

$$4a^2 - 3a - 1 = -7$$
$$4a^2 - 3a = -6$$

Multiply both sides by 4 so that the coefficient of a^2 will be 1.

$$a^2 - \frac{3}{4}a = -\frac{6}{4}$$
$$a^2 - \frac{3}{4}a = -\frac{3}{2}$$

Square half the coefficient of a and add it to both sides.

$$a^2 - \frac{3}{4}a + \frac{9}{64} = -\frac{3}{2} + \frac{9}{64}$$
$$\left(a - \frac{3}{8}\right)^2 = -\frac{96}{64} + \frac{9}{64}$$
$$\left(a - \frac{3}{8}\right)^2 = -\frac{87}{64}$$

The square root of −87/64 is not a real number, so there is no real number solution.

15. $h = -16t^2 + 32t + 50$

Let h = 30 and solve for t (which must have a positive value since it represents a number of seconds).

$$30 = -16t^2 + 32t + 50$$
$$16t^2 - 32t - 20 = 0$$

Divide both sides by 16.

$$t^2 - 2t - \frac{20}{16} = 0$$
$$t^2 - 2t = \frac{5}{4}$$

Half of −2 is −1, and $(-1)^2 = 1$. Add 1 to both sides of the equation.

$$t^2 - 2t + 1 = \frac{5}{4} + 1$$
$$(t - 1)^2 = \frac{9}{4}$$

$t - 1 = \sqrt{\frac{9}{4}}$ or $t - 1 = -\sqrt{\frac{9}{4}}$

$t - 1 = \frac{3}{2}$ or $t - 1 = -\frac{3}{2}$

$t = 1 + \frac{3}{2}$ or $t = 1 - \frac{3}{2}$

$t = \frac{5}{2} = 2\frac{1}{2}$ or $t = -\frac{1}{2}$

Reject the negative value of t. The object will reach a height of 30 feet after 2 1/2 seconds.

16. $d = 2t^2 - 5t + 2$

Let d = 14 and solve for t.

$$14 = 2t^2 - 5t + 2$$
$$0 = 2t^2 - 5t - 12$$

Divide both sides by 2.

$$t^2 - \frac{5}{2}t - 6 = 0$$

$$t^2 - \frac{5}{2}t = 6$$

$$\frac{1}{2}\left(-\frac{5}{2}\right) = -\frac{5}{4} \text{ and } \left(-\frac{5}{4}\right)^2 = \frac{25}{16}.$$

Add 25/16 to both sides of the equation.

$$t^2 - \frac{5}{2}t + \frac{25}{16} = 6 + \frac{25}{16}$$

$$\left(t - \frac{5}{4}\right)^2 = \frac{49}{16}$$

$$t - \frac{5}{4} = \sqrt{\frac{49}{16}} \quad \text{or} \quad t - \frac{5}{4} = -\sqrt{\frac{49}{16}}$$

$$t - \frac{5}{4} = \frac{7}{4} \quad \text{or} \quad t - \frac{5}{4} = -\frac{7}{4}$$

$$t = \frac{12}{4} = 3 \quad \text{or} \quad t = -\frac{2}{4} = -\frac{1}{2}$$

Reject the negative value of t. The projectile will be 14 feet from the ground after 4 seconds.

17. $3k^2 + 3k + 3 = 0$

This equation matches the standard form, so substitute a = 3, b = 2, and c = 3 into the quadratic formula.

$$k = \frac{-2 \pm \sqrt{2^2 - 4(3)(3)}}{2(3)}$$

$$= \frac{-2 \pm \sqrt{4 - 36}}{6}$$

$$= \frac{-2 \pm \sqrt{-32}}{6}$$

Since $\sqrt{-32}$ is not a real number, there is no real number solution.

18. $$x(5x - 1) = 1$$
$$5x^2 - x = 1$$
$$5x^2 - x - 1 = 0$$

Use the quadratic formula with a = 5, b = -1, and c = -1.

$$x = \frac{-b \pm \sqrt{b^2 - 4ac}}{2a}$$

$$= \frac{-(-1) \pm \sqrt{(-1)^2 - 4(5)(-1)}}{2(5)}$$

$$= \frac{1 \pm \sqrt{1 + 20}}{10}$$

$$= \frac{1 \pm \sqrt{21}}{10}$$

The solutions are

$$\frac{1 + \sqrt{21}}{10} \text{ and } \frac{1 - \sqrt{21}}{10}.$$

19. $$2p^2 + 8 = 4p + 11$$
$$2p^2 - 4p - 3 = 0$$

Use the quadratic formula with a = 2, b = -4, and c = -3.

$$p = \frac{-(-4) \pm \sqrt{(-4)^2 - 4(2)(-3)}}{2(2)}$$

$$= \frac{4 \pm \sqrt{16 + 24}}{4}$$

$$= \frac{4 \pm \sqrt{40}}{4}$$

$$= \frac{4 \pm \sqrt{4 \cdot 10}}{4}$$

$$= \frac{4 \pm 2\sqrt{10}}{4}$$

$$= \frac{2(2 \pm \sqrt{10})}{2(2)}$$

$$= \frac{2 \pm \sqrt{10}}{2}$$

The solutions are

$$\frac{2 + \sqrt{10}}{2} \text{ and } \frac{2 - \sqrt{10}}{2}.$$

20. $-4a^2 + 7 = 2a$

Rewrite the equation in standard form.

$$-4a^2 - 2a + 7 = 0$$

Then $a = -4$, $b = -2$, and $c = 7$. Substitute these values into the quadratic formula.

$$a = \frac{-(-2) \pm \sqrt{(-2)^2 - 4(-4)(7)}}{2(-4)}$$

$$= \frac{2 \pm \sqrt{4 + 112}}{-8}$$

$$= \frac{2 \pm \sqrt{116}}{-8} = \frac{2 \pm 2\sqrt{29}}{-8}$$

$$= \frac{2(1 \pm \sqrt{29})}{2(-4)} = \frac{1 \pm \sqrt{29}}{-4}$$

The solutions are

$$\frac{-1 + \sqrt{29}}{4} \text{ and } \frac{-1 - \sqrt{29}}{4}.$$

21. $\quad \frac{1}{4}x^2 = 2 - \frac{3}{4}x$

$$\frac{1}{4}x^2 + \frac{3}{4}x - 2 = 0$$

Multiply both sides by the common denominator, 4.

$$4\left(\frac{1}{4}x^2 + \frac{3}{4}x - 2\right) = 4(0)$$

$$x^2 + 3x - 8 = 0$$

Use the quadratic formula with $a = 1$, $b = 3$, and $c = -8$.

$$x = \frac{-3 \pm \sqrt{3^2 - 4(1)(-8)}}{2(1)}$$

$$= \frac{-3 \pm \sqrt{9 + 32}}{2}$$

$$= \frac{-3 \pm \sqrt{41}}{2}$$

The solutions are

$$\frac{-3 + \sqrt{41}}{2} \text{ and } \frac{-3 - \sqrt{41}}{2}.$$

22. The correct statement of the quadratic formula is

$$x = \frac{-b \pm \sqrt{b^2 - 4ac}}{2a}.$$

To state that

$$x = -b \pm \frac{\sqrt{b^2 - 4ac}}{2a}$$

is not equivalent because the $-b$ term should be above the fraction bar rather than be a separate term from the fraction.

For Exercise 23-28, see the graphs in the answer section of the textbook.

23. $y = -3x^2$

Let $x = 0$ to get

$$y = -3 \cdot 0^2 = 0;$$

the y-intercept is $(0, 0)$.

Let $y = 0$ to get $0 = -3x^2$, which implies $x = 0$; the x-intercept is $(0, 0)$.

In $y = -3x^2$, $a = -3$, $b = 0$, and $c = 0$.

The x-value of the vertex is

$$x = -\frac{b}{2a} = -\frac{0}{2(-3)} = 0,$$

so the vertex is $(0, 0)$.

Make a table of ordered pairs.

x	y
-2	-12
-1	-3
0	0
1	-3
2	-12

Plot these five ordered pairs and connect them with a smooth curve.

24. $y = x^2 - 2x + 1$

Let $x = 0$ to get
$$y = 0^2 - 2(0) + 1 = 1;$$
the y-intercept is (0, 1).
Let $y = 0$ and solve for x.
$$0 = x^2 - 2x + 1$$
$$0 = (x - 1)(x - 1)$$
$$x - 1 = 0 \quad \text{or} \quad x - 1 = 0$$
$$x = 1 \quad \text{or} \quad x = 1$$

The x-intercept is (1, 0).
In $y = x^2 - 2x + 1$, $a = 1$, $b = -2$, and $c = 1$.
The x-value of the vertex is
$$x = -\frac{b}{2a} = -\frac{-2}{2(1)} = 1.$$

The y-value of the vertex is
$$y = 1^2 - 2(1) + 1 = 0,$$
so the vertex is (1, 0).
Make a table of ordered pairs.

x	y
-1	4
0	1
1	0
2	1
3	4

Plot these five ordered pairs and connect them with a smooth curve.

25. $y = -x^2 + 5$

Let $x = 0$ to get
$$y = -0^2 + 5 = 5;$$
the y-intercept is (0, 5).
Let $y = 0$ and solve for x.
$$0 = -x^2 + 5$$
$$x^2 = 5$$
$$x = \sqrt{5} \quad \text{or} \quad x = -\sqrt{5}$$
$$x \approx 2.2 \quad \text{or} \quad x \approx -2.2$$

The x-intercepts are approximately (2.2, 0) and (-2.2, 0).
In $y = -x^2 + 5$, $a = -1$, $b = 0$, and $c = 5$.
The x-value of the vertex is
$$x = -\frac{b}{2a} = -\frac{0}{2(-1)} = 0,$$
so the vertex is (0, 5).
Make a table of ordered pairs.

x	y
-2.2	0
-2	1
-1	4
0	5
1	4
2	1
2.2	0

Plot these seven ordered pairs and connect them with a smooth curve.

26. $y = -x^2 + 2x + 3$

Let $x = 0$ to get
$$y = -0^2 + 2(0) + 3 = 3;$$
the y-intercept is (0, 3).

Let $y = 0$ and solve for x.

$$0 = -x^2 + 2x + 3$$
$$x^2 - 2x - 3 = 0$$
$$(x - 3)(x + 1) = 0$$
$$x - 3 = 0 \text{ or } x + 1 = 0$$
$$x = 3 \text{ or } x = -1$$

The x-intercepts are $(3, 0)$ and $(-1, 0)$.

In $y = -x^2 + 2x + 3$, $a = -1$, $b = 2$, and $c = 3$.

The x-value of the vertex is

$$x = -\frac{b}{2a} = -\frac{2}{2(-1)} = 1.$$

The y-value of the vertex is

$$y = -1^2 + 2(1) + 3 = 4,$$

so the vertex is $(1, 4)$.

Make a table of ordered pairs.

x	y
-1	0
0	3
1	4
2	3
3	0

Plot these five ordered pairs and connect them with a smooth curve.

27. $y = x^2 + 4x + 2$

Let $x = 0$ to get

$$y = 0^2 + 4(0) + 2 = 2;$$

the y-intercept is $(0, 2)$.

Let $y = 0$ and solve for x.

$$0 = x^2 + 4x + 2$$
$$x^2 + 4x + 2 = 0$$
$$x^2 + 4x = -2$$
$$x^2 + 4x + 4 = -2 + 4$$
$$(x + 2)^2 = 2$$

$$x + 2 = \sqrt{2} \quad \text{or} \quad x + 2 = -\sqrt{2}$$
$$x = -2 + \sqrt{2} \quad \text{or} \quad x = -2 - \sqrt{2}$$
$$x \approx -.6 \quad \text{or} \quad x \approx -3.4$$

The x-intercepts are approximately $(-.6, 0)$ and $(-3.4, 0)$.

In $y = x^2 + 4x + 2$, $a = 1$, $b = 4$, and $c = 2$.

The x-value of the vertex is

$$x = -\frac{b}{2a} = -\frac{4}{2(1)} = -2.$$

The y-value of the vertex is

$$y = (-2)^2 + 4(-2) + 2 = -2,$$

so the vertex is $(-2, -2)$.

Make a table of ordered pairs.

x	y
-4	2
-3.4	0
-3	-1
-2	-2
-1	-1
0	2
.6	0

Plot these seven ordered pairs and connect them with a smooth curve.

28. $y = (x + 4)^2$

Let $x = 0$ to get

$$y = (0 + 4)^2 = 16;$$

the y-intercept is $(0, 16)$.

Let $y = 0$ and solve for x.

$$0 = (x + 4)^2$$

$x + 4 = 0$ implies $x = -4$.

The x-intercept is $(-4, 0)$.

In $y = (x + 4)^2 = x^2 + 8x + 16$, $a = 1$, $b = 8$, and $c = 16$.

The x-value of the vertex is

$$x = -\frac{b}{2a} = -\frac{8}{2(1)} = -4.$$

The y-value of the vertex is

$$y = (-4 + 4)^2 = 0^2 = 0,$$

so the vertex is $(-4, 0)$.

Make a table of ordered pairs.

x	y
-6	4
-5	1
-4	0
-3	1
-2	4
0	16

Plot these six ordered pairs and connect them with a smooth curve.

29. $(2t - 1)(t + 1) = 54$

Rewrite the equation in standard form.

$$2t^2 + t - 1 = 54$$
$$2t^2 + t - 55 = 0$$

Solve this equation by factoring.

$$(2t + 11)(t - 5) = 0$$
$$2t + 11 = 0 \quad \text{or} \quad t - 5 = 0$$
$$2t = -11 \quad \text{or} \quad t = 5$$
$$t = -\frac{11}{2} \quad \text{or} \quad t = 5$$

The solutions are $-11/2$ and 5.

30. $(2p + 1)^2 = 100$

Solve by the square root method.

$$2p + 1 = \sqrt{100} \quad \text{or} \quad 2p + 1 = -\sqrt{100}$$
$$2p + 1 = 10 \quad \text{or} \quad 2p + 1 = -10$$
$$2p = 9 \quad \text{or} \quad 2p = -11$$
$$p = \frac{9}{2} \quad \text{or} \quad p = -\frac{11}{2}$$

The solutions are $9/2$ and $-11/2$.

31. $(k + 2)(k - 1) = 3$

Rewrite the equation in standard form.

$$(k + 2)(k - 1) = 3$$
$$k^2 + k - 2 = 3$$
$$k^2 + k - 5 = 0$$

The left side cannot be factored, so use the quadratic formula with $a = 1$, $b = 1$, and $c = -5$.

$$k = \frac{-b \pm \sqrt{b^2 - 4ac}}{2a}$$
$$= \frac{-1 \pm \sqrt{1^2 - 4(1)(-5)}}{2(1)}$$
$$= \frac{-1 \pm \sqrt{1 + 20}}{2}$$
$$= \frac{-1 \pm \sqrt{21}}{2}$$

The solutions are

$$\frac{-1 + \sqrt{21}}{2} \quad \text{and} \quad \frac{-1 - \sqrt{21}}{2}.$$

32. $6t^2 + 7t - 3 = 0$

Solve by factoring.

$$(3t - 1)(2t + 3) = 0$$

$3t - 1 = 0$ or $2t + 3 = 0$
$3t = 1$ or $2t = -3$
$t = \frac{1}{3}$ or $t = -\frac{3}{2}$

The solutions are 1/3 and -3/2.

33. $2x^2 + 3x + 2 = x^2 - 2x$

Rewrite the equation in standard form.

$$x^2 + 5x + 2 = 0$$

The left side cannot be factored, so use the quadratic formula with $a = 1$, $b = 5$, and $c = 2$.

$$x = \frac{-b \pm \sqrt{b^2 - 4ac}}{2a}$$
$$= \frac{-5 \pm \sqrt{5^2 - 4(1)(2)}}{2(1)}$$
$$= \frac{-5 \pm \sqrt{25 - 8}}{2}$$
$$= \frac{-5 \pm \sqrt{17}}{2}$$

The solutions are

$$\frac{-5 + \sqrt{17}}{2} \text{ and } \frac{-5 - \sqrt{17}}{2}.$$

34. $x^2 + 2x + 5 = 7$

Rewrite the equation in standard form.

$$x^2 + 2x - 2 = 0$$

The left side cannot be factored, so use the quadratic formula with $a = 1$, $b = 2$, and $c = -2$.

$$x = \frac{-2 \pm \sqrt{2^2 - 4(1)(-2)}}{2(1)}$$
$$= \frac{-2 \pm \sqrt{4 + 8}}{2}$$
$$= \frac{-2 \pm \sqrt{12}}{2} = \frac{-2 \pm 2\sqrt{3}}{2}$$
$$= \frac{2(-1 \pm \sqrt{3})}{2} = -1 \pm \sqrt{3}$$

The solutions are

$-1 + \sqrt{3}$ and $-1 - \sqrt{3}$.

35. $m^2 - 4m + 10 = 0$

Use the quadratic formula with $a = 1$, $b = -4$, and $c = 10$.

$$m = \frac{-(-4) \pm \sqrt{(-4)^2 - 4(1)(10)}}{2(1)}$$
$$= \frac{4 \pm \sqrt{16 - 40}}{2}$$
$$= \frac{4 \pm \sqrt{-24}}{2}$$

The radical $\sqrt{-24}$ is not a real number, so the equation has no real number solution.

36. $k^2 - 9k + 10 = 0$

The left side of the equation cannot be factored, so use the quadratic formula with $a = 1$, $b = -9$, and $c = 10$.

$$k = \frac{-b \pm \sqrt{b^2 - 4ac}}{2a}$$
$$= \frac{-(-9) \pm \sqrt{(-9)^2 - 4(1)(10)}}{2(1)}$$
$$= \frac{9 \pm \sqrt{81 - 40}}{2}$$
$$= \frac{9 \pm \sqrt{41}}{2}$$

The solutions are

$$\frac{9 + \sqrt{41}}{2} \text{ and } \frac{9 - \sqrt{41}}{2}.$$

37. $(3x + 5)^2$

   $3x + 5 = \sqrt{0}$ or $3x + 5 = -\sqrt{0}$

   These two cases are really the same.

   $$3x + 5 = 0$$
   $$3x = -5$$
   $$x = -\frac{5}{3}$$

   The solution is $-5/3$.

38. $\frac{1}{2}r^2 = \frac{7}{2} - r$

   Multiply by 2 to clear fractions; then rewrite the result in standard form.

   $$\frac{1}{2}r^2 = \frac{7}{2} - r$$
   $$2\left(\frac{1}{2}r^2\right) = 2\left(\frac{7}{2} - r\right)$$
   $$r^2 = 7 - 2r$$
   $$r^2 + 2r - 7 = 0$$

   The left side does not factor, so use the quadratic formula with $a = 1$, $b = 2$, and $c = -7$.

   $$r = \frac{-2 \pm \sqrt{2^2 - 4(1)(-7)}}{2(1)}$$
   $$= \frac{-2 \pm \sqrt{4 + 28}}{2}$$
   $$= \frac{-2 \pm \sqrt{32}}{2} = \frac{-2 \pm 4\sqrt{2}}{2}$$
   $$= \frac{2(-1 \pm 2\sqrt{2})}{2} = -1 \pm 2\sqrt{2}$$

The solutions are $-1 + 2\sqrt{2}$ and $-1 - 2\sqrt{2}$.

39. $$x^2 + 4x = 1$$
    $$x^2 + 4x - 1 = 0$$

   The left side of the equation does not factor, so use the quadratic formula with $a = 1$, $b = 4$, and $c = -1$.

   $$x = \frac{-4 \pm \sqrt{4^2 - 4(1)(-1)}}{2(1)}$$
   $$= \frac{-4 \pm \sqrt{16 + 4}}{2}$$
   $$= \frac{-4 \pm \sqrt{20}}{2} = \frac{-4 \pm 2\sqrt{5}}{2}$$
   $$= \frac{2(-2 \pm \sqrt{5})}{2} = -2 \pm \sqrt{5}$$

   The solutions are $-2 + \sqrt{5}$ and $-2 - \sqrt{5}$.

40. $$7x^2 - 8 = 5x^2 + 8$$
    $$2x^2 - 16 = 0$$
    $$2x^2 = 16$$
    $$x^2 = 8$$
    $$x = \sqrt{8} \text{ or } x = -\sqrt{8}$$

   Note that $\sqrt{8} = \sqrt{4 \cdot 2} = 2\sqrt{2}$, so

   $$x = 2\sqrt{2} \text{ or } x = -2\sqrt{2}.$$

   The solutions are $2\sqrt{2}$ and $-2\sqrt{2}$.

Chapter 9 Test

1. $x^2 = 39$
   $x = \sqrt{39}$ or $x = -\sqrt{39}$

   The solutions are $\sqrt{39}$ and $-\sqrt{39}$.

2. $(y + 3)^2 = 64$
   $y + 3 = \sqrt{64}$ or $y + 3 = -\sqrt{64}$
   $y + 3 = 8$ or $y + 3 = -8$
   $y = 5$ or $y = -11$

   The solutions are 5 and -11.

3. $(4x + 3)^2 = 24$
   $4x + 3 = \sqrt{24}$ or $4x + 3 = -\sqrt{24}$

   Note that $\sqrt{24} = \sqrt{4 \cdot 6} = 2\sqrt{6}$.

   $4x + 3 = 2\sqrt{6}$ or $4x + 3 = -2\sqrt{6}$
   $4x = -3 + 2\sqrt{6}$ or $4x = -3 - 2\sqrt{6}$
   $x = \dfrac{-3 + 2\sqrt{6}}{4}$ or $x = \dfrac{-3 - 2\sqrt{6}}{4}$

   The solutions are

   $\dfrac{-3 + 2\sqrt{6}}{4}$ and $\dfrac{-3 - 2\sqrt{6}}{4}$.

4. $x^2 - 4x = 6$
   $x^2 - 4x + 4 = 6 + 4$
   $(x - 2)^2 = 10$
   $x - 2 = \sqrt{10}$ or $x - 2 = -\sqrt{10}$
   $x = 2 + \sqrt{10}$ or $x = 2 - \sqrt{10}$

   The solutions are $2 + \sqrt{10}$ and $2 - \sqrt{10}$.

5. $2x^2 + 12x - 3 = 0$
   $x^2 + 6x - \dfrac{3}{2} = 0$
   $x^2 + 6x = \dfrac{3}{2}$
   $x^2 + 6x + 9 = \dfrac{3}{2} + 9$
   $(x + 3)^2 = \dfrac{21}{2}$
   $x + 3 = \sqrt{\dfrac{21}{2}}$ or $x + 3 = -\sqrt{\dfrac{21}{2}}$

   Note that

   $\sqrt{\dfrac{21}{2}} = \dfrac{\sqrt{21}}{\sqrt{2}} = \dfrac{\sqrt{21} \cdot \sqrt{2}}{\sqrt{2} \cdot \sqrt{2}} = \dfrac{\sqrt{42}}{2}$.

   $x + 3 = \dfrac{\sqrt{42}}{2}$ or $x + 3 = -\dfrac{\sqrt{42}}{2}$
   $x = -3 + \dfrac{\sqrt{42}}{2}$ or $x = -3 - \dfrac{\sqrt{42}}{2}$
   $x = \dfrac{-6 + \sqrt{42}}{2}$ or $x = \dfrac{-6 - \sqrt{42}}{2}$

   The solutions are

   $\dfrac{-6 + \sqrt{42}}{2}$ and $\dfrac{-6 - \sqrt{42}}{2}$.

6. For a quadratic equation to have two real solutions, the quantity under the radical in the quadratic equation must be positive.

7. $2x^2 + 5x - 3 = 0$

   Use $a = 2$, $b = 5$, and $c = -3$.

   $x = \dfrac{-b \pm \sqrt{b^2 - 4ac}}{2a}$

   $= \dfrac{-5 \pm \sqrt{5^2 - 4(2)(-3)}}{2(2)}$

   $= \dfrac{-5 \pm \sqrt{25 + 24}}{4}$

   $= \dfrac{-5 \pm \sqrt{49}}{4} = \dfrac{-5 \pm 7}{4}$

$$x = \frac{-5 + 7}{4} \quad \text{or} \quad x = \frac{-5 - 7}{4}$$

$$x = \frac{2}{4} = \frac{1}{2} \quad \text{or} \quad x = \frac{-12}{4} = -3$$

The solutions are 1/2 and −3.

8. $$3w^2 + 2 = 6w$$
    $$3w^2 - 6w + 2 = 0$$

    Use $a = 3$, $b = -6$, and $c = 2$.

    $$w = \frac{-(-6) \pm \sqrt{(-6)^2 - 4(3)(2)}}{2(3)}$$
    $$= \frac{6 \pm \sqrt{36 - 24}}{6}$$
    $$= \frac{6 \pm \sqrt{12}}{6}$$
    $$= \frac{6 \pm 2\sqrt{3}}{6}$$
    $$= \frac{2(3 \pm \sqrt{3})}{2(3)} = \frac{3 \pm \sqrt{3}}{3}$$

    The solutions are

    $$\frac{3 + \sqrt{3}}{3} \text{ and } \frac{3 - \sqrt{3}}{3}.$$

9. $$4x^2 + 8x + 11 = 0$$

    Use $a = 4$, $b = 8$, and $c = 11$.

    $$x = \frac{-8 \pm \sqrt{8^2 - 4(4)(11)}}{2(4)}$$
    $$= \frac{-8 \pm \sqrt{64 - 176}}{8}$$
    $$= \frac{-8 \pm \sqrt{-112}}{8}$$

    The radical $\sqrt{-112}$ is not a real number, so the equation has no real number solution.

10. $$t^2 - \frac{5}{3}t + \frac{1}{3} = 0$$
    $$3\left(t^2 - \frac{5}{3}t + \frac{1}{3}\right) = 3(0)$$
    $$3t^2 - 5t + 1 = 0$$

    Use $a = 3$, $b = -5$, and $c = 1$.

    $$t = \frac{-(-5) \pm \sqrt{(-5)^2 - 4(3)(1)}}{2(3)}$$
    $$= \frac{5 \pm \sqrt{25 - 12}}{6}$$
    $$= \frac{5 \pm \sqrt{13}}{6}$$

    The solutions are

    $$\frac{5 + \sqrt{13}}{6} \text{ and } \frac{5 - \sqrt{13}}{6}.$$

11. $$p^2 - 2p - 1 = 0$$

    Solve by completing the square.

    $$p^2 - 2p = 1$$
    $$p^2 - 2p + 1 = 1 + 1$$
    $$(p - 1)^2 = 2$$
    $$p - 1 = \sqrt{2} \quad \text{or} \quad p - 1 = -\sqrt{2}$$
    $$p = 1 + \sqrt{2} \quad \text{or} \quad p = 1 - \sqrt{2}$$

    The solutions are $1 + \sqrt{2}$ and $1 - \sqrt{2}$.

12. $$(2x + 1)^2 = 18$$

    Solve by the square root method.

    $$2x + 1 = \sqrt{18} \quad \text{or} \quad 2x + 1 = -\sqrt{18}$$
    $$2x + 1 = 3\sqrt{2} \quad \text{or} \quad 2x + 1 = -3\sqrt{2}$$
    $$2x = -1 + 3\sqrt{2} \quad \text{or} \quad 2x = -1 - 3\sqrt{2}$$
    $$x = \frac{-1 + 3\sqrt{2}}{2} \quad \text{or} \quad x = \frac{-1 - 3\sqrt{2}}{2}$$

    The solutions are

    $$\frac{-1 + 3\sqrt{2}}{2} \text{ and } \frac{-1 - 3\sqrt{2}}{2}.$$

13.
$$(x - 5)(2x - 1) = 1$$
$$2x^2 - 11x + 5 = 1$$
$$2x^2 - 11x + 4 = 0$$

Use $a = 2$, $b = -11$, and $c = 4$ in the quadratic formula.

$$x = \frac{-(-11) \pm \sqrt{(-11)^2 - 4(2)(4)}}{2(2)}$$
$$= \frac{11 \pm \sqrt{121 - 32}}{4}$$
$$= \frac{11 \pm \sqrt{89}}{4}$$

The solutions are
$$\frac{11 + \sqrt{89}}{4} \text{ and } \frac{11 - \sqrt{89}}{4}.$$

14.
$$t^2 + 25 = 10t$$
$$t^2 - 10t + 25 = 0$$
$$(t - 5)^2 = 0$$
$$t - 5 = 0$$
$$t = 5$$

The only solution is 5.

15. $s = -16t^2 + 64t$

Let $s = 64$ and solve for t.

$$64 = -16t^2 + 64t$$
$$16t^2 - 64t + 64 = 0$$
$$t^2 - 4t + 4 = 0$$
$$(t - 2)^2 = 0$$
$$t - 2 = 0$$
$$t = 2$$

The ball will reach a height of 64 feet after 2 seconds.

16. Equation (d) $t^2 = 0$ has exactly one real number solution because 0 has a single square root.

17. Use the Pythagorean formula with legs x and $x + 2$ and hypotenuse $x + 4$.

$$a^2 + b^2 = c^2$$
$$(x)^2 + (x + 2)^2 = (x + 4)^2$$
$$x^2 + x^2 + 4x + 4 = x^2 + 8x + 16$$
$$x^2 - 4x - 12 = 0$$
$$(x - 6)(x + 2) = 0$$
$$x - 6 = 0 \text{ or } x + 2 = 0$$
$$x = 6 \text{ or } x = -2$$

Reject the negative value because x represents a length. The value of x is 6.

For Exercises 18 and 19, see the graphs in the answer section of the textbook.

18. $y = (x - 3)^2$
$y = x^2 - 6x + 9$

This is a quadratic equation in standard form with $a = 1$, $b = -6$, and $c = 9$.

The x-value of the vertex is
$$x = -\frac{b}{2a} = -\frac{-6}{2(1)} = 3.$$

The y-value of the vertex is
$$y = (3 - 3)^2 = 0,$$

so the vertex is $(3, 0)$.
Make a table of ordered pairs.

x	y
0	9
1	4
2	1
3	0
4	1
5	4
6	9

Plot these seven ordered pairs and connect them with a smooth curve.

19. $y = -x^2 - 2x - 4$

    Here, $a = -1$, $b = -2$, and $c = -4$. The x-value of the vertex is

    $$x = -\frac{b}{2a} = -\frac{-2}{2(-1)} = -1.$$

    The y-value of the vertex is

    $$y = -(-1)^2 - 2(-1) - 4 = -3,$$

    so the vertex is $(-1, -3)$.
    Make a table of ordered pairs.

x	y
-3	-7
-2	-4
-1	-3
0	-4
1	-7

    Plot these five ordered pairs and connect them with a smooth curve.

20. $(4x - 7)^2 = -4$

    It is apparent that this equation has no real solutions because the square of a real number cannot be negative.

Cumulative Review: Chapters R–9

1. $\dfrac{-4 \cdot 3^2 + 2 \cdot 3}{2 - 4 \cdot 1} = \dfrac{-4 \cdot 9 + 2 \cdot 3}{2 - 4}$

    $\phantom{\dfrac{-4 \cdot 3^2 + 2 \cdot 3}{2 - 4 \cdot 1}} = \dfrac{-36 + 6}{-2}$

    $\phantom{\dfrac{-4 \cdot 3^2 + 2 \cdot 3}{2 - 4 \cdot 1}} = \dfrac{-30}{-2} = 15$

2. $-9 - (-8)(2) + 6 - (6 + 2)$
    $= -9 - (-8)(2) + 6 - 8$
    $= -9 - (-16) + 6 - 8$
    $= -9 + 16 + 6 - 8$
    $= 7 + 6 - 8$
    $= 13 - 8$
    $= 5$

3. $|-3| - |1 - 6| = |-3| - |-5|$
    $ = 3 - 5 = -2$

4. $-4r + 14 + 3r - 7 = -r + 7$

5. $13k - 4k + k - 14k + 2k$
    $= (13 - 4 + 1 - 14 + 2)k$ *Distributive property*
    $= -2k$

6. $5(4m - 2) - (m + 7)$
    $= 20m - 10 - m - 7$
    $= 19m - 17$

7. $6x - 5 = 13$
    $6x = 18$
    $x = 3$

8. $3k - 9k - 8k + 6 = -64$
    $-14k + 6 = -64$
    $-14k = -70$
    $k = 5$

9. $2(m - 1) - 6(3 - m) = -4$
    $2m - 2 - 18 + 6m = -4$
    $8m - 20 = -4$
    $8m = 16$
    $m = 2$

10. Let x = the number of votes Hartmann received.

    Then $x + 80$ = the number of votes Sprague received.

    A total of 346 votes were cast for the two candidates, so

    $$x + (x + 8) = 346.$$

    Solve this equation.

    $$2x + 80 = 346$$
    $$2x = 266$$
    $$x = 133$$

    Hartmann received 133 votes.

11. Solve $I = prt$ for p.

    $$\frac{I}{rt} = \frac{prt}{rt} \quad \text{Divide by } rt$$

    $$\frac{I}{rt} = p \quad \text{or} \quad p = \frac{I}{rt}$$

12. Solve $P = 2L + 2W$ for L.

    $$P - 2W = 2L$$
    $$\frac{P - 2W}{2} = L$$

    or $\quad L = \dfrac{P - 2W}{2}$

For Exercises 13 and 14, see the number line graphs in the answer section of the textbook.

13. $-8m < 16$

    Divide each side by -8 and reverse the inequality symbol.

    $$\frac{-8m}{-8} > \frac{16}{-8}$$
    $$m > -2$$

To graph this solution on a number line, place an open circle at -2 and draw an arrow extending to the right.

14. $-9p + 2(8 - p) - 6 \geq 4p - 50$
    $-9p + 16 - 2p - 6 \geq 4p - 50$
    $-11p + 10 \geq 4p - 50$
    $-15p \geq -60$
    $\dfrac{-15p}{-15} \leq \dfrac{-60}{-15}$
    $\quad\quad\quad$ Divide by -15; reverse symbol
    $p \leq 4$

To graph this solution on a number line, place a solid dot at 4 and draw an arrow extending to the left.

15. $(3^2 \cdot x^{-4})^{-1} = (3^2)^{-1}(x^{-4})^{-1}$
    $= (9)^{-1} x^{(-4)(-1)}$
    $= \dfrac{1}{9} \cdot x^4 = \dfrac{x^4}{9}$

16. $\left(\dfrac{b^{-3}c^4}{b^5c^3}\right)^{-2} = (b^{-3-5}c^{4-3})^{-2}$
    $= (b^{-8}c^1)^{-2}$
    $= (b^{-8})^{-2}(c^1)^{-2}$
    $= b^{16}c^{-2}$
    $= b^{16} \cdot \dfrac{1}{c^2} = \dfrac{b^{16}}{c^2}$

17. $\left(\dfrac{5}{3}\right)^{-3} = \left(\dfrac{3}{5}\right)^3 = \dfrac{3^3}{5^3} = \dfrac{27}{125}$

18. $(5x^5 - 9x^4 + 8x^2) - (9x^2 + 8x^4 - 3x^5)$
    $= 5x^5 - 9x^4 + 8x^2 - 9x^2 - 8x^4 + 3x^5$
    $= 8x^5 - 17x^4 - x^2$

19. $(2x - 5)(x^3 + 3x^2 - 2x - 4)$

Multiply vertically.

$$\begin{array}{rrrrr} & x^3 & +\ 3x^2 & -\ 2x & -\ 4 \\ & & & 2x & -\ 5 \\ \hline & -5x^3 & -\ 15x^2 & +\ 10x & +\ 20 \\ 2x^4 & +\ 6x^3 & -\ 4x^2 & -\ 8x & \\ \hline 2x^4 & +\ x^3 & -\ 19x^2 & +\ 2x & +\ 20 \end{array}$$

20. $(5t + 9)^2 = (5t)^2 + 2(5t)(9) + (9)^2$

$= 25t^2 + 90t + 81$

21. $\dfrac{3x^3 + 10x^2 - 7x + 4}{x + 4}$

$$\begin{array}{r} 3x^2 - 2x + 1 \\ x + 4 \overline{)\, 3x^3 + 10x^2 - 7x + 4} \\ \underline{3x^3 + 12x^2} \qquad\qquad \\ -2x^2 - 7x \qquad \\ \underline{-2x^2 - 8x} \qquad \\ x + 4 \\ \underline{x + 4} \\ 0 \end{array}$$

The remainder is 0, so the answer is the quotient, $3x^2 - 2x + 1$.

22. $16x^3 - 48x^2y = 16x^2(x - 3y)$

$16x^2$ is greatest common factor

23. $16x^4 - 1$

$= (4x^2 + 1)(4x^2 - 1)$ *Difference of two squares*

$= (4x^2 + 1)(2x + 1)(2x - 1)$ *Difference of two squares*

24. $2a^2 - 5a - 3$

Use the grouping method. Look for two integers whose product is $2(-3) = -6$ and whose sum is -5.

The integers are -6 and 1.

$2a^2 - 5a - 3$

$= 2a^2 - 6a + a - 3$

$= 2a(a - 3) + 1(a - 3)$

$= (a - 3)(2a + 1)$

25. $25m^2 - 20m + 4$

Since $25m^2 = (5m)^2$, $4 = 2^2$, and $20m = 2(5m)(2)$, $25m^2 - 20m + 4$ is a perfect trinomial.

$25m^2 - 20m + 4$

$= (5m)^2 - 2(5m)(1) + 2^2$

$= (5m - 2)^2$

26. $x^2 + 3x - 54 = 0$

$(x + 9)(x - 6) = 0$

$x + 9 = 0$ or $x - 6 = 0$

$x = -9$ or $x = 6$

The solutions are -9 and 6.

27. $3x^2 = x + 4$

$3x^2 - x - 4 = 0$

$(3x - 4)(x + 1) = 0$ *Factor*

$3x - 4 = 0$ or $x + 1 = 0$

$3x = 4$

$x = \dfrac{4}{3}$ or $x = -1$

The solutions are $4/3$ and -1.

28. Let x represent the width of the rectangle. Then $2.5x$ represents the length.

Use the formula for the area of a rectangle.

$$A = LW$$
$$1000 = (2.5x)x \quad \text{Let } A = 1000$$
$$1000 = 2.5x^2$$
$$\frac{1000}{2.5} = \frac{2.5x^2}{2.5} \quad \text{Divide by 2.5}$$
$$400 = x^2$$
$$0 = x^2 - 400$$
$$0 = (x + 20)(x - 20) \quad \text{Factor}$$
$$x + 20 = 0 \quad \text{or} \quad x - 20 = 0$$
$$x = -20 \quad \text{or} \quad x = 20$$

Reject $x = -20$ since the width cannot be negative. The width is 20 meters and the length is $2.5(20) = 50$ meters.

29. $\dfrac{2}{a - 3} \div \dfrac{5}{2a - 6}$

$= \dfrac{2}{a - 3} \cdot \dfrac{2a - 6}{5} \quad$ Multiply by reciprocal of divisor

$= \dfrac{2}{a - 3} \cdot \dfrac{2(a - 3)}{5} \quad$ Factor

$= \dfrac{4(a - 3)}{(a - 3)5} \quad$ Multiply

$= \dfrac{4}{5} \quad$ Lowest terms

30. $\dfrac{1}{k} - \dfrac{2}{k - 1}$

$= \dfrac{1(k - 1)}{k(k - 1)} - \dfrac{2(k)}{k(k - 1)} \quad$ LCD $= k(k - 1)$

$= \dfrac{(k - 1) - 2k}{k(k - 1)} \quad$ Subtract numerators

$= \dfrac{-k - 1}{k(k - 1)} \quad$ Combine terms

31. $\dfrac{2}{a^2 - 4} + \dfrac{3}{a^2 - 4a + 4}$

$= \dfrac{2}{(a + 2)(a - 2)} + \dfrac{3}{(a - 2)(a - 2)} \quad$ Factor denominators

$= \dfrac{2(a - 2)}{(a + 2)(a - 2)(a - 2)}$

$+ \dfrac{3(a + 2)}{(a + 2)(a - 2)(a - 2)}$

LCD $= (a + 2)(a - 2)(a - 2)$

$= \dfrac{2(a - 2) + 3(a + 2)}{(a + 2)(a - 2)(a - 2)} \quad$ Add numerators

$= \dfrac{2a - 4 + 3a + 6}{(a + 2)(a - 2)(a - 2)} \quad$ Distributive property

$= \dfrac{5a + 2}{(a + 2)(a - 2)(a - 2)} \quad$ Combine terms

or $\dfrac{5a + 2}{(a + 2)(a - 2)^2}$

32. $\dfrac{6 + \dfrac{1}{x}}{3 - \dfrac{1}{x}}$

Multiply numerator and denominator of the complex fraction by x, which is the least common denominator for all the denominators in the complex fraction.

$\dfrac{x\left(6 + \dfrac{1}{x}\right)}{x\left(3 - \dfrac{1}{x}\right)}$

$= \dfrac{6x + x\left(\dfrac{1}{x}\right)}{3x - x\left(\dfrac{1}{x}\right)} \quad$ Distributive property

$= \dfrac{6x + 1}{3x - 1}$

33. $\frac{1}{x+3} + \frac{1}{x} = \frac{7}{10}$

Multiply each side by the least commo denominator, $10x(x+3)$.

$$10x(x+3)\left(\frac{1}{x+3} + \frac{1}{x}\right) = 10x(x+3)\left(\frac{7}{10}\right)$$

$$10x(x+3)\left(\frac{1}{x+3}\right) + 10x(x+3)\left(\frac{1}{x}\right)$$
$$= 10x(x+3)\left(\frac{7}{10}\right)$$

$$10x + 10(x+3) = 7x(x+3)$$
$$10x + 10x + 30 = 7x^2 + 21x$$
$$20x + 30 = 7x^2 + 21x$$
$$0 = 7x^2 + x - 30$$
$$0 = (7x + 15)(x - 2)$$

$7x + 15 = 0$ or $x - 2 = 0$
$7x = -15$ or $x = 2$
$x = -\frac{15}{7}$ or $x = 2$

The solutions are $-15/7$ and 2.

For Exercises 34–36, see the graphs in the answer section of the textbook.

34. $2x + 3y = 6$

Find the intercepts.
Let $x = 0$.
$$2(0) + 3y = 6$$
$$3y = 6$$
$$y = 2$$

The y-intercept is $(0, 2)$.
Let $y = 0$.
$$2x + 3(0) = 6$$
$$2x = 6$$
$$x = 3$$

The x-intercept is $(3, 0)$.
The graph is the line through the points $(0, 2)$ and $(3, 0)$.

35. $y = 3$

For any value of x, the value of y will always be 3. Three ordered pairs are $(-2, 3)$, $(0, 3)$, and $(4, 3)$. Plot these points and draw a line through them. This will be a horizontal line.

36. $2x - 5y < 10$

First, graph the boundary line $2x - 5y = 10$. If $x = 0$, then $-5y = 10$ and $y = -2$, so the y-intercept is $(0, -2)$. If $y = 0$, then $2x = 10$ and $x = 5$, so the x-intercept is $(5, 0)$. Because of the "<" sign, the line through $(0, -2)$ and $(5, 0)$ should be dashed.
Next, use $(0, 0)$ as a test point in $2x - 5y < 10$.

$$2(0) - 5(0) < 10$$
$$0 < 10 \quad True$$

Since $0 < 10$ is a false statement, shade the region on the side of the line that contains $(0, 0)$. This is the region above the line.

37. Line through (−1, 4) and (5, 2)

Use the slope formula with $(-1, 4) = (x_1, y_1)$ and $(5, 2) = (x_2, y_2)$.

$$m = \frac{y_2 - y_1}{x_2 - x_1}$$

$$= \frac{2 - 4}{5 - (-1)}$$

$$= \frac{-2}{6} = -\frac{1}{3}$$

38. Slope 2; y-intercept (0, 3)

Let m = 2 and b = 3 in slope-intercept form.

$$y = mx + b$$
$$y = 2x + 3$$

Now rewrite the equation in the form $Ax + By = C$.

$$-2x + y = 3$$
$$2x - y = -3 \quad \text{Multiply by } -1$$

39. $2x + y = -4$ (1)
    $-3x + 2y = 13$ (2)

Use the addition method. Multiply equation (1) by −2 and add the result to equation (2).

$$-4x - 2y = 8$$
$$-3x + 2y = 13$$
$$-7x = 21$$
$$x = -3$$

To find y, substitute −3 for x in equation (1).

$2x + y = -4$
$2(-3) + y = -4$
$-6 + y = -4$
$y = 2$

The solution is (−3, 2).

40. $3x - 5y = 8$ (1)
    $-6x + 10y = 16$ (2)

Use the addition method. Multiply equation (1) by 2 and add the result to equation (2).

$$6x - 10y = 16$$
$$-6x + 10y = 16$$
$$0 = 32 \quad \textit{False}$$

Since the result is a false statement, there is no solution.

41. $2x + y \leq 4$
    $x - y > 2$

Draw a solid boundary line through (2, 0) and (0, 4), and shade the side that includes the origin. Draw a dashed boundary line through (2, 0) and (0, −2), and shade the side that does not include the point (1, 1). The solution of the system of inequalities is the intersection of these two shaded half-planes. See the final graph in the answer section of the textbook.

42. The square roots of 289 are 17 and −17, because $17^2 = 289$ and $(-17)^2 = 289$.

43. $\sqrt{100} = 10$ since $10^2 = 100$ and $\sqrt{100}$ represents the positive square root.

44. $\dfrac{6\sqrt{6}}{\sqrt{5}} = \dfrac{6\sqrt{6} \cdot \sqrt{5}}{\sqrt{5} \cdot \sqrt{5}} = \dfrac{6\sqrt{30}}{5}$

45. $\sqrt[3]{\dfrac{7}{16}} = \dfrac{\sqrt[3]{7}}{\sqrt[3]{16}}$

$= \dfrac{\sqrt[3]{7} \cdot \sqrt[3]{4}}{\sqrt[3]{16} \cdot \sqrt[3]{4}}$

$= \dfrac{\sqrt[3]{28}}{\sqrt[3]{64}} = \dfrac{\sqrt[3]{28}}{4}$

46. $3\sqrt{5} - 2\sqrt{20} + \sqrt{125}$

$= 3\sqrt{5} - 2\sqrt{4 \cdot 5} + \sqrt{25 \cdot 5}$

$= 3\sqrt{5} - 2 \cdot 2\sqrt{5} + 5\sqrt{5}$

$= 3\sqrt{5} - 4\sqrt{5} + 5\sqrt{5}$

$= (3 - 4 + 5)\sqrt{5} = 4\sqrt{5}$

47. $\sqrt[3]{16a^3b^4} - \sqrt[3]{54a^3b^4}$

$= \sqrt[3]{8a^3b^3 \cdot 2b} - \sqrt[3]{27a^3b^3 \cdot 2b}$

$= \sqrt[3]{8a^3b^3} \cdot \sqrt[3]{2b} - \sqrt[3]{27a^3b^3} \cdot \sqrt[3]{2b}$

$= 2ab\sqrt[3]{2b} - 3ab\sqrt[3]{2b}$

$= (2 - 3)ab\sqrt[3]{2b} = -ab\sqrt[3]{2b}$

48. $\sqrt{x + 2} = x - 4$

$(\sqrt{x + 2})^2 = (x - 4)^2$

$x + 2 = x^2 - 8x + 16$

$0 = x^2 - 9x + 14$

$0 = (x - 7)(x - 2)$

$x - 7 = 0$ or $x - 2 = 0$

$x = 7$ or $x = 2$

Check

Let $x = 7$.

$\sqrt{x + 2} = x - 4$

$\sqrt{7 + 2} = 7 - 4$

$\sqrt{9} = 3$

$3 = 3$ True

Let $x = 2$.

$\sqrt{x + 2} = x - 4$

$\sqrt{2 + 2} = 2 - 4$

$\sqrt{4} = -2$

$2 = -2$ False

Of these two numbers, 7 checks in the original equation but 2 does not, so the only solution is 7.

49. $2a^2 - 2a = 1$

Rewrite the equation to match the form of the standard quadratic equation.

$2a^2 - 2a = 1$

$2a^2 - 2a - 1 = 0$

The left side cannot be factored, so use the quadratic formula with $a = 2$, $b = -2$, and $c = -1$.

$a = \dfrac{-b \pm \sqrt{b^2 - 4ac}}{2a}$

$a = \dfrac{-(-2) \pm \sqrt{(-2)^2 - 4(2)(-1)}}{2(2)}$

$= \dfrac{2 \pm \sqrt{4 + 8}}{4}$

$= \dfrac{2 \pm \sqrt{12}}{4} = \dfrac{2 \pm 2\sqrt{3}}{4}$

$= \dfrac{2(1 \pm \sqrt{3})}{4} = \dfrac{1 \pm \sqrt{3}}{2}$

The solutions are $\dfrac{1 + \sqrt{3}}{2}$ and $\dfrac{1 - \sqrt{3}}{2}$.

50. $y = x^2 - 4$

Find the y-intercept.

Let $x = 0$; then $y = -4$. The y-intercept $(0, -4)$.

Find the x-intercepts.

$$0 = x^2 - 4 \qquad \text{Let } y = 0$$
$$0 = (x - 2)(x + 2) \quad \text{Factor}$$
$$x - 2 = 0 \quad \text{or} \quad x + 2 = 0$$
$$x = 2 \quad \text{or} \quad x = -2$$

The x-intercepts are $(2, 0)$ and $(-2, 0)$.

The x-value of the vertex is

$$x = -\frac{b}{2a} = \frac{0}{2(1)} = 0.$$

The point $(0, -4)$ is both the vertex and the y-intercept.

Find some other points on the graph.

Let $x = 1$; then
$$y = (1)^2 - 4 = -3.$$

Let $x = -1$; then
$$y = (-1)^2 - 4 = -3.$$

Plot the points $(0, -4)$, $(2, 0)$, $(-2, 0)$, $(1, -3)$, and $(-1, -3)$ and draw the parabola. See the graph in the answer section of the textbook.

NOTES

NOTES

NOTES

NOTES